T0227298

# HALOGENATED-ORGANIC-CONTAINING WASTES

# HALOGENATED-ORGANIC CONTAINING WASTES

## Treatment Technologies

by

**N. Surprenant, T. Nunno,
M. Kravett, M. Breton**

Alliance Technologies Corporation
Bedford, Massachusetts

NOYES DATA CORPORATION
Park Ridge, New Jersey, U.S.A.

Copyright © 1988 by Noyes Data Corporation
Library of Congress Catalog Card Number 88-17016
ISBN: 0-8155-1178-7
ISSN: 0090-516X
Printed and bound in the United Kingdom

Published in the United States of America by
Noyes Data Corporation
Mill Road, Park Ridge, New Jersey 07656

Transferred to Digital Printing, 2011

Library of Congress Cataloging-in-Publication Data

Halogenated-organic-containing wastes : treatment technologies / by N.
    Surprenant . . . [et al.] .
        p.  cm. -- (Pollution technology review, ISSN 0090-516X : no.
    157)
        Bibliography:  p.
        Includes index.
        ISBN 0-8155-1178-7 :
        1. Organohalogen compounds--Handbooks, manuals, etc. 2. Organic
    wastes--Handbooks, manuals, etc. I. Surprenant, N.F. II. Series.
    TD812.5.O73H35     1988
    628.5'4--dc19               88-17016
                            CIP

# Foreword

This book provides information describing alternative technologies to land disposal for halogenated-organic-containing wastes. Emphasis is placed on presenting performance data for proven technologies; however, information dealing with emerging technologies is also presented.

The halogenated-organic constituents covered are 78 RCRA-listed, halogenated-organic compounds not classified as solvents, dioxins, or polychlorinated biphenyls. An estimated 24.2 million gallons of these wastes were generated in 1981. Of these, 3.2 million gallons were land disposed. The Resource Conservation and Recovery Act (RCRA), as amended by the Hazardous and Solid Waste Amendments (HSWA) of 1984, prohibits the continued placement of these wastes in or on the land (with certain exceptions). Treatment alternatives, thus, must be found for those wastes exceeding 1,000 ppm of halogenated organics.

The treatment technologies discussed in the book include biological treatment as well as physical, chemical and thermal treatments. Each treatment system, plus solidification/fixation processes for residuals, is described as follows: (1) process description, including design and operating parameters, pretreatment requirements, and post-treatment of residuals; (2) performance data available from bench, pilot and full-scale studies; (3) cost of treatment; and (4) current status of the process.

Approaches to identifying and selecting appropriate technologies for specific halogenated-organic-compound-bearing waste streams are also covered.

The information in the book is from *Technical Resource Document: Treatment Technologies for Halogenated Organic Containing Wastes,* prepared by N. Surprenant, T. Nunno, M. Kravett, and M. Breton of Alliance Technologies Corporation for the U.S. Environmental Protection Agency, December 1987.

The table of contents is organized in such a way as to serve as a subject index and provides easy access to the information contained in the book.

Advanced composition and production methods developed by Noyes Data Corporation are employed to bring this durably bound book to you in a minimum of time. Special techniques are used to close the gap between "manuscript" and "completed book." In order to keep the price of the book to a reasonable level, it has been partially reproduced by photo-offset directly from the original report and the cost saving passed on to the reader. Due to this method of publishing, certain portions of the book may be less legible than desired.

## ACKNOWLEDGMENTS

The authors would like to thank Harry M. Freeman, the Hazardous Waste Engineering Research Laboratory Work Assignment Manager, for his assistance and support throughout the program. The authors also extend thanks to other members of the HWERL staff for their assistance and to the many industrial representatives who provided design, operating, and performance data for the waste treatment technologies.

## NOTICE

The materials in this book were prepared as accounts of work sponsored by the U.S. Environmental Protection Agency. The information has been reviewed by the Hazardous Waste Engineering Research Laboratory of the U.S.E.P.A. and approved for publication. Approval does not signify that the contents necessarily reflect the views and policies of the U.S. E.P.A. or the Publisher. On this basis the Publisher assumes no responsibility nor liability for errors or any consequences arising from the use of the information contained herein.

Mention of trade names or commercial products does not constitute endorsement or recommendation for use by the Agency or the Publisher. Final determination of the suitability of any information or product for use contemplated by any user, and the manner of that use, is the sole responsibility of the user. The book is intended for informational purposes only. The reader is warned that caution must always be exercised when dealing with hazardous chemicals, wastes, or processes such as those involving halogenated organics; and expert advice should be obtained before implementation is considered.

# Contents and Subject Index

# 1. Introduction

Section 3004 of the Resource Conservation and Recovery Act (RCRA), as amended by the Hazardous and Solid Waste Amendments of 1984 (HSWA), prohibits the continued placement of RCRA-regulated hazardous wastes in or on the land, including placement in landfills, land treatment areas, waste piles, and surface impoundments (with certain exceptions for surface impoundments used for the treatment of hazardous wastes). The amendments specify dates by which these prohibitions are to take effect for specific hazardous wastes as shown in Table 1.1. After the effective date of a prohibition, wastes may only be land disposed if: (1) they comply with treatment standards promulgated by the Agency that minimize short-term and long-term threats arising from land disposal; or (2) the Agency has approved a site-specific petition demonstrating, to a reasonable degree of certainty, that there will be no migration from the disposal unit for as long as the waste remains hazardous. In addition, the statute authorizes the Agency to extend the effective dates of prohibitions for up to 2 years nationwide if it is determined that there is insufficient alternative treatment, recovery or disposal capacity.

PURPOSE

This Technical Resource Document (TRD) for halogenated organic wastes identifies recovery and treatment alternatives to land disposal for these wastes and provides performance data and other technical information needed to assess potentially applicable alternatives. This document is one of a series of documents designed to assist regulatory agency and industrial personnel in meeting the land disposal bans promulgated by the 1984 RCRA Amendments.

TABLE 1.1.   SCHEDULING FOR PROMULGATION OF REGULATIONS BANNING
LAND DISPOSAL OF SPECIFIED HAZARDOUS WASTES

| Waste category | Effective date* |
|---|---|
| ● Dioxin containing waste | 11/8/86 |
| ● Solvent containing hazardous wastes numbered F001, F002, F003, F004, F005 | 11/8/86 |
| ● California List | |
| -Liquid hazardous wastes, including free liquids associated with any solid or sludge containing: | |
|    - Free or complex cyanides at $\geq$1,000 mg/L<br>   - As $\geq$500 mg/L<br>   - Cd $\geq$100 mg/L<br>   - Cr$^{+6}$ $\geq$500 mg/L<br>   - Pb $\geq$500 mg/L<br>   - Hg $\geq$20 mg/L<br>   - Ni $\geq$134 mg/L<br>   - Se $\geq$100 mg/L<br>   - Tl $\geq$130 mg/L | 7/8/87 |
| -Liquid hazardous wastes with: | |
|    - pH $\leq$2.0<br>   - PCBs $\geq$50 ppm | 7/8/87 |
|    - Hazardous wastes containing halogenated organic compounds in total concentration $\geq$1,000 mg/kg | 7/8/87 |
| ● Other listed hazardous wastes (§§261.31 and 32), for which a determination of land disposal prohibition must be made: | |
|    - One-third of wastes<br>   - Two-thirds of wastes<br>   - All wastes | 8/8/88<br>6/8/89<br>5/8/90 |
| ● Hazardous wastes identified on the basis of characteristics under Section 3001 | 5/8/90 |
| ● Hazardous wastes identified or listed after enactment | Within 6 months |

*Not including underground injection.

Although emphasis is placed on performance data, technical factors affecting
the performance of recovery/treatment alternatives, including restrictive
waste characteristics, are discussed to assist in the evaluation of options to
land disposal.  Cost data are also presented to assist in the evaluation and
ultimate selection of treatment/recovery technologies.

## DOCUMENT ORGANIZATION AND CONTENT

The following section (Section 2) will identify the nonsolvent
halogenated wastes of concern, including the constituents of concern for each
specific RCRA waste code designation.  Available information concerning waste
stream characteristics, generation, and management practices will also be
provided in Section 2.  Remaining sections (Sections 3 through 10) will
discuss, respectively; pretreatment, recovery practices, and all potentially
applicable physical, chemical, biological, incineration, other thermal
treatment processes and approaches to land disposal of residuals.  Each
treatment process will be reviewed with regard to the following four factors:

1.   Process description, including design and operating parameters,
     pretreatment requirements, and post-treatment of residuals;

2.   Performance data available from bench, pilot, and full-scale studies;

3.   Cost of treatment; and

4.   Present status of the process.

A final section (Section 11) will provide approaches to identifying and
selecting appropriate technologies for halogenated organic compound bearing
waste streams.  Although emphasis is placed on technical performance, cost
data will also be presented and discussed to assist in process selection.

# 2. Waste Characteristics, Generation, and Management

IDENTIFICATION AND CHARACTERIZATION OF HALOGENATED ORGANIC WASTES AND THEIR
CONSTITUENTS

The halogenated organic wastes addressed in this TRD are listed by RCRA
code in Table 2.1.  The wastes consist of RCRA D code pesticide wastes, listed
because of EP toxicity; specific K code process wastes; and P and U code
wastes containing specific halogenated organic compounds (HOCs).  As noted in
the table, many of the K code process wastes contain halogenated organic
compounds that are used as solvents.  Although these waste streams will
generally be amenable to the recovery/treatment processes discussed in this
TRD, a more detailed and relevant discussion of the properties and management
of halogenated solvent compounds can be found in the solvent TRD.[1]

A total of 78 nonsolvent halogenated organic compounds are constituents
of concern within the 120 waste streams listed in Table 2.1.  These
constituents are identified by chemical compound name in Table 2.2 for the D,
U, and P Codes comprising the halogenated organic waste category.
Constituents found in the specific process waste stream K codes, along with
their concentrations, are listed in Table 2.3.  The data in Table 2.3 were
assembled in the Reference 2 study and were compiled from four previous EPA
sponsored programs.[3-6]

An examination of Table 2.3 shows that the composition and constituent
concentrations of many of the specific waste streams is subject to wide
variability both within and among the four programs shown in the table.  A
fairly detailed analysis of the data provided by the four data sources can be
found in Reference 7.  As noted in References 2 and 7, the waste characterization efforts are subject to a great deal of uncertainty.  Very little
sampling data were available to characterize the waste streams, and estimates

TABLE 2.1.  RCRA-LISTED WASTES CONTAINING HALOGENATED ORGANIC COMPOUNDS (HOCs)

| Waste category | Total number listed in Part 261 | Total number containing HOCs (%) | Listing of specific hazardous waste codes containing one or more HOCs |
|---|---|---|---|
| DOXX | 17 | 6 (35) | D012, D013, D014, D015, D016, D017 |
| KXXX | 76 | 27 (36) | K001, K009[a], K010[a], K015[a], K016[a], K017[a], K018[a], K019[a], K020[a], K021[a], K028[a], K029[a], K030[a], K032[a], K033, K041, K042[a], K043, K073[a], K085[a], K095[a], K096[a], K097, K098, K099, K105[a] |
| PXXX | 107 | 23 (21) | P004, P017, P023, P024, P025, P026, P027, P028, P033, P035, P036, P037, P043, P050, P051, P057, P058, P059, P060, P090, P095, P118, P123 |
| UXXX | 233 | 64 (26) | U006, U017, U020, U023, U024, U025, U026, U027, U029, U030, U033, U034, U035, U036, U038, U039, U041[a], U042, U043, U044[a], U045[a], U046[a], U047, U048, U049, U060, U061, U062, U066, U067, U068, U072, U073, U081, U082, U097, U127, U128, U129, U130, U132, U138, U142, U150, U156[a], U158, U183, U184, U185, U192, U207, U212, U224, U230, U231, U232, U233, U235, U237, U240, U242, U243, U246, U247 |
| Totals | 433 | 120 (28) | |

[a]Contains or represents a specific halogenated organic compound addressed in the solvent TRD.[1]

Source:  Reference 2.

TABLE 2.2.   HALOGENATED ORGANIC COMPOUNDS IN THE RCRA D, P, AND U WASTE CODES

| | | | |
|---|---|---|---|
| D012 | Eldrin | D015 | Toxaphene |
| D013 | Lindane | D016 | 2,4-D |
| D014 | Methoxychlor | D017 | 2,4,5-TP (silvex) |
| | | | |
| P004 | Eldrin | P043 | Vinyl chloride |
| P017 | Bromoacetone | P050 | Endosulfan |
| P023 | Chloroacetaldehyde | P051 | Endrin |
| P024 | p-chloroadniline | P057 | Fluoroacetamide |
| P025 | Indomethacin | P058 | Fluoracetic acid (Na salt) |
| P026 | 1-(o-chlorophenyl)thiourea | P059 | Heptachloc |
| P027 | 3-chloroprionitrile | P060 | Isodrin |
| P028 | Benzyl chloride | P090 | Pentachlorophenol |
| P033 | Cyanogen chloride | P095 | Phosgene |
| P035 | 2,4-D | P118 | Trichloromethanethiol |
| P036 | Dichlorophenylarsine | P123 | Toxaphene |
| P037 | Dieldrin | | |
| | | | |
| U006 | Acetyl chloride | U073 | 3,3-dichlorobenzidene |
| U017 | Benzal chloride | U081 | 2,4-dichlorophenol |
| U020 | Benzenesulfonyl chloride | U082 | 2,6-dichlorophenol |
| U023 | Benzotrichloride | U097 | Dimethylcarbomyl chloride |
| U024 | Bis(2-chloroethoxy) methane | U127 | Hexachlorobenzene |
| U025 | Bis(2-chloroethyl) ether | U128 | Hexachlorobutadiene |
| U026 | N,N-Bis(2-chloroethyl) naphthyl amine | U129 | Lindane |
| U027 | Bis(2-chloroisopropyl) ether | U130 | Hexachloropentadiene |
| U039 | Bromomethane | U132 | Hexachlorophene |
| U030 | 4-Bromophenyl phenyl ether | U138 | Methyl iodine |
| U033 | Carbonyl fluoride | U142 | Kepone |
| U034 | Trichloroacetaldehyde | U150 | Melphalan |
| U035 | Chlorambucil | U156 | Methyl chlorocarbonate |
| U036 | Chlordane | U158 | 4,4'-methylene Bis(2-chloroaniline) |
| U038 | Chlorobenzilate | U183 | Pentachlorobenzene |
| U039 | p-chloro-m-cresol | U184 | Pentachloroethane |
| U041 | 1-chloro-2,3-epoxy propane | U185 | Pentadhlororonitrobenzene |
| U042 | 2-chloroethyle vinyl ether | U192 | Pronamide |
| U043 | Vinyl chloride | U207 | 1,2,4,5-tetrachlorobenzene |
| U044 | Chloroform | U212 | 2,3,4,6-tetrachlorophenol |
| U045 | Chloromethane | U224 | Toxaphene |
| U046 | Chloromethyl methyl ether | U230 | 2,4,5-trichlorophenol |
| U047 | 2-chloronaphthalene | U231 | 2,4,6-trichlorophenol |
| U048 | 2-chlorophenol | U232 | 2,4,5-trichlorophenoxy acetic acid |
| U049 | 4-chloro-o-toluidine HCl | U233 | Silvex (2,4,5-TP) |
| U060 | DDD | U235 | Tris(2,3-dibromopropyl) phosphate |
| U061 | DDT | U237 | Uracil mustard |
| U062 | Diallate | U240 | 2,4-D Salts & Esters |
| U066 | 1,2-dibromo-3-chloropropane | U242 | Pentachlorophenol |
| U067 | 1,2-dibromoethane | U243 | Hexachloropropene |
| U068 | Dibromomethane | U246 | Nyanogen bromide |
| U072 | p-dichlorobenzene | U247 | Methoxychlor |

TABLE 2.3.   CONSTITUENT CONCENTRATIONS IN K TYPE HALOGENATED PROCESS WASTES

| RCRA hazardous waste number | Description | Halogenated organic compound constituents | Constituent concentrations given in the following sources of waste characterization data (mg/kg): | | | |
|---|---|---|---|---|---|---|
| | | | W-E-T Model Report[3] | MITRE WP81W00-65[4] | MITRE WP83000065[5] | JRR Associates[6] |
| K001 | Wastewater sludge from wood preserving operations using creosote and/or pentachlorophenol | Pentachlorophenol | 780 | NA | $1.0 \times 10^4$ | |
| | | 2-Chlorophenol | | NA | | |
| | | p-chloro-m-cresol | | NA | | |
| | | Trichlorophenols | | NA | | |
| | | Tetrachlorophenols | | NA | | |
| | | Benzo(h)fluoranthene | | NA | | |
| | | 2,4-Dichlorophenol | | NA | | |
| | | 2,4,6-Trichlorophenol | | NA | | |
| | | 2,3,7,8-Tetrachlorodibenzo-p-dioxin (TCDD) | | NA | | |
| | | Pentachlorodibenzo-p-dioxins | | NA | | |
| | | Hexachlorodibenzo-p-dioxins | | NA | | |
| | | Octachlorodibenzo-p-dioxins | | NA | | |
| | | Tetrachlorodibenzofurans | | NA | | |
| | | Pentachlorodibenzofurans | | NA | | |
| | | Hexachlorodibenzofurans | | NA | | |
| | | Heptachlorodibenzofurans | | NA | | |
| | | Octachlorodibenzofurans | | NA | | |
| | | Heptachlorodibenzo-p-dioxins | | NA | | |
| | | Higher chlorophenols | | NA | | |
| | | Tetrachloroethylene | | NA | | |
| | | Chloroform | 3,000 | NA | | |
| K009 | Distillation bottoms from the production of acetaldehyde from ethylene | Acetyl chloride | | 5,000 | | |
| | | Chloral | | 1,000 | | |
| | | Chloroacetaldehyde | | | | |
| | | Chloroform | 3,000 | <5,000 | | |
| | | Methylene chloride | | <5,000 | | |
| | | Methyl chloride | | | | |
| K010 | Distillation bottoms from the production of acetaldehyde from ethylene | Acetyl chloride | $11.9 \times 10^4$ | $1.0 \times 10^5$ | | |
| | | Chloral | $2.0 \times 10^4$ | $6.8 \times 10^4$ | | |
| | | Chloroacetaldehyde | | $11.1 \times 10^4$ | | |
| | | Chloroform | | $<4.0 \times 10^4$ | | |
| | | Methylene chloride | | $<4.0 \times 10^4$ | | |
| | | Methyl chloride | | $<4.0 \times 10^4$ | | |

(continued)

TABLE 2.3 (continued)

| RCRA hazardous waste number | Description | Halogenated organic compound constituents | Constituent concentrations given in the following sources of waste characterization data (mg/kg): | | | |
|---|---|---|---|---|---|---|
| | | | W-E-T Model Report[3] | MITRE WP81W004654[4] | MITRE WP83000655[5] | JRB Associates[6] |
| K015 | Distillation bottoms from the distillation of benzyl chloride | Benzyl chloride | 5,300 | $1.0 \times 10^5$ | NA | |
| | | Monochlorobenzene | | | NA | $1 \times 10^5 - 5 \times 10^5$ |
| | | Benzotrichloride | 3,400 | $5.0 \times 10^5$ | $9.0 \times 10^4$ | $1 \times 10^5 - 5 \times 10^5$ |
| | | Benzal chloride | | $8.0 \times 10^5$ | $8.3 \times 10^5$ | $1 \times 10^5 - 5 \times 10^5$ |
| | | Trichlorobenzene | 4,100 | $<5.0 \times 10^4$ | NA | $1 \times 10^5 - 5 \times 10^5$ |
| | | Tetrachlorobenzene | | $<5.0 \times 10^4$ | NA | |
| | | Pentachlorobenzene | | $<5.0 \times 10^4$ | NA | |
| | | Hexachlorobenzene | 410 | $<5.0 \times 10^4$ | NA | |
| K016 | Heavy ends or distillation residues from the production of carbon tetrachloride | Hexachloroethane | $2.0 \times 10^5$ | $1.6 \times 10^5$ | $1.5 \times 10^5$ | $1 \times 10^5 - 5 \times 10^5$ |
| | | Hexachlorobenzene | $4.0 \times 10^4$ | | | $1 \times 10^5 - 5 \times 10^5$ |
| | | Hexachlorobutadiene | $4.0 \times 10^4$ | $3.1 \times 10^4$ | $4.99 \times 10^4$ | $1 \times 10^5 - 5 \times 10^5$ |
| | | Carbon tetrachloride | | | $1.5 \times 10^5$ | $1 \times 10^5 - 5 \times 10^5$ |
| | | Tetrachloroethylene | $2.0 \times 10^5$ | $1.7 \times 10^5$ | $1.5 \times 10^5$ | $>5.0 \times 10^5$ |
| | | Tetrachloroethane | | | | $1 \times 10^5 - 5 \times 10^5$ |
| | | Trichloroethylene | | | | $1 \times 10^5 - 5 \times 10^5$ |
| | | Chlorobenzenes | | | | $100 - 1 \times 10^5$ |
| | | Chloropropanes | | | | $100 - 1 \times 10^5$ |
| | | Chloroform | | | | $<1,000$ |
| | | Pentachloroethane | | | | $1 \times 10^4 - 1 \times 10^5$ |
| | | Heavy chlorinated tars | | $2.8 \times 10^4$ | 410 | |
| | | Dichlorobiphenyl | | | | |
| | | Chloropropenes | | | | $1 \times 10^4 - 5 \times 10^5$ |
| K017 | Heavy ends (still bottoms) from the purification column in the production of epichlorohydrin | Epichlorohydrin | $2.0 \times 10^4$ | $2.0 \times 10^4$ | | |
| | | bis(chloromethyl) ether | | $7.0 \times 10^4$ | | $<100 - >5 \times 10^5$ |
| | | bis(2-chloroethyl) ether | | $7.0 \times 10^4$ | | |
| | | Trichloropropane | | $7.0 \times 10^5$ | | |
| | | Dichloropropanols | | $1.0 \times 10^5$ | | |
| | | Chlorinated aliphatics | | NA | | |
| | | 2-chloroallyl alcohol | | 5,000 | | $<100 - 1 \times 10^4$ |
| | | 1-Chloro-2,3-hydroxypropane | | 2,000 | | $100 - 1 \times 10^5$ |
| | | Propyl chlorides | | | | $1,000 - 1 \times 10^5$ |
| | | Isopropyl chlorides | | | | $1,000 - 1 \times 10^5$ |
| | | 1,2-Dichloropropane | $1 \times 10^4$ | 1,000 | | $1,000 - 1 \times 10^5$ |
| | | Dichloropropyl ethers | | | | $<100 - >5 \times 10^5$ |

(continued)

**TABLE 2.3 (continued)**

| RCRA hazardous waste number | Description | Halogenated organic compound constituents | Constituent concentrations given in the following sources of waste characterization data (mg/kg): | | | |
|---|---|---|---|---|---|---|
| | | | W-E-T Model Report [3] | MITRE WP81W004465 [4] | MITRE WP83000655 [5] | JRB Associates [6] |
| K018 | Heavy ends from the fractionation column in ethyl chloride production | Ethylene Dichloride (1,2-Dichloroethane) | $1.1 \times 10^5$ | | | $1 \times 10^5 - 5 \times 10^5$ |
| | | Trichloroethylene | $3.2 \times 10^5$ | $3.2 \times 10^5$ | | $1 \times 10^5 - 5 \times 10^5$ |
| | | Hexachlorobutadiene | $2.15 \times 10^5$ | | | |
| | | Hexachlorobenzene | $2.15 \times 10^5$ | | | |
| | | Heavy chlorinated organics | | | $7.0 \times 10^5$ | $<100 - >5 \times 10^5$ |
| | | Ethyl chloride | | $4.3 \times 10^5$ | | $1 \times 10^4 - 1 \times 10^5$ |
| | | Dichloroethylenes | | $7.0 \times 10^4$ | | $100 - 5 \times 10^5$ |
| | | Chlorinated butanes | | $2.2 \times 10^5$ | | $100 - 5 \times 10^5$ |
| | | Chloroform | | | | |
| K019 | Heavy ends from the distillation of ethylene dichloride in ethylene dichloride production | Ethylene dichloride | $2.06 \times 10^5$ | $3.0 \times 10^5$ | | $>5.0 \times 10^5$ |
| | | 1,1,1-Trichloroethane | $2.42 \times 10^5$ | $1.9 \times 10^5$ | $2.82 \times 10^5$ | $100 - 1,000$ |
| | | 1,1,2-Trichloroethane | | $1.9 \times 10^5$ | | |
| | | 1,1,2,2-Tetrachloroethane | $1.21 \times 10^5$ | $1.9 \times 10^5$ | $1.25 \times 10^5$ | $1 \times 10^4 - 1 \times 10^5$ |
| | | 1,1,1,2-Tetrachloroethane | $1.21 \times 10^5$ | $1.9 \times 10^5$ | $1.25 \times 10^5$ | $1 \times 10^4 - 1 \times 10^5$ |
| | | Hexachloroethane | | | | $100 - 1,000$ |
| | | Trichloroethylene | | | | $1 \times 10^2 - 1 \times 10^5$ |
| | | Tetrachloroethylene | | | | $1,000 - 1 \times 10^4$ |
| | | Pentachloroethane | | | | $1 \times 10^4 - 1 \times 10^5$ |
| | | Carbon tetrachloride | | | | $1 \times 10^4 - 1 \times 10^5$ |
| | | Chloroform | | | | |
| | | Vinyl chloride (chloroethylene) | | | | |
| | | Vinylidene chloride (1,1-Dichloroethylene) | | | | |
| | | 1-Chloro-2-bromoethane | | | | |
| | | 1-Chloro-4-ethylhexane | | | | |
| | | 1-Chloroethylbenzene | | | | |
| | | 1,2,3-chloro-1,3-butadiene | | | $1 \times 10^4$ | |
| | | 1,3-Dichlorobenzene | | | $1 \times 10^4$ | |
| | | 1,4-Dichlorobenzene | | | | |
| | | Dichlorohexadiene | | | $2 \times 10^4$ | |
| | | 1,2-Dichloropropane | | | $1,000$ | |
| | | 1,3-Dichloropropane | | | | |
| | | 1-Methyl-2-chlorobenzene | | | | |
| | | 1-Methyl-3-chlorobenzene | | | | |
| | | Chlorobenzene | | | $7$ | |
| | | 1,2,3-Trichloropropane | | | $8.8 \times 10^4$ | |

(continued)

**TABLE 2.3 (continued)**

| RCRA hazardous waste number | Description | Halogenated organic compound constituents | Constituent concentrations given in the following sources of waste characterization data (mg/kg): | | | |
|---|---|---|---|---|---|---|
| | | | W-E-T Model Report[3] | MITRE WP81W00-65[4] | MITRE WP83000065[5] | JRB Associates[6] |
| K019 (cont.) | | 1,1,2-Trichloropropane | | | | |
| | | Heavy chlorinated tars | | | 12[5] | |
| | | 1,4-Dichloro-2-butene | | | | |
| | | Ethylidene chloride (1,1-Dichloro-ethane) | | | $2.18 \times 10^5$ | |
| | | $C_4$ chlorinated organics | | | | $1 \times 10^4 - 1 \times 10^5$ |
| | | $C_6$ chlorinated organics | | | | $1 \times 10^4 - 1 \times 10^5$ |
| K020 | Heavy ends from the distillation of vinyl chloride in vinyl chloride monomer production | Ethylene dichloride | $2.42 \times 10^5$ | $2.25 \times 10^5$ | | $1 \times 10^5 - 5 \times 10^5$ |
| | | 1,1,1-Trichloroethane | $2.72 \times 10^5$ | 8,000 | $2.5 \times 10^5$ | $1 \times 10^5 - 5 \times 10^5$ |
| | | 1,1,2-Trichloroethane | | $2.8 \times 10^5$ | | $1,000 - 1 \times 10^4$ |
| | | 1,1,2,2-Tetrachloroethane | | $5.0 \times 10^5$ | | |
| | | 1,1,1,2-Tetrachloroethane | $2.11 \times 10^5$ | $3.0 \times 10^5$ | | |
| | | Trichloroethylene | | 2,000 | | |
| | | Tetrachloroethylene | | 9,000 | | $1,000 - 1 \times 10^4$ |
| | | Carbon tetrachloride | | $1.0 \times 10^4$ | 2,800 | $100 - 1,000$ |
| | | Chloroform | | $1.0 \times 10^4$ | 2,800 | $1 \times 10^4 - 5 \times 10^5$ |
| | | Vinyl chloride | 2,000 | $1 \times 10^4$ | 2,800 | $1 \times 10^5 - 5 \times 10^5$ |
| | | Vinylidene chloride | | $1 \times 10^4$ | | |
| | | Heavy chlorinated tars | | $1 \times 10^4$ | 5,600 | $100 - 1,000$ |
| | | 1-Chlorobutane | | 3,000 | | $1 \times 10^5 - 5 \times 10^5$ |
| | | 1,2-Dichlorobutane | | 5,000 | | |
| | | Dichlorobutenes | | $1.8 \times 10^4$ | | |
| | | Chlorobenzene | | $1.0 \times 10^4$ | | <100 |
| | | 1,2-Dichlorohexane | | 6,000 | | |
| | | 2-Chloroethanol | | 3,000 | | |
| | | 1,4-Dichlorobutane | | 7,000 | | |
| | | Pentachloroethane | | 5,000 | | |
| | | Hexachloroethane | | 4,000 | | |
| | | 1,2,3-Trichlorobutane | | 9,000 | $3.79 \times 10^4$ | $100 - 1,000$ |
| | | 1,2,3-Trichloropropane | | 8,000 | | |
| | | Bis(2-chloroethyl) ether | | $2.0 \times 10^4$ | $4.71 \times 10^4$ | $1,000 - 1 \times 10^4$ |
| | | 1,2,4-Trichlorobutane | | $2.0 \times 10^4$ | | |
| | | Freons (soluble and insoluble material) | | $1.0 \times 10^4$ | | |
| | | Toxaphene | | | | |
| | | Trichlorophenol | | | $2.03 \times 10^5$ | $1 \times 10^5 - 5 \times 10^5$ |
| | | 1,1-Dichloroethane | | | $4.71 \times 10^4$ | |
| | | Chlorobutadienes | | | | |

(continued)

TABLE 2.3 (continued)

| RCRA hazardous waste number | Description | Halogenated organic compound constituents | Constituent concentrations given in the following sources of waste characterization data (mg/kg): | | | |
|---|---|---|---|---|---|---|
| | | | W-E-T Model Report[3] | MITRE WP81M00465[4] | MITRE WP83000065[5] | JRB Associates[6] |
| K021 | Aqueous spent antimony catalyst waste from fluoro-methanes production | Carbon tetrachloride | 1,000 | | | |
| | | Chloroform | | | | |
| K028 | Spent catalyst from the hydrochlorinator reactor in the production of 1,1,1-trichloroethane | 1,1,1-Trichloroethane | $3.97 \times 10^5$ | $1.0 \times 10^4$ | | $100 - 1 \times 10^4$ |
| | | Vinyl chloride | | $1.0 \times 10^4$ | | |
| | | Acetyl chloride | $2.73 \times 10^5$ | | | |
| | | 1,1,2-Tetrachloroethane | $3.16 \times 10^5$ | | | |
| | | Pentachloroethane | $1.4 \times 10^4$ | | | |
| | | Dichloroethanes | | | | $1 \times 10^5 - 5 \times 10^5$ |
| K029 | Waste from the product steam stripper in the production of 1,1,1-trichloroethane | Ethylene dichloride | $8.3 \times 10^5$ | $8.3 \times 10^5$ | | |
| | | 1,1,1-Trichloroethane | $1.7 \times 10^5$ | $1.7 \times 10^5$ | | |
| | | Vinyl chloride | | $1.0 \times 10^5$ | | |
| | | Vinylidene chloride | | $1.0 \times 10^5$ | | |
| | | Chloroform | | $1.0 \times 10^5$ | | |
| K030 | Column bottoms or heavy ends from the combined production of trichloroethylene and tetrachloroethylene | Hexachlorobenzene | $2.0 \times 10^5$ | $2.0 \times 10^5$ | | |
| | | Hexachlorobutadiene | $3.38 \times 10^5$ | $3.38 \times 10^5$ | | |
| | | Hexachloroethane | | $4.0 \times 10^4$ | | |
| | | 1,1,1,2-Tetrachloroethane | $6.3 \times 10^4$ | $6.3 \times 10^4$ | | |
| | | 1,1,2,2-Tetrachloroethane | $2.3 \times 10^5$ | $2.3 \times 10^5$ | | |
| | | Ethylene dichloride | | 6,000 | | |
| | | Tetrachloroethylene | | $4.5 \times 10^4$ | | |
| | | Trichloroethylene | | $4.5 \times 10^4$ | | |
| | | Pentachloroethane | | $3.3 \times 10^4$ | | |
| | | 1,1-Dichloroethane | | NA | | |
| | | Chlorobenzenes | | NA | | |
| | | Chloroethanes | | NA | | |
| | | Chlorobutadienes | | NA | | |
| | | 1,1,2-Trichloroethane | | NA | | |
| | | Tars | | NA | | |
| K032 | Wastewater treatment sludge from the production of chlordane | Hexachlorocyclopentadiene | | NA | | |
| | | Chlorinated cyclic compounds | | NA | | |

(continued)

## TABLE 2.3 (continued)

| RCRA hazardous waste number | Description | Halogenated organic compound constituents | Constituent concentrations given in the following sources of waste characterization data (mg/kg): | | | |
|---|---|---|---|---|---|---|
| | | | W-E-T Model Report[3] | MITRE MP81W004654[4] | MITRE MP83000655[5] | JRB Associates[6] |
| K033 | Wastewater and scrub water from the chlorination of cyclopentadiene in the production of chlordane | Hexachlorocyclopentadiene<br>Chlorinated cyclic compounds<br>Hexachlorobenzene<br>Pentachlorobenzene | | NA<br>NA | | $7.14 \times 10^{2}$<br><br>$3.57 \times 10^{5}$<br>$7.14 \times 10^{2}$ |
| K034 | Filter solids from the filtration of hexachloro-cyclopentadiene in the production of chlordane | Hexachlorocyclopentadiene<br>Hexachlorobenzene<br>Pentachlorobenzene<br>Dichlorobenzenes, N.O.S. | | | | $2.0 \times 10^{5}$<br>$1.0 \times 10^{5}$<br>$1.0 \times 10^{5}$ |
| K041 | WW treatment sludges from the production of toxaphene | Toxaphene | $1.0 \times 10^{4}$ | $1.0 \times 10^{4}$ | | |
| K042 | Heavy ends or distillation residues from the distillation of tetrachloro-benzene in the production of 2,4,5-T | Hexachlorobenzene<br>o-Dichlorobenzene<br>Chlorobenzene | | NA<br>NA<br>NA | | |
| K043 | 2,6-Dichlorophenol waste from the production of 2,4-D | 2,4-Dichlorophenol<br>2,6-Dichlorophenol<br>2,4,6-Trichlorophenol<br>Chlorophenol polymers | | NA<br>NA<br>NA<br>NA | | |
| K073 | Chlorinated hydrocarbon waste from the purification step of the diaphragm cell process using graphite anodes in chlorine production | Chloroform<br>Carbon tetrachloride<br>Hexachloroethane<br>Trichloroethane<br>Tetrachloroethylene<br>Dichloroethylene<br>1,1,2,2-Tetrachloroethane<br>Pentachloroethane | $7.4 \times 10^{5}$<br>$1.1 \times 10^{5}$ | $7.4 \times 10^{5}$<br>$1.1 \times 10^{5}$<br>$8.0 \times 10^{4}$<br>$1.0 \times 10^{4}$<br>6,000<br>3,000<br>5,000<br>$1.3 \times 10^{4}$ | $7.4 \times 10^{5}$<br>$1.1 \times 10^{5}$<br>$8.0 \times 10^{4}$<br>$1.0 \times 10^{4}$<br><br><br><br>$1.3 \times 10^{4}$ | |

(continued)

## TABLE 2.3 (continued)

| RCRA hazardous waste number | Description | Halogenated organic compound constituents | W-F-T Model Report[3] | MITRF WP81M004654[4] | MITRE WP83000655[5] | JRB Associates[2] |
|---|---|---|---|---|---|---|
| | | | | | Constituent concentrations given in the following sources of waste characterization data (mg/kg): | |
| K085 | Distillation or fraction- ation column bottoms from the production of mono- chlorobenzene | Dichlorobenzenes, N.O.S. | | | | $1 \times 10^4 - 5 \times 10^5$ |
| | | Trichlorobenzenes, N.O.S. | | | | $1 \times 10^4 - 1 \times 10^5$ |
| | | Tetrachlorobenzenes, N.O.S. | | NA | $1.58 \times 10^5$ | $100 - 1 \times 10^5$ |
| | | Pentachlorobenzenes | | NA | $1.58 \times 10^5$ | |
| | | Hexachlorobenzene | $1.0 \times 10^5$ | NA | $1.58 \times 10^5$ | |
| | | Benzyl chloride | | NA | $1.58 \times 10^5$ | |
| | | Chlorotoluene | | NA | $1.93 \times 10^5$ | $1,000 - 5 \times 10^5$ |
| | | Monochlorobenzene | | NA | $1.58 \times 10^5$ | |
| | | Toxaphene | | NA | | |
| | | PCBs | | | | $1,000 - 1 \times 10^7$ |
| | | Chlorobenzophenones | | | | $100 - 1,000$ |
| K095 | Distillation bottoms from the production of 1,1,1- trichloroethane | 1,1,1-Trichloroethane | $2.4 \times 10^5$ | $2.7 \times 10^5$ | | |
| | | 1,1,1,2-Tetrachloroethane | $2.7 \times 10^5$ | $3.7 \times 10^5$ | | |
| | | 1,1,2,2-Tetrachloroethane | | $1.0 \times 10^4$ | | |
| | | Pentachloroethane | | $4.0 \times 10^5$ | | |
| | | 1,1,2-Trichloroethane | | | | |
| K096 | Heavy ends from the heavy ends column from the produc- tion of 1,1,1-trichloro- ethane | Ethylene dichloride | $1.67 \times 10^5$ | $1.7 \times 10^5$ | | |
| | | 1,1,1-Trichloroethane | $2.71 \times 10^5$ | $2.2 \times 10^5$ | | |
| | | 1,1,2-Trichloroethane | $2.71 \times 10^5$ | $2.9 \times 10^5$ | | |
| | | 1,1-Dichloroethane | | $<1.0 \times 10^4$ | | |
| | | Tetrachloroethane | $1.4 \times 10^5$ | $1.5 \times 10^5$ | | |
| | | Hexachloroethane | $1.4 \times 10^5$ | $1.5 \times 10^5$ | | |
| K097 | Vacuum stripper discharge from chlorinator in produc- tion of chlordane | Chlordane | $3.0 \times 10^4$ | | | |
| | | Heptaclor | | | | |
| K098 | Untreated process waste- water from the production of toxaphene | Toxaphene | $1.0 \times 10^4$ | | | |

(continued)

**TABLE 2.3 (continued)**

| RCRA hazardous waste number | Description | Halogenated organic compound constituents | Constituent concentrations given in the following sources of waste characterization data (mg/kg): | | | |
|---|---|---|---|---|---|---|
| | | | W-E-T Model Report[3] | MITRE WP81W00465[4] | MITRE WP83000065[5] | JRB Associates[6] |
| K099 | Untreated wastewater from the production of 2,4-D | 2,4-Dichlorophenol | | | | $100 - 1.0 \times 10^4$ |
| | | 2,4,6-Trichlorophenol | | | | <100 |
| | | Dichlorophenols | | | | <100 |
| | | TCDD | | | | <100 |
| | | TCDD isomers | | | | <100 |
| | | TCDF isomers | | | | <100 |
| | | Chlorophenols | | | | $1,000 - 1 \times 10^4$ |
| K105 | Separated aqueous stream from the reactor product washing step in the production of chlorobenzenes | Chlorobenzene | 80 | | | |
| | | 1,2-Dichlorobenzene | 50 | | | |
| | | 2,4,6-Trichlorophenol | | | | |
| | | 1,4-Dichlorobenzene | 50 | | | |

NA = constituent listed, concentration not available.

N.O.S. = not otherwise specified.

Source:  Reference 2.

were based on survey studies of limited scope and reliability.  EPA is in the process of updating the data base to provide more reliable information related to the characteristics of the specific waste streams and the quantities generated.

## CLASSIFICATION OF HAZARDOUS ORGANIC WASTES

Several approaches have been proposed for classifying wastes in a manner that would provide meaningful insight into the applicability of treatment technologies.[1,2,11-13]  Although many of these approaches were devised for solvents, including halogenated solvents, they are for the most part, equally applicable to other halogenated organics.[1]

In an attempt to provide a meaningful classification system, the halogenated organic compounds were grouped in Reference 2 by their chemical structure and functionality.  However, the Reference 2 study members and others involved in selection of treatment technologies have concluded that other factors are generally more important in assessing treatability.  The halogen content and the physical nature of the waste matrix were identified as key factors.  Table 2.4 lists the halogenated organic compounds of concern by their physical state (gaseous, liquid, or solid) at 25°C in order of their halogen content.  Other factors which are useful in assessing the applicability of recovery/treatment technologies for these halogens (e.g. vapor pressure, functionality, solubility, octanol water partition coefficients, and heat of combustion) are provided in Appendix A.

## HALOGENATED ORGANIC WASTE GENERATION AND MANAGEMENT

As shown in Table 2.5, estimates of the quantities of wastes generated obtained from the above sources of information (i.e., those used to characterize waste stream composition) and EPA's National Survey Data Base differ widely.[2]  The data shown is for the specific waste stream codes (K type) and for the six pesticide codes (D012 through D017).  Little or no information is available concerning the U and P waste codes for specific halogenated organic constituents.

TABLE 2.4.   WASTE CATEGORIZATION BASED ON PHYSICAL STATE

| RCRA waste code | Compound name | Molecular formula | Molecular weight | Halogen content (% by weight) |
|---|---|---|---|---|
| Gaseous Compounds (@25°C) | | | | |
| U043 | Vinyl chloride | $C_2H_3Cl$ | 62.5 | 57 Cl |
| U033 | Carbonyl fluoride | $CF_2O$ | 66 | 58 F |
| P033 | Cyanogen chloride | CClN | 61.5 | 58 Cl |
| U045 | Methyl chloride | $CH_3Cl$ | 50.5 | 70 Cl |
| P095 | Carbonyl chloride | $CCl_2O$ | 98.9 | 72 Cl |
| U029 | Methyl bromide | $CH_3BR$ | 9.5 | 84 Br |
| Liquid Compounds (@25°C) | | | | |
| P043 | Diisopropyl fluorophosphate | $C_6H_{14}FO_3P$ | 184 | 10 F |
| U062 | Diallate | $C_{10}H_{17}Cl_2NOS$ | 270.2 | 13 Cl |
| U020 | Benzene sulfonyl chloride | $C_6H_5ClO_2S$ | 176.6 | 20 Cl |
| U038 | Ethyl-4,4'-dichlorobenzilate | $C_{16}H_{14}Cl_2O_3$ | 325.2 | 22 Cl |
| U048 | 2-Chlorophenol | $C_6H_5ClO$ | 128.6 | 28 Cl |
| P028 | Benzyl chloride | $C_7H_7Cl$ | 126.6 | 28 Cl |
| U030 | 1-Bromo-4-phenoxy benzene | $C_{12}H_9B_2O$ | 249 | 32 Br |
| P036 | Dichlorophenyl arsive | $C_6H_5AsCl_2$ | 222.9 | 32 Cl |
| U097 | Dimethyl carbamoyl chloride | $C_3H_6ClNO$ | 107.6 | 33 Cl |
| U042 | 2-Chloroethylvinyl ether | $C_4H_7ClO$ | 106.6 | 33 Cl |
| U041 | Epichlorohydrin | $C_3H_3ClO$ | 92.5 | 38 Cl |
| U156 | Methyl chlorocarbonate | $C_2H_3ClO_2$ | 94.5 | 38 Cl |
| P027 | 3-Chloropropionitrile | $C_3H_4ClN$ | 89.5 | 40 Cl |
| U024 | Bis(2-chloroethoxy) methane | $C_5H_{10}Cl_2O_2$ | 173.1 | 41 Cl |
| U027 | Bis(2-chloroisopropyl) ether | $C_6H_{12}Cl_2O$ | 171.1 | 42 Cl |
| U046 | Chloromethoxymethane | $C_2H_5ClO$ | 80.5 | 44 Cl |
| U017 | Benzal chloride | $C_7H_6Cl_2$ | 161 | 44 Cl |
| P023 | Chloroacetaldehyde | $C_2H_3ClO$ | 78.5 | 45 Cl |
| U006 | Acetyl chloride | $C_2H_3Cl$ | 98.9 | 45 Cl |
| U025 | Bis(2-chloroethyl) ether | $C_4H_8Cl_2O$ | 143 | 50 Cl |

(continued)

TABLE 2.4 (continued)

| RCRA waste code | Compound name | Molecular formula | Molecular weight | Halogen content (% by weight) |
|---|---|---|---|---|
| U023 | Benzotrichloride | $C_7H_5Cl_3$ | 195.5 | 54 Cl |
| P017 | Bromoacetone | $C_3H_5Be$ | 137 | 58 Br |
| U235 | Tris(2,3-dibromopropyl) phosphate | $C_9H_{15}Br_6PO_4$ | 697.7 | 69 Br |
| U034 | Trichloroacetaldehyde | $C_2HCl_3O$ | 147.4 | 72 Cl |
| U130 | Hexachlorocyclopendadiene | $C_5Cl_6$ | 272.8 | 78 Cl |
| U128 | Hexachlorobutadiene | $C_4Cl_6$ | 260.8 | 82 Cl |
| U066 | 1,2-Dibromo-3-chloropropane | $C_3H_5Br_2Cl$ | 236.4 | 68 Br/15 Cl |
| U067 | Ethylene dibromide | $C_2H_4Br_2$ | 187.9 | 85 Cl |
| U184 | Pentachloroethane | $C_2HCl_5$ | 202.3 | 88 Cl |
| U138 | Methyl iodide | $CH_3I$ | 142 | 89 I |
| U068 | Methylene bromide | $CH_2Br_2$ | 173.9 | 92 Br |
| Solid Compounds (@25°C) | | | | |
| P025 | Indomethacin | ----- | --- | 10 Cl |
| P058 | Fluoracetic acid (Na salt) | $C_2H_2FNaO_2$ | 137 | 19 F |
| P026 | o-(1-chlorophenyl) thiourea | $C_7H_7ClN_2$ | 187 | 19 Cl |
| U047 | 2-Chloronaphthalene | $C_{10}H_7Cl$ | 162.6 | 22 Cl |
| U150 | Melphalan | $C_{13}H_{18}Cl_2N_2O_2$ | 305 | 23 Cl |
| U035 | Chlorambucil | $C_{14}H_{19}Cl_2NO_2$ | 304.2 | 23 Cl |
| U039 | p-chloro-m-cresol | $C_7H_7ClO$ | 142.6 | 25 Cl |
| P057 | Fluoroacetamide | $C_2H_4FNO$ | 77 | 25 F |
| U026 | Chlornaphazine | $C_{14}H_{15}Cl_2N$ | 268.2 | 26 Cl |
| U158 | 4,4'-Methylene-bis-2-chloroaniline | $C_{13}H_{12}Cl_2N$ | 267.2 | 27 Cl |
| P024 | p-chloroaniline | $C_6H_6ClN$ | 127.6 | 28 Cl |
| U192 | Pronamide | $C_{12}H_{11}Cl_2NO$ | 256.1 | 28 Cl |
| U237 | Uracil mustard | $C_8H_{11}Cl_2N_3O_2$ | 252.1 | 28 Cl |
| U073 | 3,3'-dichlorobenzidine | $C_{12}H_{10}Cl_2N_2$ | 253.1 | 28 Cl |
| D014, U247 | Methoxychlor | $C_{16}H_{15}Cl_3O_2$ | 345.7 | 31 Cl |

(continued)

TABLE 2.4 (continued)

| RCRA waste code | Compound name | Molecular formula | Molecular weight | Halogen content (% by weight) |
|---|---|---|---|---|
| D016, P035 | 2,4-D | $C_8H_6Cl_2O_3$ | 221 | 32 Cl |
| D017, U233 | 2,4,5-TP | $C_9H_7Cl_3O_3$ | 269.5 | 40 Cl |
| U232 | 2,4,5-T | $C_8H_6Cl_3O_3$ | 255.5 | 42 Cl |
| U060 | DDD | $C_{14}H_{10}Cl_4$ | 320.1 | 44 Cl |
| U082 | 2,6-Dichlorophenol | $C_6H_4Cl_2O$ | 163 | 44 Cl |
| U081 | 2,4-Dichlorophenol | $C_6H_4Cl_2O$ | 163 | 44 Cl |
| U061 | DDT | $C_{14}H_9Cl^5$ | 354.5 | 50 Cl |
| U132 | Hexachlorophene | $C_{13}H_6Cl_6O_2$ | 406.9 | 52 Cl |
| P050 | Endosulfan | $C_9H_6Cl_6O_3S$ | 406.9 | 52 Cl |
| U231 | 2,4,6-Trichlorophenol | $C_6H_3Cl_3O$ | 197.5 | 54 Cl |
| U230 | 2,4,5-Trichlorophenol | $C_6H_3Cl_3O$ | 197.5 | 54 Cl |
| P037 | Dieldrin | $C_{12}H_8Cl_6O$ | 380.9 | 56 Cl |
| D012, P051 | Endrin | $C_{12}H_8Cl_6O$ | 380.9 | 56 Cl |
| P060 | Isodrin | $C_{12}H_8Cl_6$ | 365 | 58 Cl |
| P004 | Aldrin | $C_{12}H_8Cl_6$ | 365 | 58 Cl |
| U185 | Pentachloronitrobenzene | $C_6Cl_5NO_2$ | 295.4 | 60 Cl |
| U212 | 2,3,4,5-Tetrachlorophenol | $C_6H_2Cl_4O$ | 231.9 | 61 Cl |
| U207 | 1,2,4,5-Tetrachlorobenzene | $C_6H_2Cl_4$ | 215.9 | 66 Cl |
| P059 | Heptachlor | $C_{10}H_5Cl_7$ | 373.4 | 67 Cl |
| P090, U242 | Pentachlorophenol | $C_6HCl_5O$ | 266.4 | 67 Cl |
| U036 | Chlordane | $C_{10}H_6Cl_8$ | 409.8 | 69 Cl |
| D015, P123, U224 | Toxaphene | $C_{10}H_{10}Cl_8$ | 413.8 | 69 Cl |
| U183 | Pentachlorobenzene | $C_6HCl_5$ | 250.3 | 71 Cl |
| U142 | Kepone | $C_{10}Cl_{10}O$ | 490.7 | 72 Cl |
| D013, U129 | Lindane | $C_6H_6Cl_6$ | 290.9 | 73 Cl |
| U127 | Hexachlorobenzene | $C_6Cl_6$ | 284.8 | 75 Cl |

TABLE 2.5.    WASTE QUANTITY DATA FOR HALOGENATED PROCESS WASTES

| | National Survey data base[8] | W-E-T Model/ICF[3] | MITRE[4] | MITRE[5] | JRB[6] |
|---|---|---|---|---|---|
| | (Waste Quantity from the Following Sources (MT/YR) | | | | |
| K001 | 10,980 | 35,700 | 86,709 | 56 | – |
| K009 | | 399,500 | 565,000 | 45 | – |
| K010 | | 26,300 | 25,000 | 9 | – |
| K015 | | 8,380 | 50 | – | – |
| K016 | | 1,600 | 3,200 | 3,628 | 23,817 |
| K017 | | 6,360 | 8,599 | – | 7,403,100 |
| K018 | | 35,400 | 35,000 | 2,834 | 6,858 |
| K019 | | 80,300 | 8,192 | 15,900 | 22,499 |
| K020 | | 52,700 | 125,795 | 2,504 | 27,515 |
| K021 | | 270 | – | – | – |
| K028 | | 610 | 580 | – | 1,201 |
| K029 | | 1,300 | 1,240 | – | – |
| K030 | | 48,400 | 61,299 | – | – |
| K032 | 43,000 | – | – | – | – |
| K033 | | – | – | 14 | – |
| K034 | | – | – | 3 | – |
| K035 | | – | 27,000 – 52,000 | – | – |
| K041 | | 5,000 | 17,000 | – | – |
| K042 | | – | – | – | – |
| K043 | | – | – | – | – |
| K073 | | 340 | 125,000 | 1,361 | – |
| K085 | | 4,500 | 6,615 | 57 | 435 |
| K095 | | 35,500 | – | – | – |
| K096 | | 3,200 | 5,090 | – | – |
| K097 | | – | – | – | – |
| K098 | | – | – | – | 13,377 |
| K099 | | – | – | – | – |
| K105 | | 520 | – | – | – |
| D012 | | – | – | – | – |
| D013 | | – | – | 2,268 | – |
| D014 | 14,435 | – | – | – | – |
| D015 | | – | – | – | – |
| D016 | | – | – | – | – |
| D017 | | – | – | – | – |

Source:    Adapted from Reference 2.

The National Survey data are roughly comparable to more recent data developed by Westat for the Office of Solid Waste[9] and reported in a recent review of treatment technologies for nonsolvent halogenated organics.[10] Westat estimated that the maximum quantity of various subgroups of halogenated organic wastes generated is as follows:

| Halogenated Organic subgroup | Estimated maximum quantity generated[9] ($10^6$ gal/yr) |
|---|---|
| Pesticides (D wastes) | 7.6 |
| Specific processes (K wastes) | 12.5 |
| Single constituents (U and P wastes) | 4.1 |
| Total: | 24.2 |

National survey estimates of treatment process utilization for nonsolvent halogenated organic wastes is 31.2 million gallons of halogenated organic wastes per year. The estimate compares well with the above estimate of 24.2 million gallons generated per year, recognizing that some double counting exist for wastes that are handled in multiple processes. An estimated 3.1 million gallons (roughly 10 percent) of the 31.2 million gallons treated are land disposed.[9] A summary of existing waste treatment technologies that could be applied to the treatment of these land disposed wastes is provided in Table 2.6. As indicated, the applicability of treatment technologies was determined largely by the physical form of the wastes. Other factors related to constituents properties and cost will play roles in selecting the best treatment for a specific waste.

Despite uncertainties in the data base, it is important to note that the quantities of halogenated organic wastes generated and managed are not large. Halogenated solvent waste generation was estimated at about 2,600 million gallons/year,[1] roughly 100 times greater than the 31 million gallons/year estimate for nonsolvent halogenated organic wastes provided by Westat.[9]

TABLE 2.6. SUMMARY OF EXISTING WASTE TREATMENT TECHNOLOGIES

| Waste category | Land disposed waste volume (gal/yr) | Existing treatment technology | Comment |
|---|---|---|---|
| High chlorine content KXXX wastes | 1,673,977 (liquid) | Liquid injection incineration/waste blending/caustic scrubbing | ~4,000 Btu/lb |
| | 612,291 (solid) | Rotary kiln incineration | ~4,000 Btu/lb |
| Halogenated aqueous KXXX wastes | 0 | Filtration/steam stripping/carbon adsorption | |
| Halogenated aqueous sludge KXXX wastes | 23,970 | Waste blending/liquid injection incineration | ~4,000 Btu/lb |
| | | Rotary kiln incineration | ~4,000 Btu/lb |
| Halogenated high inorganic KXXX liquid wastes | 128 | Rotary kiln incineration with high efficiency scrubber | ~1,000 Btu/lb |
| | | Solidification/land disposal | |
| Halogenated potential gases | 0 | Liquid injection incineration/caustic scrubbing | Unknown Btu content |
| Halogenated potential solids | 68,216 | Rotary kiln incineration with caustic scrubbing | Assumed ~4,000 Btu/lb |
| Other halogenated organics with inorganic solids | 759,274 | Rotary kiln incineration with caustic scrubbing | Assumed ~1,000 Btu/lb |
| Total | 3,137,860 | | |

Source:  Reference 1.

REFERENCES

1.   Breton, Marc, et al.  Technical Resource Document; Treatment Technologies
     for Solvent-Containing Wastes.  Prepared for U.S. EPA, HWERL, Cincinnati,
     Ohio under Contract No. 68-03-3243, Work Assignment No. 2.  August 1986.

2.   Arienti, Mark, et al.  Technical Assessment of Treatment Alternatives for
     Wastes Containing Halogenated Organics.  Prepared for U.S. EPA, OSW,
     Washington, D.C. under Contract No. 68-01-6871, Work Assignment No. 9.
     October 1984.

3.   ICF, Incorporated.  The RCRA Risk-cost Analysis Model Phase III Report.
     Submitted to the U.S. EPA, Office of Solid Waste, Economic Analysis
     Branch.  March 1, 1984.

4.   MITRE Corporation.  Composition of Selected Hazardous Waste Streams.
     Working Paper WP-81W00465.  November 1981.

5.   MITRE Corporation.  Composition of Hazardous Waste Streams Currently
     Incinerated.  Working Paper WP-8300065.  April 1981.

6.   Letter Report from J. Harris, JRB Associates, to M. Scott, ENVIRON.
     May 21, 1984.

7.   Roeck, D., et al.  Assessment of Wastes Containing Halogenated Organic
     Compounds and Current Disposal Practices.  Report prepared for U.S. EPA,
     OSW, Washington, D.C. under Contract No. 68-01-6871, Work Assignment
     No. 2.  October 1984.

8.   Westat.  National Survey of hazardous Waste Generators and Treatment
     Storage and Disposal Facilities Regulated Under RCRA in 1981.  Office of
     Solid Waste, U.S. EPA.  1983.

9.   Dietz, S., et al.  National Survey of Hazardous Waste Generators and
     Treatment, Storage and Disposal Facilities Regulated under RCRA in 1981.
     Prepared by Westat, Inc. for U.S. EPA, Office of Solid Waste.  April 1984.

10.  Turner, R. J.  Treatment Technologies for Hazardous Wastes, Part V:
     Nonsolvent Halogenated Organics JAPCA.  June 1986.

11.  Allen, C. C., and B. L. Blaney.  Techniques for Treating Hazardous Waste
     to Remove Volatile Organic Constituents.  JAPCA, Vol. 35, No. 8.
     August 1985.

12.   Blaney, B. L.   Alternative Techniques for Managing Solvent Wastes.
      Journal of the Air Pollution Control Association, 36(3):   275-285.
      March 1986.

13.   Engineering-Science.   Supplemental Report on the Technical Assessment of
      Treatment Alternatives for Waste Solvents.   Washington, D.C.:   U.S.
      Environmental Protection Agency.   1985.

# 3. Pretreatment

Most of the halogenated organic waste streams require some sort of pretreatment before they are introduced to the final treatment process. Pretreatment is needed to modify restrictive waste stream characteristics that affect process performance. Generally some sort of phase separation will be required to remove either solid materials that can adversely affect process efficiency or operation (e.g., through plugging of fuel atomization nozzles or packed bed adsorption towers) or an aqueous phase that can drastically lower waste stream Btu values.

Frequently, phase separation permits a significant volume reduction, particularly if the hazardous component is present to a significant extent in only one of the phases. Furthermore, by concentrating the hazardous portion of the stream, sequential processing steps may be accomplished more readily. Phase-separation processes usually are mechanical, inexpensive and simple, and can be applied to a broad spectrum of wastes and waste components.

Emulsions are generally very difficult to separate. Heating, cooling, change of pH, salting out, centrifugation, API separators, and other techniques may all be tried, but there is no accurate way to predict separation. Appropriate methods can only be developed empirically for any given waste stream.[1]

Conceptually, the simplest separation process is sedimentation, or gravity settling. The output streams will consist of a sludge and a decantable supernatant liquid possibly containing both organic and aqueous fractions. A closely related process and the phase separation technique in most common use is filtration. Centrifugation is essentially a high gravity sedimentation process whereby centrifugal forces are used to increase the rate of particle settling.

The basic concept in all the above processes for settable slurries is to get the solid phases or water to separate from the organic phase, through the use of gravitational, centrifugal or hydrostatic forces. Such forces generally do not act on colloidal suspended particles.

The simplest and most commonly used colloidal separation process is flocculation. Ultrafiltration, another possible separation technology, has many industrial applications, but has yet to demonstrate its full potential.

The major phase separation desired in the handling of sludges is dewatering. Vacuum or press filtration are the processes in most common use. Some research has been done on simple freezing, but the process is not well developed and the work that has been done is not promising.

Sludges and slurries (colloidal or separable) in which the liquid phase is volatile may be treated by either evaporation or distillation. Solar evaporation is very commonly used, however, the impact of the emission of volatile organics to the atmostphere should be considered. Engineered evaporation or distillation systems would normally be operated if recovery of the liquid is desired.

In the case of halogenated organic waste treated by incineration or other thermal destruction technologies, blending the waste stream with other organic compounds may be required to adjust heat content or reduce halogen content to meet a prescribed specification. Blending and/or heating may also be required to adjust viscosity to ensure proper atomization of a waste treated by liquid injection incineration.

In addition to phase separation and/or modification of the waste properties through mixing or heating, some sort of treatment may be required to separate components. Component separation can be achieved physically by, for example, distillation of volatiles or use of solvent extraction processes, or chemically by neutralization or precipitation. Pretreatment options are numerous and must be tailored to the waste stream and the process used for final treatment.

Extensive descriptions of pretreatment technologies (e.g., demulsification, filtration, centrifugation, sedimentation, flotation, evaporation, size reduction, neutralization and precipitation) can be found in the technical literature, including references cited in the following discussions of recovery/treatment processes. These discussions of alternative technologies have been structured to include considerations of pretreatment requirements.

Although pretreatment processes are generally simple, low cost operations, their impact must be considered in any assessment and ultimate selection of alternative technologies.

## REFERENCES

1.  Berkowitz, J. B., et al.  Unit Operations for Treatment of Hazardous Industrial Wastes.  Noyes Data Corporation; Park Ridge, New Jersey.  1978.

# 4. Waste Minimization Processes and Practices

Waste minimization consists of two distinct aspects of hazardous waste management: source reduction and recycling/reuse. Source reduction refers to preventive measures taken to reduce the volume or toxicity of hazardous wastes generated at a facility; recycling/reuse refers to procedures and processes aimed at the recovery of generated wastes or their reuse, e.g., as a fuel.

Very little is known about the extent of waste minimization practices undertaken by generators or reprocessors of halogenated organic waste streams. However, despite the fact that data sources generally fail to draw a distinction between halogenated solvents and halogenated organic compounds, it is unlikely that halogenated organic wastes are reprocessed offsite to the extent that are halogenated solvent wastes. This can be attributed to two reasons. First, many of the halogenated organics are higher molecular weight materials that are not as amenable to recovery by distillation or stripping practices as are the lower molecular weight halogenated solvents. Second, far less halogenated organic wastes are generated than are halogenated solvents, and as in the case of pesticides, generation often occurs in a dispersive pattern. Thus, waste collection and establishment of processing routines is more difficult. Onsite waste recovery activities are also probably less frequent for the halogenated organic industry than for the solvent sector. As noted, recovery is generally more difficult. Incineration of distillation bottoms containing many of the high molecular weight halogenated organics would appear to be the primary method of disposal, especially if the bottoms are not suitable for process recovery.

## 4.1   SOURCE REDUCTION

Source reduction is defined as any onsite activity which reduces the
volume and/or hazard of waste generated at a facility.  Source reduction
represents a preventive approach to hazardous waste management, since the
reduction of hazardous waste volume reduces problems associated with waste
handling, treatment, disposal, and liability.  Source reduction practices may
impact all aspects of industrial processes generating hazardous wastes,
including raw materials, equipment, and products.  A primary motivation for
plants to implement certain source reduction practices is the potential
economic benefit they may accrue.  These economic benefits increase as
restrictions on waste management practices become more stringent.

Waste source reduction practices vary widely from plant to plant,
reflecting the variability of industrial processes and waste characteristics.
In general, source reduction practices may be classified as follows:

- Raw material substitution;

- Product reformulation;

- Process redesign/modification; and

- Waste segregation.

The extent to which these activities are practiced is unknown, although
certain mandatory actions resulting from regulatory restrictions or outright
bans on the use of certain pesticides and other halogenated compound
intermediates have occured within the halogenated organic chemical process
industries.  Examples of waste minimization practices also include equipment
modernization, e.g., use of new distillation equipment to maximize overhead
and bottom product separation.  Another approach to source reduction based on
waste segregation has been reported by the North Carolina Department of
Natural Resources and Community Development.[1]  A North Carolina firm
formulating pesticide mixtures was able to reduce their hazardous waste
generations by over 45,000 pounds per year by installing bag houses on each
product line and returning collected dust to the product line as raw material
feed.  The company realized a 10 month payback period on an initial investment
of $96,000 for each baghouse.

Review of documented case studies on source reduction indicates that these practices have been applied in more instances to solvent wastes than any other waste type.  Additionally, it appears that source reduction practices are used more frequently for chlorinated solvents, especially 1,1,1-trichloroethane and methylene chloride, although similar practices may be applicable to other halogenated organics.  With respect to cost savings, the data appear to indicate that source reduction of large generating sources may yield annual savings of tens of thousands of dollars.  Savings in the hundreds of thousands of dollar may even be possible if source reduction practices eliminate the need for unit operations such as air pollution controls.

Regulatory trends appear to be moving toward the promotion of source reduction at sites generating hazardous wastes.  The EPA has recently proposed requirements that generators certify institution of hazardous waste reduction programs (51 FR 10177, March 24, 1986).  This would involve the institution at generator sites of programs to reduce the volume or toxicity of hazardous wastes to a degree determined by the generator to be economically practicable.  Generators must also certify that their current method of management is the most practicable method available to minimize present and future threats to human health and the environment.

Three states currently have established source reduction/pollution prevention programs:  North Carolina, Minnesota, and Massachusetts.  In addition, Tennessee has established a "pilot program", and Kentucky, California, Maryland, and Washington have programs currently in development.  These programs vary from State to State but, in general, include information exchange, technical assistance, and economic incentives to companies to encourage development of their program.

The reader is referred to other EPA and state publications[2-13] and the solvent TRD (Section 5)[14] which address approaches to source reduction.  Although most of the literature is focused on solvent reduction, the same general technological principles will apply.

## 4.2  RECYCLING/REUSE

According to EPA guidance issued on January 4, 1985, "recycling" was defined as practices in which wastes are:  1) reclaimed, or 2) reused.  A reclaimed waste is one which is processed or treated through some means to purify it for subsequent reuse, or to recover specific constituents for reuse. Reused wastes are those which serve directly as feedstocks, without any treatment. Recycling of wastes may be done by either the original generator or other firms.

The recovery of halogenated organic constituents from their waste materials can be accomplished in several ways. The preferred recovery method is determined by both constituent and waste matrix characteristics. Volatile materials can be recovered by evaporation/fractionation processes. Solids can be recovered as residues from evaporative processes. Organics can be selectively extracted from liquids (generally aqueous) or solids by an organic solvent. Very little is known about the extent of recovery practices as applied to halogenated organic waste streams. However, an examination of available data for the specific K type waste streams shown previously in Table 2.3 indicate that constituent concentrations are high enough to warrant application of recovery processes. Many of the components are low molecular weight halogenated organic solvents which can be volatilized leaving the higher molecular weight components as principal constituents of the distillation bottoms. Conceivably, some of these bottom constituents are suitable for recycle. In other cases, further evaporation or destruction by incineration may be preferred.

The use of recovery processes is generally highly dependent on the value of the products which can be recovered or produced. For example, the viability of the chlorinolysis process is highly dependent on the demand for its product, carbon tetrachloride. If the demand for this material is low, then the price will probably be low and the process will not be economically viable.

In addition, recovery processes are dependent on the feasibility of disposing of the waste material. In the past, when land disposal of wastes was relatively inexpensive, recovery processes were rarely considered. However, with increasing land disposal costs and a potential ban on land disposal of some wastes, recovery processes become economically feasible and/or necessary.

Distillation, extraction, and steam stripping are possible unit recovery processes. Conceptual systems designs which utilize these unit processes are described below and summarized in Table 4.1 for some K type halogenated organic wastes.[15]

TABLE 4.1.   RECOVERY PROCESSES SUMMARY

| Waste code | Waste stream in | Recovery process | Recovered products | Waste stream out | Cost (1977 cost) |
|---|---|---|---|---|---|
| KO30 | Hexachlorobutadiene 77% <br> Chlorobenzenes 7% <br> Chloroethanes 3% <br> Chlorobutadiene 3% <br> Tars and others 10% <br><br> 12,000 KKG/yr | (1) Steam stripper <br><br> (2) Steam stripper | Chlorobenzenes } <br> Chloroethanes } <br> Chlorobutadienes } <br> 1560 KKg/yr recycled to process <br><br> Hexachlorobutadiene 9240 KKg/yr to be sold | Tars <br> Others 1200 KKg/yr <br><br> Sent to landfill | Capital cost: $731,900 <br><br> Net annual cost: -$4,536,600 |
| KO17 | Epichlorohydrin 80 KKg/yr <br> Dichlorohydrin 440 KKg/yr <br> Chloroethers 560 KKg/yr <br> Trichloropropane 2800 KKg/yr <br> Tars, Resins, and Others 120 KKg/yr | (1) Solvent extraction <br><br> (2) Evaporation <br><br> (3) Distillation | Epichlorohydrin 80 KKg/yr <br><br> Dichlorohydrin 440 KKg/yr To process <br><br> Trichloropropane 2470 KKg/yr To process | Trichloropropane 330 KKg/yr <br> Chloroethers 560 KKg/yr <br> Others 120 KKg/yr <br><br> Chemical land disposal | Capital cost: $243,500 <br><br> Net annual cost: $2,000 |
| KO19 | 1,2-dichloroethane 23% <br> 1,1,2-trichloroethane 38% <br> 1,1,2-tetrachloroethane 38% <br> Tars 1% <br><br> 1,400 KKg/yr | (1) Distillation <br><br> (2) Dehydrochlorination | Ethylene dichloride 284 KKg/yr To process <br><br> 1,1-dichloroethylene 310 KKg/yr Storage <br><br> 1,1,2-trichloroethylene 333 KKg/yr Storage | Calcium hydroxide 122 KKg/yr <br> Tars 14 KKg/yr <br> Water 1422 KKg/yr <br> Calcium chloride 318 KKg/yr <br> Other 249 KKg/yr | Capital cost: $223,400 <br><br> Net annual cost: -$1,200 |

(continued)

## TABLE 4.1 (continued)

| Waste code | Waste stream in | Recovery process | Recovered products | Waste stream out | Cost (1977 cost) |
|---|---|---|---|---|---|
| K021 | Antimony penta-chloride (SbCl$_5$) 18.0 KKg/yr<br>CCl$_4$ 4.9 KKg/yr<br>Organics 0.8 KKg/yr | (1) Dechlorination utilizing 7.9 KKg/yr of ethylene trichloride | Antimony pentachloride 18.0 KKg/yr | Tars 3.1 KKg/yr<br>Chemical landfill | Capital cost: $368,000<br><br>Net annual cost: $100,000 |
| | | (2) Filtration | CCl$_4$ 4.4 KKg/yr<br>To process | | |
| | | (3) Chlorination utilizing 1.3 KKg/yr of antimony trichloride and 4.7 KKg/yr of chlorine | Ethylene pentachloride 12.1 KKg/yr<br>Storage | | |
| | | (1) ⎫<br>(2) ⎬ Distillation<br>(3) ⎭ columns | Freons – To process<br>CCl$_4$ 4.9 KKg/yr<br>To process<br>SbCl$_5$ 18.0 KKg/yr<br>To process | Tars 0.08 KKg/yr<br>To chemical landfill | Capital cost: $24,000<br><br>Net annual cost: $8,450 |

Source: Reference 15.

## Alternative Treatment Process For K030 Waste--Heavy Ends from the Production of Perchloroethylene

This recovery scheme involves steam stripping of the waste stream to separate out the light ends which consist of chlorobenzenes, chloroethanes, and chlorobutadienes. These light ends are then recycled as feed stock components to the perchloroethylene chlorinator.

The bottoms from this stream stripping operation consist mainly of hexachlorobutadiene along with tars and other heavy materials. This stream would subsequently be fractionated in a second stripping column to produce a pure hexachlorobutadiene stream overhead while the bottoms consist of tarry materials. The recovered hexachlorobutadiene could be sold for use as a liquid phase chlorination medium, and the tarry material would have to be incinerated.

Assuming a throughput of 12,000 KKg of heavy ends per year, 9,240 KKg of hexachlorobutadiene could be recovered and sold. In addition, more than 1,500 KKg/yr of "lights" are recovered and can be recycled into the production process.

## Alternative Treatment Process for K017 Waste--Heavy Ends from the Purification Column in the Production of Perchloroethylene

This recovery process involves extracting epichlorohydrin and dichlorohydrin from the waste stream using water and recycling the dried epichlorohydrin and dichlorohydrin back to the process.

The raffinate from the extractor would contain mainly trichloropropane and chloroethers. This stream could be fractionated to recover trichloropropane overhead. The bottoms would contain some trichloropropane and the chloroethers. This stream would have to be disposed of by incineration.

For a waste volume of 4,000 metric tons per year, this treatment scheme should be able to reduce the amount of waste material to 1,000 metric tons. Assuming 80 percent effectiveness of the treatment process, 315 KKg of epichlorohydrin and 2,000 KKg of trichloropropane would be recovered.

## Alternative Treatment Process For K019 Waste--Heavy Ends from the Distillation of Ethylene Dichloride in Ethylene Dichloride Production

This recovery process involves distillation of the ethylene dichloride remaining in the heavy ends using a thin film evaporator. The distillate, consisting mainly of ethylene dichloride, would be returned to the process. Other recovery steps involve dehydrochlorination with a slurry of calcium hydroxide. The bottoms from the distillation column would contain 1,1,2-trichloroethane and 1,1,1,2-tetrachloroethane. In the first dehydrochlorination reactor, the trichloroethane would be converted to 1,1-dichloroethylene at a temperature of 50°C. In the second dehydrochlorination reactor, the tetrachloroethane is converted to 1,1,2-trichloroethylene at 100°C.

A typical plant producing 136,000 MT per year of vinyl chloride monomer would generate 1,400 KKg/yr of ethylene dichloride still bottoms. Utilizing the recovery scheme detailed above, it would be possible to recover over 300 metric tons of 1,1-dichloroethylene, which would be used in the production of copolymeric materials and 1,1,1-trichloroethane. In addition, 280 metric tons of ethylene dichloride and 330 metric tons of trichloroethylene would be recovered.

## Alternative Treatment Process for K021 Waste--Spent Reactor Catalyst from Fluorocarbon Manufacture

This waste stream consists largely of antimony pentachloride ($SbCl_5$) contaminated with carbon tetrachloride and other organics, as shown in Table 4.1. In the past, this waste has been landfilled. Two processes for recovery of the catalyst have been proposed. The first one involves dechlorinating the antimony pentachloride in the presence of ethylene trichloride, resulting in the precipitation of antimony trichloride. The reaction is:

$$SbCl_5 + EtCl_3 \longrightarrow SbCl_3 + EtCl_5$$

Subsequently, the $SbCl_3$ is filtered off and rechlorinated to form antimony pentachloride which is returned to the fluorocarbon process. The filtrate is separated by distillation into carbon tetrachloride, ethylene trichloride, ethylene pentachloride and still bottoms. Each of these materials can be sold or reused in the process except for the still bottoms. These would be drummed and treated.

The second method for recovery of the spent catalyst involves distillation of the waste stream into several components: light ends (predominately freons), carbon tetrachloride, antimony pentachloride, and still bottoms. The antimony pentachloride is recovered and returned to the hydrofluorinator, and the lights and the carbon tetrachloride are reused in the fluorocarbon manufacturing process.

## 4.3  WASTE EXCHANGE

Reuse of wastes may be accomplished either by the generator itself, or through sales to a commercial processor. Marketing of wastes for reuse is often facilitated through use of waste exchanges. Waste exchanges are institutions which serve as brokers of wastes or clearing houses for information on wastes available for reuse. In some waste exchanges, potential buyers of wastes are brought into contact with generators, while other waste exchanges accept or purchase wastes from a generator for sales to other users. Waste exchanges are considered by EPA to be of great potential value in future waste management because, through waste exchanges, recycling practices may be increased.

In general, the "exchangeability" of a waste is enhanced by higher concentration and purity, quantity, availability, and higher offsetting disposal costs. Some of the limitations to waste exchangeability are the high costs and other difficulties associated with transportation and handling, costs of purification or pretreatment required, and, in certain cases, the effect on process or product confidentiality. In general, waste exchange involves transfer of products from large, continuous processors to small, batch processors, or products from high purity processors (e.g., pharmaceutical manufacturers) to low purity processors (e.g., paint manufacturers).

Waste exchanges are operated by both private firms and public organizations.  Several public waste exchanges are listed below:

- California Waste Exchange (California);

- Canadian Waste Materials Exchange (Ontario);

- Chemical Recycle Information Program (Texas);

- Colorado Waste Exchange (Colorado);

- Georgia Waste Exchange (Georgia);

- Great Lakes Regional Waste Exchange (Michigan);

- Industrial Materials Exchange Service (Illinois);

- Industrial Waste Information Exchange (New Jersey);

- Inter-Mountain Waste Exchange (Utah);

- Louisville Area Waste Exchange (Kentucky);

- Midwest Industrial Waste Exchange (Missouri);

- Montana Industrial Waste Exchange (Montana);

- Northeast Industrial Waste Exchange (New York);

- Piedmont Waste Exchange (North Carolina);

- Southern Waste Information Exchange (Florida);

- Techrad (Oklahoma);

- Tennessee Waste Exchange (Tennessee);

- Virginia Waste Exchange (Virginia);

- Western Waste Exchange (Arizona); and

- World Association for Safe Transfer and Exchange (Connecticut).

The following is a list of the private material exchanges currently in business:

- Zero Waste Systems, Inc. (California);

- ICM Chemical Corporation (Florida);

- Environmental Clearinghouse Organization - ECHO (Illinois);

- American Chemical Exchange - ACE (Illinois);

- Peck Environmental Laboratory, Inc. (Maine);

- New England Materials Exchange (New Hampshire);

- Alkem, Inc. (New Jersey);

- Enkarn Research Corporation (New York);

- Ohio Resource Exchange - ORE (Ohio); and

- Union Carbide Corporation (in-house operation only; West Virginia).

## REFERENCES

1.  North Carolina Department of Natural Resources and Community
    Development.  Accomplishments of N.C. Industries, Case Summaries.
    January 1986.

2.  Minnesota Waste Management Board.  Hazardous Waste Management Report.
    1983.

3.  Huisingh, D., et al.  Proven Profit from Pollution Prevention.
    Conference draft.  The Institute for Local Self-Reliance, Washington,
    DC.  July 1985.

4.  Roeck, D.R., et al.  GCA Technology Division, Inc.  Hazardous Waste
    Generation and Source Reduction in Massachusetts.  Bedford, MA.  Contract
    No. 84-198, MA Dept. of Env. Mgt., Bureau of Solid Waste Disposal, June
    1985 (Draft).

5.  Kohl, Jerome, P. Moses, and B. Triplett.  Managing and Recycling
    Solvents:  North Carolina Practices, Facilities and Regulation.
    Industrial Extension Services, School of Engineering, North Carolina
    State University, Raleigh, NC.  December 1984.

6.  Versar, Inc. National Profiles for Recycling/A Preliminary Assessment,
    Draft.  EPA Contract No.  68-01-7053, U.S. EPA Waste Treatment Branch.
    July 1985.

7.  Hobbs, B., and R.R. Hall, GCA Technology Division, Inc. Study of Solvent
    Reprocessors.  Bedford, MA.  EPA Contract No.  68-01-5960 (Draft), U.S.
    EPA, Office of Chemical Control, January 1982.

8.  Radminsky, Jan, et al.  Department of Health Services, Alternative
    Technology and Policy Development Section.  Alternative Technology for
    Recycling and Treatment of Hazardous Wastes, Second Biennial Report.
    California, July 1984.

9.  Engineering-Science.  Supplemental Report on the Technical Assessment of
    Treatment Alternatives for Waste Solvents.  Washington, DC. U.S.
    Environmental Protection Agency.  1985.

10. Noll, K.D., et al., Illinois Institute of Technology.  Recovery, Reuse
    and Recycle of Industrial Waste.  Chicago, IL.  EPA-600/2-83-114, U.S.
    EPA/ORD, Cincinnati, OH.  November 1983.

11.  Tierney, D.R., and T.W. Hughes, Monsanto Research Corp.  Source
     Assessment:  Reclaiming of Waste Solvents, State of the Art.
     EPA-600/2-78-004f, 66 pp.  April 1978.

12.  Center for Environmental Management, Tufts University, Medford, MA.
     Waste Reduction:  The Ongoing Saga.  Conference, League of Women Voters
     of Massachusetts and U.S. EPA.  Woods Hole, MA.  June 4-6, 1986.

13.  Center for Environmental Management, Tufts University, Medford, MA.
     Waste Reduction:  The Untold Story.  Conference, League of Women Voters
     of Massachusetts and U.S. EPA.  Woods Hole, MA.  June 19-21, 1985.

14.  Breton, M. et al., Technical Resources Document.  Solvent-Bearing
     Wastes.  Final Report submitted by GCA Technology Treatment Technologies
     for Division to EPA, HWERL, Cincinnati, Ohio under Contract
     No. 68-03-3243, Work Assignment No. 2, September 1986.

15.  Arienti, M., et al.  Technical Assessment of Treatment Alternatives for
     Wastes Containing Halogenated Organics.  Report prepared for U.S. EPA,
     OSW, under Contract No. 68-01-6871, Work Assignment No. 9.  October 1984.

# 5. Physical Treatment Technologies

The physical treatment processes discussed in this halogenated organic waste TRD are:

5.1 Distillation,

5.2 Evaporation (thin film evaporators),

5.3 Steam Stripping,

5.4 Liquid-Liquid Extraction,

5.5 Carbon Adsorption, and

5.6 Resin Adsorption.

These physical processes generally do not result in the destruction of waste constituents and provide opportunities for recovery of valuable components. The first three processes are largely dependent on the volatility of the constituents to affect evaporation from the waste matrix. These technologies are considered more applicable to halogenated solvents which are generally lower in molecular weight and more volatile than the halogenated organic compounds considered in this document. However, as noted in Section 2, many of the K type waste streams considered here contain halogenated solvents in addition to other halogenated organic constituents. Thus, application of technologies such as distillation, evaporation, and steam stripping will produce an overhead fraction consisting of low molecular weight halogenated solvents and a bottom fraction consisting of higher molecular weight organics.

Liquid-liquid extraction (solvent extraction) is widely used in the chemical process industry but has not yet been extensively employed for treatment of hazardous wastes.  It may possibly be utilized to advantage to separate components that can not be separated by processes based on differential volatilization.  Extraction processes are applicable to both aqueous and organic matrices, although partition coefficients are greatest for aqueous/organic combinations.

Adsorption is highly applicable to most of the high molecular weight halogenated organic compounds considered in this TRD.  It is usually used to remove small quantities of organic contaminants from aqueous waste streams although it is sometimes used as a pretreatment followed by a biological finishing process.

Some level of residual contamination of the waste stream(s) can be expected for the other physical treatment processes.  These processes will require subsequent post-treatment (e.g., incineration of distillation bottoms, adsorption/biological treatment of steam stripper aqueous condensate) in order to meet surface water discharge (e.g., NPDES, POTW) or land disposal requirements.

The physical treatment processes (and the other treatment processes discussed in following sections) are considered within the framework of four major areas; i.e., 1) process description including pretreatment and post-treatment  requirements, 2) demonstrated performance in field and laboratory, 3) cost of treatment, and 4) overall status of the technology. Discussions of the various types of treatment processes applicable to halogenated organic waste streams are contained in Sections 5.0 through 10.0. These are followed by a review section (Section 11.0) that addresses approaches to identifying potential treatment processes or process combinations that are likely to meet treatment requirements in the most cost effective manner.

5.1  DISTILLATION

Distillation is primarily a recovery operation that is most applicable to the low boiling organic components addressed in the solvent TRD.  However, distillation is also applied to other segments of the organic chemical industry, including halogenated organic compound waste streams of interest to this TRD.  As noted in Section 2, many of these specific waste streams are K type wastes which contain both halogenated organics and halogenated solvents.  Depending upon their relative volatilities, these mixed wastes can be recovered as overhead constituents or as overhead and bottom fractions. For the specific K type waste streams shown in Section 2, the latter case is most likely.  As discussed in the examples of recovery processes for selected K type waste streams (Section 4), the bottoms stream can be further purified by a subsequent fractionation, or used directly as feedstock to a production process.  Lacking value as a feedstock raw material, incineration is usually the preferred ultimate disposal option.

Distillation is often useful as an auxilliary to other treatment processes.  For example, it may be used to separate mixtures of solvent and solute resulting from liquid extraction processes or from wash solvents used to regenerate resin adsorbents.  Because it is such an important process it is discussed in some detail below.  However, the reader is referred to the solvent TRD,[1] Perry's Chemical Engineers' Handbook[2] and other references for additional detail concerning design and performance.[3-10]

5.1.1  Process Description

Distillation is a separation technique that operates on the principle of differential volatility.  More volatile constituents can be enriched or separated by heating and volatilizing from less volatile constituents. Distillation systems fall into one of four general categories.

1.   Batch distillation;

2.   Continuous distillation;

3.   Batch fractionation; and

4.   Continuous fractionation.

Fractionation is distinguished from distillation through the use of multiple distillation steps in tray or packed towers to separate two or more volatile components of the waste feed into distinct fractions.  In batch distillation, the system is charged with a given quantity of a waste (generally a liquid although some solids can be tolerated if minimum heat and mass transfer levels can be achieved) and heated indirectly with steam or oil.  Coils or the vessel wall act as the heat transfer surface.  Heating continues until a predetermined fraction of the volatile components is removed, as indicated by the time/temperature profile of the charge within the still.  At this point, heating is discontinued and the bottom product (i.e., less volatile residues) are removed.

For the halogenated organic K type waste streams, the bottoms may represent the halogenated constituent which can be returned to the process as feedstock.  In other cases, further fractionation may be required to recover a waste product of sufficient quality for reuse or sale.  In many cases, it will be expeditious to maintain a residue that can be handled readily and/or incinerated; i.e., waste that is pumpable and has Btu value.  In any event, it must be recognized that complete removal of volatiles will not be feasible. Attempts to reduce volatile content to low levels can result in compound destruction, equipment fouling and excessive operating costs.  Optimal removal efficiencies must balance the benefits of overhead recovery against the costs of equipment maintenance and the cost benefits of still bottom disposal or reuse.

Continuous distillation functions much the same as batch distillation, except that continuous charging and bottoms removal results in steady-state operation.  Alternatively, waste feed or removal of the bottoms product can take place at specific intervals resulting in semi-continuous operation.  Continuous bottoms removal generally implies that the stream is a pumpable liquid.

In response to economic incentives to recycle and minimize wastes, several small stills have been recently developed which yield high percent recovery as a result of design features such as removable liners for bottoms disposal.  Many can be operated under vacuum with batch bottoms removal to approach nearly 100 percent recovery.  Although aimed at the solvent recovery market, they can be applied to halogenated organic compounds provided the vapor pressure of the compound is sufficient under operating conditions and thermal decomposition is not a factor.

The separation of mixtures requires multiple distillations or fractionation since adequate separation of fluids with similar boiling ranges is not achievable in a single stage.  Fractionation column designs include the use of multiple trays or packing to maximize surface area so that rising vapors are intimately contacted with falling condensate (reflux).  These columns can be operated as batch or continuous units.  Figure 5.1.1 shows basic process schematics for batch and continuous fractionation systems.[11]

Fractional distillation is a very well developed industrial technology that has been studied extensively for many years, particularly by the petrochemical industry.  Vapor-liquid equilibrium data provide the basis for evaluating the feasibility of fractionation.  A McCabe-Thiele diagram, which graphically illustrates the fractionation process, is shown in Figure 5.1.2. The figure shows the minimum number of theoretical column stages required to effect separation; i.e., 100 percent efficiency at total reflux.  In certain cases, relative composition differences between the liquid and vapor phase might be so small as to render the separation economically unfeasible.  The vapor-liquid equilibrium (VLE) data must be evaluated to determine separational feasibility.  It is evident from studying the McCabe-Thiele diagram (Figure 5.1.2) that if the vapor and liquid composition lines are close enough, many equilibrium stages will be required.  Detailed discussions of fractional distillation theory and practice, including the use of well-developed models for predicting separation, can be found in chemical engineering texts such as the section on distillation in Kirk-Othmer's Encyclopedia of Chemical Technology,[10] the literature sources cited therein, and in Perry's Chemical Engineers' Handbook.[2]

In continuous fractionation, feed is constantly charged to the column at a point which provides the specified top and bottoms product.  The section of the tower above the feed point is the rectifying or enriching section, and the section below the feed point is the stripping section.  A reboiler is connected to the bottom of the fractionation tower to provide the reboil heat needed for added reflux and better fractionation of complex mixtures.  Batch fractionation differs in that the charge is introduced at the bottom of the tower.  As a result, it is possible to obtain a distillate of high purity, but recovery of volatile component(s) must proceed in a step-wise manner.  As the more volatile constituent is taken off, the reflux and thus, energy consumption, must be increased to maintain overhead product purity.  Eventually, the

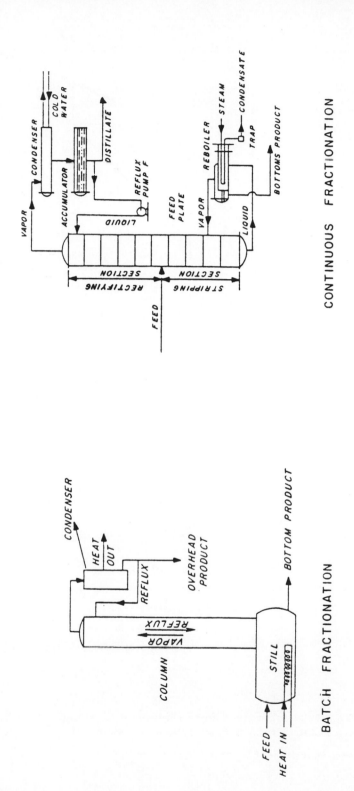

Figure 5.1.1.  Basic schematic for batch and continuous fractionation systems.

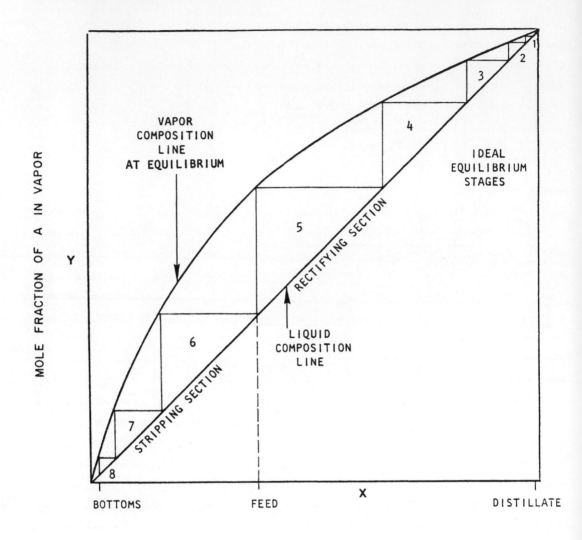

Figure 5.1.2.   McCabe-Thiele diagram for distillation.
(Total reflux = no products taken off.)

Source:   Reference 10.

overhead quality decreases to the point where it must be removed and stored separately until the next most volatile component(s) is enriched enough to generate a useful product.  However, batch fractionation does permit handling of wastes with higher dissolved solids content which would foul the stripping zone surfaces in a continuous feed column.

5.1.1.1  Restrictive Waste Characteristics--

<u>Batch Distillation</u>

Batch distillation equipment has undergone considerable development in recent years, particularly that of interest to the small quantity generator. Almost any liquid waste can be processed in commercially available equipment, including high solids, ignitable, and potentially explosive mixtures.  However, there are some restrictions which affect safety and product purity that should be considered.  Although, from the standpoint of safety, they are less critical for halogenated organics which are generally nonignitable under normal operating conditions.

Typically, at least 40°C between autoignition point and boiling point is required for safe distillation of high purity mixtures.  Autoignition temperatures for most volatile compounds can be found in the National Fire Protection Association's "Fire Protection Guide for Hazardous Materials."[12] Heptane is generally considered to have the lowest autoignition point (399°F) of commonly recovered organics.[9]  For this reason (to minimize risk of explosion), many manufacturers design units which are limited to a maximum temperature of 365°F.

While many small stills are explosive proof, the potential for explosion should be evaluated, especially for highly volatile mixtures.  Autoignition temperatures of ignitable wastes may pose particular handling problems in distillation units.  Some of the particular waste streams addressed in this document can contain constituents with low autoignition points.

Most halogenated organics are nonflammable but are susceptible to thermal decay.  Thermal decay occurs at low temperatures relative to autoignition points.  For low autoignition point and low decomposition temperature compounds, vacuum operation is an option that is available in many of the

currently marketed batch stills.  Additional capital and operating expenses
are at least partially offset by the reduced boiling temperature and higher
potential recovery rates.

Solids content, which is a critical limitation for continuous
distillation and fractionation processes, is less critical for batch
distillation or continuous feed with batch bottoms removal.  In particular,
jacketed and immersion heated boilers with provisions for easy bottoms removal
are capable of achieving high recovery rates.  Residual low boiling components
remain in the bottoms produced from these operations, but potential volatile
component removal efficiencies are usually greater than 90 percent, regardless
of initial solid constituent concentration.  In many cases, the distillation
occurs over a long enough period of time such that potentially reduced mass
diffusion and heat transfer do not greatly reduce the overall effectiveness of
the recovery process.

## Continuous Distillation

Continuous distillation of wastes is subject to the same constraints that
apply to batch distillation in terms of operating temperature, i.e., autoignition temperature, boiling point, and thermal decay.  However, the
distinction between batch and continuous bottoms removal greatly affects the
applicability and achievable performance for certain waste types.

Continuous feed, batch bottoms removal units are essentially the same as
batch units in terms of processing capabilities.  Continuous feed units have
greater capital costs, since they do require some additional control features
(i.e., level control), but are more automated thereby requiring less labor.
These units are preferred for high throughput applications where labor costs
become a higher fraction of total costs than capital equipment outlays.

The continuous feed, continuous bottoms removal units are different in
that the achievable recovery is controlled by the ultimate disposition of the
bottoms.  Continuous bottoms removal implies that the bottoms remain fluid.
In cases where bottoms become highly viscous, ultimate recovery is limited by
equipment processing constraints.  These materials are more optimally
processed in thin-film evaporators which are capable of achieving high
recovery rates and throughputs in a shorter period of time (see Section 5.2).

Scraped-surface stills also increase the ultimate recovery but by a smaller amount. Thus, continuous bottoms recovery stills are best suited to recovering wastes that are non-fouling and that remain pumpable after separation.

## Batch Fractionation

Fractionation is a multi-stage process used for separating volatile mixtures when the value of the pure component products justifies the additional expenses associated with separation. In the case of solvents, pure components can be sold for 80 to 90 percent of the virgin price whereas blends typically sell for only 50 to 60 percent of virgin prices.[13] Batch fractionation can handle a higher solids content waste form relative to continuous fractionation since these materials do not come into contact with the packing or trays. However, the quantity and nature of the solids in the waste may become a limitation depending upon the design of the heat transfer unit. Excessive fouling may interfere with heat transfer resulting in higher energy costs, reduced throughput, and additional labor. Agitated units are available to reduce the potential problems due to fouling.

## Continuous Fractionation

As opposed to batch fractionation, continuous fractionation is reserved for materials which are essentially void of dissolved and suspended solids. The feed enters at a mid-point in the column where it comes into intimate contact with tray or packing surfaces. Labor costs associated with cleaning these units justify pretreatment in either a distillation or evaporation unit. Also, the bottoms product will have to be treated to remove the nonvolatile hazardous constituents.

5.1.1.2 Operating Parameters and Design Criteria--

Distillation and fractionation processes are based on the evaporation and condensation of constituents. Operating parameters of critical importance for all units are process temperature and pressure. Higher operating pressures are routinely used for low boiling point constituents to avoid the use of energy intensive refrigeration to achieve condensation.  This is particularly true for fractionation processes which requires greater reboiler and condenser duties as a result of occurring reflux.  Additionally, distillation units, particularly small batch units capable of processing up to 60 gallons per hour, are often equipped with vacuum capability.  High temperatures combined with low pressure makes high recoveries possible, even for high boiling constituents.

Other parameters that must be considered include batch time, viscosity (flow and mass transfer), reflux ratio, and the location of feed introduction to fractionation columns.  Batch time is chosen based on economics, desired recovery, and restrictive waste characteristics.  Certain units may be susceptible to fouling or viscosity problems.  High viscosity wastes are best treated in agitated thin-film evaporators (see Section 5.2).

Reflux rate and feed tray location are process variables strictly applicable to fractionation, with feed tray designation applicable specifically to continuous fractionation.  Reflux rate or ratio is set based on the economic evaluation of product purity versus utility costs.  High reflux ratios produce higher purity products but are more energy intensive. In addition, the high internal flow rates often establish the need for either larger units, or lower throughputs.  Optimal feed location in a column will be at the point of intersection of the rectifying and stripping operating lines.[2]

The reader is referred to numerous chemical engineering texts, including Perry[2] and Kirk-Othmer[10], for more extensive discussions of distillation equipment operation and design.  The solvent TRD also includes an appendix describing design features and cost of small, commercially available distillation equipment.[9]

5.1.1.3  Treatment Combinations--

Distillation pretreatment options consist of filtration, centrifugation and other physical means to separate solids from the liquid stream, and decanting to separate gross sediment and immiscible fluids.

Post-treatment methods used for overhead product include further refinement through separation of mixtures into pure components.  This is performed to enhance the value of the recovered overhead or to meet product

purity specifications.  Fractionation is usually performed on low molecular
weight organics (e.g., solvents) which have already been separated from
nonvolatile constituents.  It is also a typical regeneration technique for
solvents used in liquid-liquid extraction.  In cases where the distillate
product consists of two phases, decanting is a typical post-treatment
procedure.  If water is soluble in the overhead to an extent which exceeds
product purity specifications, the overhead stream will undergo some form of
drying to remove the water.  Commonly employed drying methods include
molecular sieve, calcium chloride, ionic resin adsorption, and caustic
extraction.

Post-treatment options for bottoms products depend on the physical form
and cost of the material.  Approximately two-thirds of recycled solvent
bottoms generated by commercial recyclers are used as is or blended with
higher Btu products for use as a fuel.[1]  For blends containing expensive
halogenated constituents, there are cases where liquids are added prior to
distillation to promote maximum removal of the halogenated constituents by
keeping the bottoms fluid.  Other post-treatment options consist of further
solid-liquid separation for organic liquids and sludges, organic removal or
extraction processes for aqueous wastes, and thermal destruction techniques.

Solid-liquid separations can be accomplished by physical means such as
centrifugation, filtration, decanting and extraction.  Dilute aqueous wastes
can be treated through air or steam stripping, carbon or resin adsorption, or
biological treatment as described in other sections.

## 5.1.2  Demonstrated Performance

A number of studies have been cited by EPA as demonstrating the
feasibility of solvent recovery by distillation processes.[14]  Although
residual solvent concentrations in treated wastes were always high enough to
require additional treatment, all these studies demonstrated significant
economic benefits.  A discussion of these studies is provided in the solvent
TRD.  Unfortunately, very little information was found in the literature
pertaining to the treatment of nonsolvent halogenated wastes.  However, 13 of
the 27 specific K type wastes represent distillation bottoms.  An examination
of the constituent concentration data previously shown in Table 2.3 would

suggest that further recovery of both halogenated solvents and compounds could be achieved by a subsequent bottoms redistillation or treatment by other technologies.  Four alternative treatment procedures have been identified in Reference 15, three of which identified distillation as a key technology for waste streams K017, K019, K021, and K030.  EPA is undertaking work to further examine alternatives for these and other specific K type waste streams containing halogenated organics.

### 5.1.3  Cost of Treatment

Cost data developed for solvent recovery applications is addressed in the solvent TRD and references cited therein.  As noted in the solvent TRD, the Naval Energy and Environmental Support Activity (NEESA) has assessed the costs of small to moderate sized batch stills and larger continuous stills, with results reported in Reference 16.  NEESA developed the following methodology to evaluate the economics of a solvent reclamation program.

Data Requirements:

D - Cost of distillation unit ($)

E - Recovery efficiency of still (decimal fraction)

I - Cost of installation ($)

M - Cost of additional labor (varies with size and operational requirements of still; assume $0 for under 3,500 gallons per year, $2,500 between 3,500 and 13,000 gallons per year, and $7,500 for over 13,000 gallons per year)

S - Cost of halogenated organic ($/gallon)

U - Cost of utilities in reclaiming halogenated organic ($/gallon)

V - Volume of halogenated waste generated (gallons)

W - Waste disposal costs for halogenated waste and still bottoms ($/gallon)

(A/P:i:n) - Appropriate capital recovery factor to evaluate payback period.

These parameters, modified to represent halogenated organics rather than solvents, may be combined to form an equation representing the parameter interactions at specific payback periods as follows:

$$V[ES+(1-E)W] - (A/P:i:n)(D+I) - UV - M = 0$$

Using this expression it is possible to calculate break-even volumes for various applications and payback periods.

NEESA has estimated the costs of reclamation for various types of solvents, including mineral solvents. The example for mineral solvents is reproduced below to illustrate the magnitude of costs one might expect for the recovery of halogenated organic wastes. Many of these wastes, like mineral spirits, tend to have relatively low cost, low volatility, and high boiling points (mineral spirits boiling range is 150 to 200°C). These factors make cost-effective recovery high volume dependent because of the more expensive reclamation equipment required and the low value of the recovered organic. Payback periods were calculated using the cost equation and the assumptions listed below.

Assumptions:

    Cost of Still (D):   $ 8,600 for under 3,500 gallons per year
            $17,700 for under 3,500-13,000 gallons per year
            $45,000 for 13,000-60,000 gallons per year

    Installation Cost (I):   1.5 D

    Recovery Efficiency (E): 0.95

    Additional Manpower (M): See original assumptions

    Organic Constituent Cost (S): $1.75 per gallon

    Utility Costs (U):  $0.05 per gallon

    Waste Disposal Costs (W):    $2.00 per gallon, $3.00 per gallon

    Discount Factor (i):    10 percent

Due to the high sensitivity to waste disposal costs for the small (15 gpd) and medium (55 gpd) size distillation units, economic analyses were conducted using three disposal costs. For the large distillation unit (250 gpd) where

the waste disposal cost is less influential, the economic analysis was
conducted at the median disposal cost of $2.00 per gallon.  NEESA originally
assumed a $0.25 per gallon utility cost.  Case studies conducted by GCA and
others report utility costs to be on the order of $0.05 per gallon.  Working
with the basis provided by NEESA, payback period curves were generated for the
reduced utility costs in as shown Figures 5.1.3 to 5.1.5.

## 5.1.4  Overall Status

### 5.1.4.1  Availability/Application--

Distillation and fractionation are two of the most established unit
operations.  Many firms provide design services and manufacture equipment.
The Chemical Engineers Equipment Buyers' Guide provides a comprehensive list
of suppliers.  In December 1985, the Naval Energy and Environmental Support
Activity (NEESA), presented the results of a package distillation equipment
manufacturer's survey.[9]  The document provides detailed price, option
(including safety features), and specification information for products from
17 different manufacturers of small batch and continuous units ($\leq$60 gallons
per hour).  Table 5.1.1 lists some of the general specifications of the stills
that were considered.  Appendix B of the solvent TRD[1] provides more detailed
information including capital purchase information.

### 5.1.4.2  Environmental Impact--

The environmental impact of distillation processes for halogenated
organic waste recovery should be minimal.  Apart from questions related to the
disposal of still bottoms, other emissions, including air emissions from the
condenser, do not appear to be significant.

### 5.1.4.3  Advantages and Limitations--

Distillation would appear to be a key technology for the recovery or
minimization of many halogenated organic wastes.  Based on well understood
principles, its implementation is relatively straightforward and its economic
benefits can be appreciable.

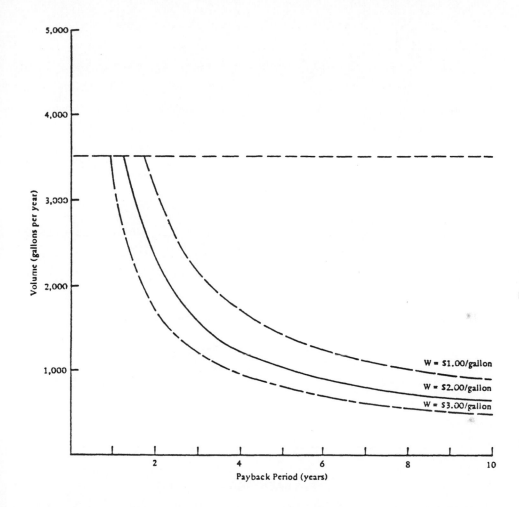

Figure 5.1.3.    Reclamation of cold cleaning solvents via small
batch stills (15 gpd).

Source:    Reference 16.

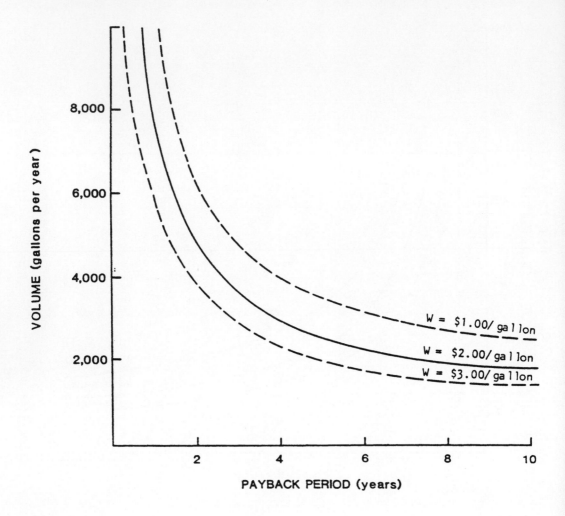

Figure 5.1.4.    Reclamation of cold cleaning solvents via medium
batch stills (55 gpd).

Source:    Reference 16.

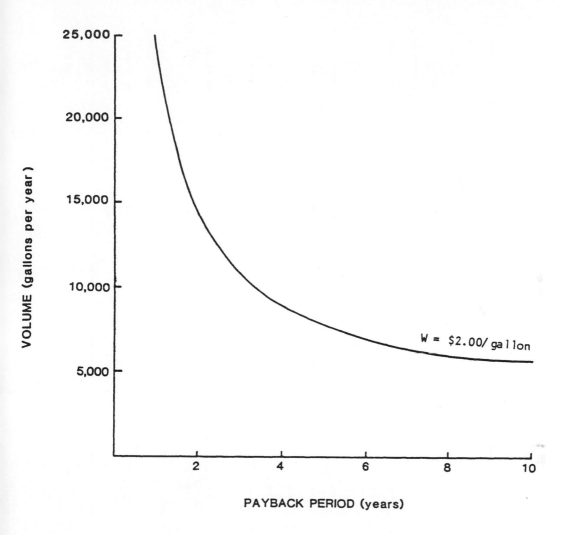

Figure 5.1.5.   Reclamation of cold cleaning solvents via a large
continuous still (250 gpd).

Source:   Reference 16.

TABLE 5.1.1. COMMERCIALLY AVAILABLE SOLVENT STILLS

| Manufacturer | Solvent types | Maximum b.p.(F) | Explosion proof | Vacuum available | Water separator | Heating | Cooling |
|---|---|---|---|---|---|---|---|
| Alt. Resource Mgmt | All | 500 | X | X | X | Electric/Steam | Water |
| Baron-Blakeslee | Halogenated | 350 | | | X | Electric/Steam | Water/Refrig |
| Branson | Halogenated | 350 | | | X | Electric/Steam | Water/Refrig |
| Brighton | All | 500 | X | X | X | Hot Oil/Steam | Water |
| Cardinal | Halogenated | 350 | X | | | Electric | Refrig |
| DCI | All | 500 | X | X | X | Dir. Steam Injection | Water |
| Disti | All | 500 | X | X | X | Hot Oil/Steam | Water |
| Finish Engineering | All | 500 | X | X | | Electric/Steam | Water |
| Hoyt | All | 350 | X | | | Hot Oil | Water |
| Lenape | Halogenated | 350 | | | X | Electric | Water/Refrig |
| Phillips | Halogenated | 350 | | | X | Electric/Steam/Gas | Water |
| Progressive Recovery | All | 500 | X | X | | Hot Oil/Steam | Water |
| Ramco | Halogenated | 350 | | | | Electric/Steam | Water/Air |
| Recyclene | All | 420 | X | | | Hot Oil | Water |
| Unique Industries | Halogenated | 350 | | | X | Electric/Steam/Gas | Water/Refrig |
| Venus | All | 210 | X | | | Electric | Water/Refrig |
| Westinghouse | Halogenated | 350 | | | X | Electric | Water/Refrig |

Source: Reference 9.

## REFERENCES

1.  Breton, M., et al.  Technical Resource Document Treatment Technologies for Solvent Containing Wastes.  Final Report prepared for U.S. EPA, HWERL, Cincinnati, Ohio, September 1986.

2.  Perry, R.H., et al.  Chemical Engineers' Handbook, Sixth Edition. McGraw-Hill, 1984.

3.  Gavlin, A. Benagali and W. Lagdon.  Case Studies of Solvent Recovery Via Continuous Processing.  AIChE Symposium Series, 76(192): 46-50, 1980. Paper presented at September Symposium of AIChE Annual Meeting, New York, NY, November 1977.

4.  Peters, M.S. and K.D. Timmerhaus, Plant Design and Economics for Chemical Engineers, McGraw-Hill, 1980.

5.  Shulka, H.M., and R.E. Hicks, Water General Corp.  Process Design Manual for Stripping of Organics.  EPA-600/2-84-139, U.S. EPA/IERL, Cincinnati, Ohio, August 1984.

6.  Yeshe, P.  Low-Volume, Wet-Scrap Processing.  Chem. Eng. Progress, 80(9): 33-36, September 1984.

7.  GCA, Waste Minimization Case Studies for EPA HWERL, Contract 68-03-3243, 1986.

8.  Higgens, T.E. CH2M Hill, Reston, Va.  Industrial Processes to Reduce Generation of Hazardous Waste at DOD Facilities, Phase 2 Report, Evaluation of 18 Case Studies.  July 15, 1985.

9.  Naval Energy and Environmental Support Activity, Assessment of Solvent Distillation Equipment, NEESA 20.3 - 012, December 1985.

10.  Kirk-Othmer Encyclopedia of Chemical Technology.  John Wiley & Sons, New York, NY, 3rd Edition.  1978.

11.  Allen, C., et al.  Field Evaluations of Hazardous Waste Pretreatment as an Air Pollution Control Technique.  Report prepared for U.S. EPA, ORD, Cincinnati, Ohio under Contract No. 68-02-3992.  April 1985.

12.  National Fire Protection Association, Fire Protection Guide for Hazardous Materials, 9th edition, 1986.

13.  Horsak, R.D., et al., Pace Company Consultants and Engineers, Inc.
     Solvent Recovery in the United States: 1980 - 1990.  Houston, TX.
     Prepared for Harding Lawson Associates, January 1983.

14.  U.S. EPA Office of Solid Waste.  Background Document for Solvents to
     Support 40 CFR Part 268 Land Disposal Restrictions.  Volume II.  Analysis
     of Treatment and Recycling Technologies for Solvents and Determination of
     Best Available Demonstrated Technology.  U.S. EPA Public Docket.
     January 1986.

15.  Arienti, M., et al.  Technical Assessment of Treatment Alternatives for
     Waste Containing Halogenated Organics.  Final Report by GCA Technology
     Division to OSW under Contract No. 68-01-6871, Work Assignment No  9,
     October 1984.

16.  Nelson, W.L., Naval Energy and Environmental Support Activity (NEESA),
     In-House Solvent Reclamation, Port Hueneme, CA.  NEESA 20.3-012,
     October 1984.

5.2   EVAPORATION PROCESSES

Available equipment designs used for evaporation/distillation include
simple stills, flash evaporators, forced circulation evaporators, and falling
film and agitated thin film evaporators.  All designs are capable of
concentrating nonvolatile components of waste mixtures.  However, in cases
where fouling, foaming, high viscosity, thermal degradation, or other factors
present potential operational problems, agitated thin film evaporators (ATFEs)
provide the most versatile service.[1]  They also represent the most
effective, high volume evaporation equipment which is capable of reducing
viscous wastes to low residual organics.[2]  This is a direct result of high
mass transfer rates achieved through turbulence.  For these reasons, a
majority of large commercial solvent recycling companies use ATFEs as
indicated by several industry surveys.

The emphasis in this section will be on the agitated thin film designs
because of their widespread use and applicability in reclaiming wastes which
are too viscous or otherwise too difficult to recover in conventional
distillation equipment.  Although used extensively for recovery of solvents,
they will also be applicable to the recovery of many halogenated organic
wastes.

5.2.1   Process Description

Liquid waste is fed to the top of ATFEs where longitudinal blades mounted
on a motor driven rotor maintain the waste against the heat transfer surface;
i.e., the inside wall of the cylindrical vessel.  This surface is enclosed in
a heating jacket which usually employs steam or hot oil as the heating medium
(temperatures up to 650°F).

The agitation and liquid film are maintained by the blades as they move
along the heat transfer surface.  The blade tips typically travel at 30 to
40 feet per second at clearances of 0.007 to 0.10 inches which creates high
turbulence (see Figure 5.2.1).  This facilitates efficient heat and mass
transfer, shortens required waste residence time, and creates a degree of
mixing that maintains solids or heavy molecular weight solutes in a manageable
suspension without fouling the heat transfer surface.  Mass diffusivities in

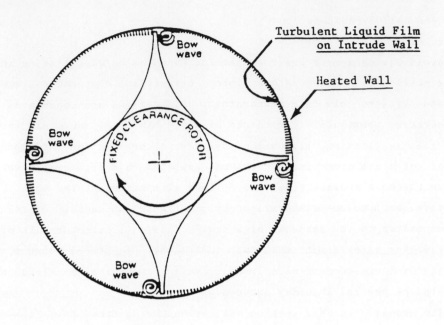

Figure 5.2.1.   Cross section of agitated thin film evaporator.

Source:   The LUWA Corporation
          Bulletin EV-24, 1982.
          Reference 3.

ATFEs can be increased by 1,000 to 10,000 times over nonagitated designs.[2] To further promote separation, the unit is usually operated under vacuum conditions which permits lower temperature processing of thermally unstable mixtures.

Finally, thermodynamic properties of the waste and ATFE operating pressure set a limit on ultimate recovery imposed by vapor-liquid equilibrium constraints. Material will boil when its vapor pressure reaches the operating pressure of the ATFE. Waste vapor pressure for miscible fluids is equal to the sum of the partial pressure of each volatile species. Partial pressure of each component is, in turn, equal to its molar concentration multiplied by the pure component vapor pressure and a constant which is dependent on the ideality of the solution. Thus, operating pressure and partial pressure determine the minimum attainable (i.e., equilibrium) concentration of each volatile species in solution. High separation efficiency will be associated with low system pressure, high pure component vapor pressure, high activity coefficient, high Henry's Law constant, and low solubility (solubility decreases with temperature). Theoretically, very high separations can be achieved for highly volatile compounds in systems of low liquid phase miscibility.

Ultimate recovery will depend on the extent to which equilibrium is achieved which will be limited by diffusive resistance to mass transfer and residence time in the system. For viscous wastes, economical recovery is limited by waste viscosity as a result of decreasing diffusivity of volatile compounds through the waste as its viscosity increases. As diffusivity decreases, resistance to overall mass transfer into the gas phase increases. However, this effect is less pronounced in ATFEs relative to other evaporator designs due to high turbulence and large exposed waste surface area.

Residual solvent concentrations below 1,000 ppm have been routinely achieved and a concentration below 100 ppm can be achieved if conditions are optimal. However, except in unusual circumstances (e.g., immiscible fluids), the sole use of ATFE cannot be expected to yield residual concentrations of volatiles in the low ppm range.

5.2.1.1  Pretreatment and Post-Treatment Requirements--

A schematic of an ATFE and associated pretreatment and post-treatment options is shown in Figure 5.2.2.  As shown, the pretreatment techniques most commonly applied to wastes undergoing ATFE are a distillation/evaporation process aimed at removal of light ends, oil or suspended solids, or a dissolved solids concentration process.  The most cost-effective application of an ATFE is in treating viscous wastes which are generally not amenable to treatment using other, less expensive evaporation processes.  In many cases, the source of these wastes will be bottoms products from conventional evaporation/distillation processes.  These processes have been described previously (Section 5.1).

Constraints on acceptable waste viscosities differ between manufacturers and specific unit types.  In general, specially designed ATFEs can process wastes with viscosities up to 1,000,000 cps.  This is shown in Table 5.2.1 and Figure 5.2.3 which present equipment selection guides for evaporator products based on waste viscosity.[1,4]

Post-treatment methods are also identified in Figure 5.2.2.  These basically include further refinement of the overhead product through water removal or separation of product mixtures, and further recovery or disposal of bottom products.

The recovered overhead may be used as is or further purified through decanting, dehydration or fractionation.  In cases where the waste feed is a mixture of organic compounds, separation by fractional distillation is sometimes justified by the increased value of the separated components.

Further recovery of bottoms from ATFE treatment of organic wastes is generally not considered practically achievable in liquid handling equipment. However, in some cases a drum dryer, centrifuge or other solids handling equipment might be employed depending on the nature of the waste.  The residue is often solidified and landfilled, incinerated, or burned as fuel if the Btu value, chlorine content, ash content and viscosity are within required specifications.  In some cases the bottoms may be suitable for process reuse.

Further treatment of aqueous ATFE bottoms will generally be required to remove remaining volatiles and other contaminants.  Candidate technologies include steam and air stripping, carbon adsorption, or biological treatment if toxic contaminant concentrations are low.

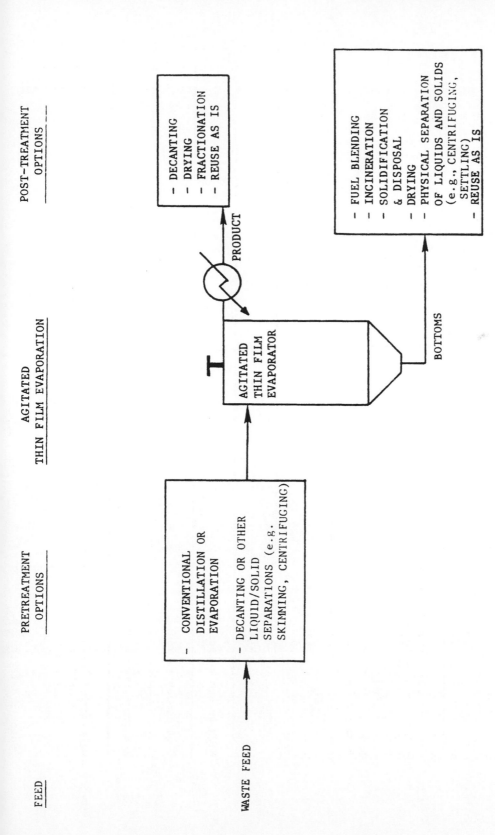

Figure 5.2.2.  Treatment train using an agitated thin film evaporator.

TABLE 5.2.1.  KEYS TO SELECTING KONTRO THIN FILM EVAPORATORS

|  | Horizontal Agitated Thin Film Evaporator | Vertical Agitated Thin Film Evaporator | High Viscosity ATFE (Film Truder) | Conventional Evaporator |
|---|---|---|---|---|
| **Liquid Property** | | | | |
| Low viscosity 10,000 cps or less | ++ | ++ | - | - |
| Medium viscosity up to 50,000 cps | ++ | + | - | X |
| High viscosity up to 1,000,000 cps | X | X | ++ | X |
| Slurry | + | + | - | X |
| High vacuum evaporation | ++ | ++ | ++ | ++ |
| High concentration | ++ | - | - | - |
| Residence time control | ++ | + | - | X |

Note:  ++ = Particularly suitable
     + = Suitable
     - = Usable in special cases
    X = Unusable

Source:  Reference 4.

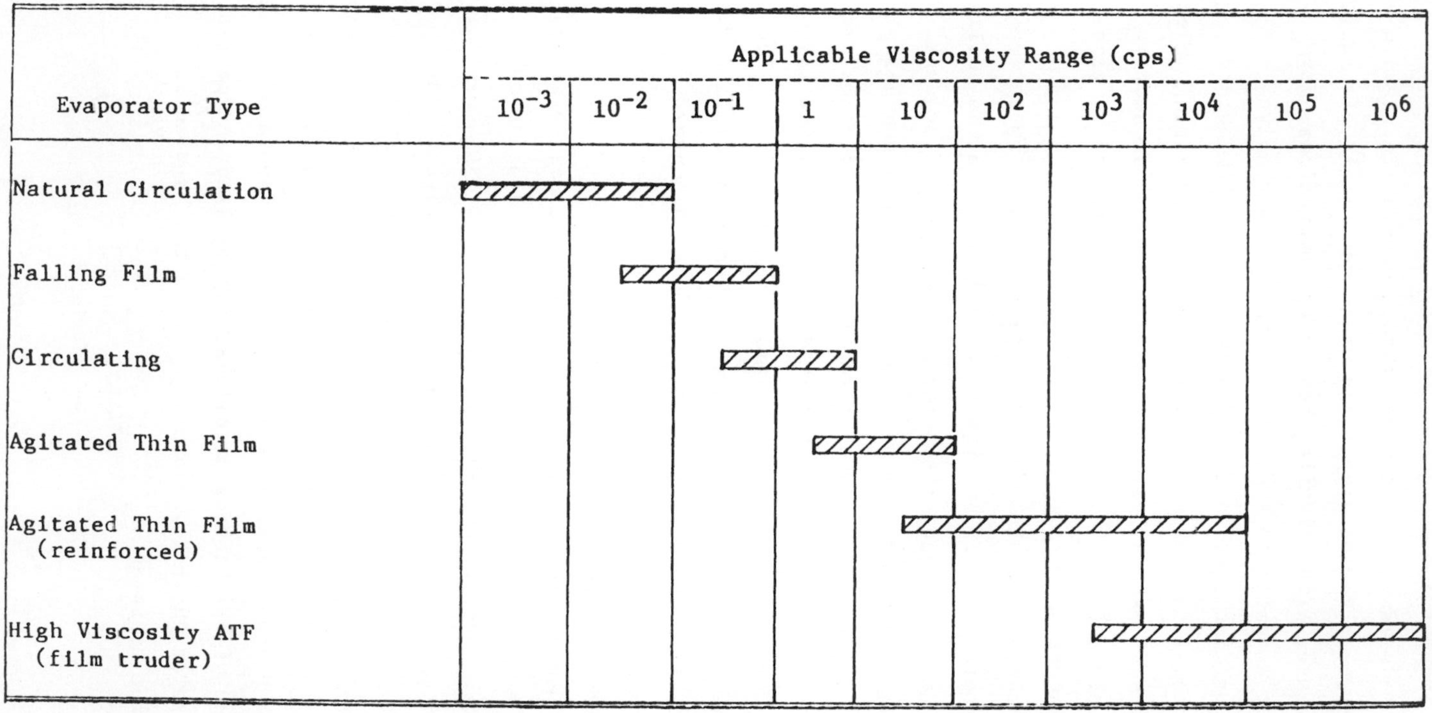

Figure 5.2.3.    Selection of LUWA Evaporators based on waste viscosity.

Source:  Reference 1.

5.2.1.2  Operating Parameters and Design Criteria--

ATFEs are suited to treatment of concentrated, nonvolatile organic wastes contaminated with water or other more volatile organics.  It is also suitable for treating aqueous wastes with volatile organic concentrations above 5 percent which are not amenable to treatment using less expensive conventional evaporation/distillation technologies.  Typically, waste viscosity (feed or bottoms) is the restrictive waste characteristic which results in adoption of ATFE as the preferred technology.  Consequently, pretreatment requirements are less stringent and may be limited to gross solid removal or waste concentration through decanting.

Operating system temperature and pressure are limited by waste type and equipment design.  Temperature must be higher (0 to 30°F) than the boiling point of the material which is to be recovered as the overhead product, and sufficiently high to maintain a minimum waste viscosity.  Maximum design temperature (less than 650°F) may be further restricted by explosion limits or by the decomposition temperature of the recoverable materials; the latter is of particular concern for some halogenated organics (Section 5.1).  Operation at low pressure reduces the temperature required to reach the boiling point. It also enables higher recovery rates to be achieved, as discussed previously.  The lower limit of pressure is restricted by cost and equipment design.  Typical system pressures range from 2 to 760 mm Hg.

For a given flow and desired recovery, an ATFE system has to be designed to produce a specific evaporation rate.  Evaporation rate depends on temperature and pressure as discussed above, heat transfer surface area, waste type and heat transfer coefficient.  Figure 5.2.4 and 5.2.5 show the relationship between heat transfer surface area, waste type, and evaporation rate for a Cherry-Burrel  ATFE (Turba-film processor).[5]  Figure 5.2.6 shows the same relationship, based on unit area of heat transfer surface, for a LUWA ATFE.[3]  As shown, high heat transfer area is required to evaporate solvents from aqueous wastes and highly viscous materials; e.g., waxes and pastes. Aqueous wastes, because of their low viscosity, are often handled in conventional evaporation equipment.

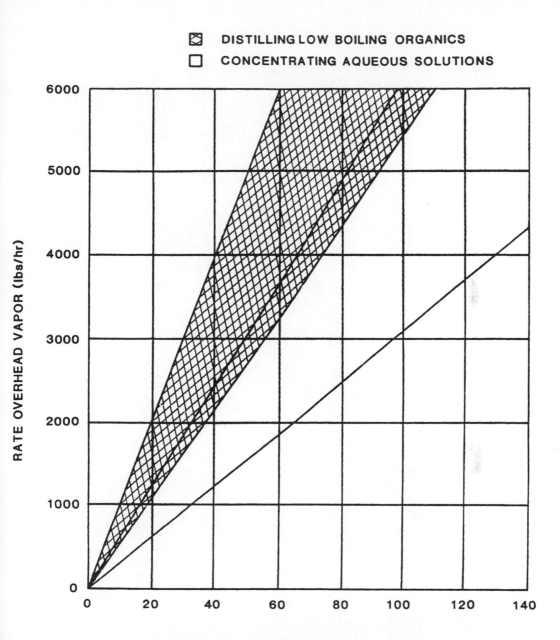

Figure 5.2.4. Required heat transfer surface area for distilling low boiling organics and concentrating aqueous solutions.*

*Source: Cherry-Burrell, Reference 5.

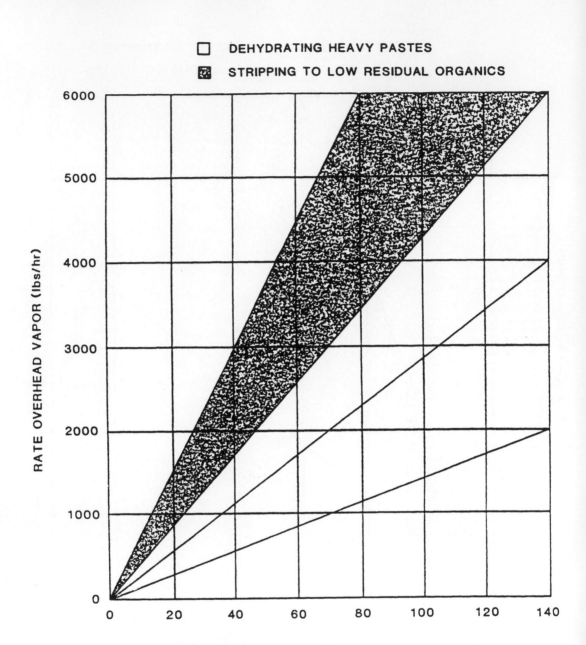

Figure 5.2.5.   Required heat transfer surface area for dehydrating heavy pastes and stripping wastes to low residual organics.*

*Source:   Cherry-Burrell, Reference 5.

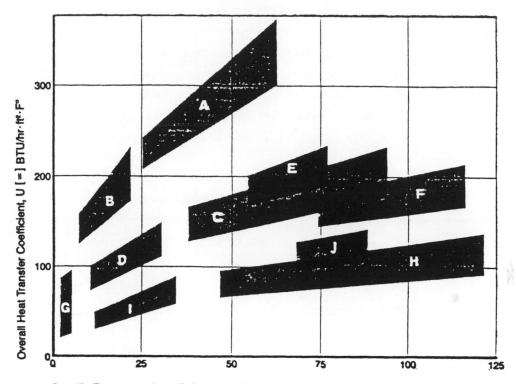

Heating Medium:   dry and saturated steam

A    Concentration of aqueous solutions
B    Dehydration of organics
C    Distillation of organics
D    Stripping of low boilers from organics
E    Reboiler service
F    Solvent reclaiming
G    Deodorization

Heating Medium:  Dowtherm or hot oil

H    Distillation of high-boiling organics
I    Stripping of high boilers
J    Reboiler service

Figure 5.2.6.    Heat transfer and evaporation rates in LUWA
Thin Film Evaporators.
Source:   Reference 3.

Typical ATFE system operating data are summarized in Table 5.2.2.[6]  As shown, heat transfer surface area, steam consumption and cooling water requirement vary directly with overhead recovery rate.  However, electricity requirement drops per unit of overhead quantity recovered thus providing a slight drop in unit operating costs as capacity increases.  Throughput of ATFEs is generally high.  If wastes are generated in low quantities, package distillation units capable of handling high solids waste may be more economical (see Section 5.1).

TABLE 5.2.2.   TYPICAL AGITATED THIN FILM EVAPORATOR DESIGN CHARACTERISTICS

| Overhead recovery (gal/hr) | Heat transfer surface area ($ft^2$) | Utilities[a] | | | System dimensions L x W x H, (ft) |
| | | Heating Btu/hr (1,000) | Cooling water (gpm) | Electricity (KW) | |
|---|---|---|---|---|---|
| 40  | 4.2  | 79  | 5  | 1.5 | 4 x 6 x 10 |
| 85  | 8.8  | 168 | 11 | 1.5 | 4 x 6 x 11.5 |
| 130 | 13.4 | 251 | 16 | 3.5 | 4 x 6 x 13 |
| 240 | 25   | 474 | 30 | 3.5 | 5 x 8 x 11.5 |
| 500 | 51.2 | 989 | 63 | 4   | 5 x 8 x 14 |

[a]Based on an average latent heat of vaporization of 175 Btu/lb and preheating feed by 200°F.

Source:  Reference 6.

Commercially available evaporator equipment design parameters are summarized as follows.[2]  Evaporators range from 1 to over 400 square feet of heat transfer surface with liquid throughput ranging as high as 250 lb/hr/$ft^2$.  Overhead to bottoms splits for lightly contaminated fluids can be as high as 100 to 1 with controlled residence times of up to 100 seconds.  Blade tip speeds of nonscraping designs average 30 to 40 ft/sec while scraping blades average 5 to 10 ft/sec.  Some units come equipped with variable clearance while scraping blades are typically spring mounted or maintain contact with the wall as a result of centrifugal force.  Operating temperatures range up to 650°F and pressure ranges from 2 mm Hg to atmospheric.

## 5.2.2    Demonstrated Performance

Actual performance data from commercially operated units, conducted in accordance with EPA Quality Assurance/Quality Control requirements, are limited to EPA sponsored studies of halogenated solvents as reported by GCA (1986)[7] and the Research Triangle Institute (1984).[2] These data, and associated cost data are present in the solvent TRD.[8] None of the units studied demonstrated a capability of reducing volatile concentration in the bottoms product to low ppm levels. These data are applicable to halogenated organic wastes provided adjustments are made for differences in the factors such as relative volatilities.

## 5.2.3    Cost of Treatment

Cost estimates obtained by GCA[7] and RTI[2] during their case studies to assess solvent recovery using ATFEs were discussed in the solvent TRD. As noted in the solvent TRD, good agreement was obtained for the various cost estimates developed by the two studies and another estimate developed by the Pace Company.[9] These estimates are applicable to halogenated organic wastes provided volatilities, recoveries, viscosities and other factors remain comparable. Generally somewhat higher costs can be anticipated for the higher molecular weight, less volatile halogenated organics due to increased energy or lower throughput.

GCA estimated an hourly cost of operation of $77.98 as shown in Table 5.2.3. Assuming an annual recovery of 940,000 gallons of waste solvent and a capital recovery factor of 17.5 percent, total cost per gallon ranged from $1.12 to $2.08 depending upon the composition of the feed, the product recovery, and the time of processing. RTI determined processing costs for an ATFE unit at about $1 per gallon when organic is stripped as the overhead product and $1.50 per gallon when water is stripped overhead with the organic becoming the bottom product. Pace estimated costs of $0.85 per gallon with the major cost difference resulting from a higher projected recovery rate and higher assumed bottoms disposal costs. Costs based on feed rate were almost identical as were the capital cost estimates of about $300,000 for a 5 square meter heat transfer surface area unit. Capital cost estimates developed by Pace are shown in Table 5.2.4 for ATFE units of different size.

TABLE 5.2.3.   HOURLY COSTS OF LUWA THIN FILM EVAPORATOR

| Expenses | Cost ($/hour) |
|---|---|
| Fuel | 8.00 |
| Auxiliary Chemicals | 0.72 |
| Electricity | 2.60 |
| Laboratory | 9.37 |
| Operating Labor | 15.76 |
| Maintenance Labor | 14.58 |
| Spare Parts (Repair and Maintenance) | 12.69 |
| Regulatory Compliance | 6.30 |
| Insurance | 0.69 |
| Capital Depreciation | 7.27 |
| Total | 77.98 |

In addition to hourly cost, the generator pays $0.26/gallon to dispose of still bottoms to an incinerator.

Source:   Reference 7.

TABLE 5.2.4.   CAPITAL COST RECOVERY COMPONENTS FOR ONSITE ATFE
RECOVERY SYSTEMS

| Nominal feed rate (gph) | Thin film evaporator | | |
|---|---|---|---|
| | 100 | 350 | 500 |
| Process Equipment | 126,000 | 212,900 | 264,900 |
| Tanks | 18,500 | 37,000 | 74,000 |
| Subtotal (1) | 144,500 | 249,900 | 338,900 |
| Engineering, Electrical, Instrumentation (20% of (1)) | 28,900 | 50,000 | 67,800 |
| Subtotal (2) | 173,400 | 299,900 | 406,700 |
| Contingency 15% of (2) | 26,000 | 45,000 | 61,000 |
| TOTAL | 199,400 | 344,900 | 467,700 |

Source:   Reference 9.

### 5.2.4.  Overall Status of Process

#### 5.2.4.1  Availability--

ATFEs are widely used in a number of industries (e.g, the solvent recovery industry) due to their unique ability to process viscous wastes relative to other evaporation/distillation technologies.  Evaporators and accessory equipment can be obtained from a number of manufacturers in various sizes and configurations.  Ten firms are identified as suppliers of the ATFE in the 1986 edition of McGraw-Hill's Chemical Engineering Equipment Buyers' Guide.[10]  Major producers include Blaw-Knox (Buffalo, NY), Cherry Burrell (Louisville, KY), LUWA (Charlotte, NC), Kontro (Orange, MA), Pfaudler (Rochester, NY) and Artisan Industries (Waltham, MA).[2]

#### 5.2.4.2  Application--

Evaporators can be used to recover solvents and other volatile organics from both organic and aqueous waste streams provided the treated waste does not exceed viscosity limits imposed by the system design (see Table 5.2.1) operating temperature.  Excessive solids content will increase viscosity and foul heat transfer surfaces.  Therefore, some pretreatment for gross solids removal may be required.

In practice, recovery to low residual organics is limited by viscosity due to increased resistance to mass transfer.  This resistance is partially offset in an ATFE due to the high turbulence generated in the vessel. Recovery is also limited by operating pressure since this pressure determines the equilibrium concentration of volatiles remaining in the waste.  Finally, recovery of organics from waste streams may not be economical unless the recoverable organic content is greater than 6 to 8 percent.[2]  A rough cost analysis based on raw material purchase and disposal costs supports this figure.  However, wastes with recoverable concentrations of as little as 3 to 5 percent may be profitably recovered when processed in existing, high volume onsite facilities which are underutilized.

ATFEs are likely to find more widespread use relative to conventional distillation equipment.  Land disposal restrictions and limitations on halogen content in supplemental fuels will compel recyclers to pursue higher recovery rates when processing halogenated organic wastes.

5.2.4.3  Environmental Impacts--

In selecting evaporators as a recovery technology for volatiles it should be recognized that, except in isolated cases, further treatment of the bottoms stream will be required to meet EPA land disposal or NPDES discharge requirements.  Air emissions from the overhead condenser have been identified as a potentially significant source of VOC emissions by RTI.[2]  VOC concentration averaged 41.1 and 34.4 mg/L at the vacuum pump outlet at two units tested.  However, no estimates of total release of emissions were provided.  Emission rate would be greatest during process start-up.  It would increase if air was leaking into the system, noncondensible gases were being formed, or if the condenser became overloaded.  Vacuum pump emissions controls should be examined as a potential additional cost since treatment requirements (e.g., carbon adsorbers) may be necessary to avoid adverse environmental impacts.

5.2.4.4  Advantages and Limitations--

Evaporation as a means of recovering useful halogenated organics is a common unit operation used by a variety of industries.  It also finds application in removing water or other volatiles from viscous, non-volatile fluids with recovery value.  The ATFE unit's most significant advantage over other recovery processes is its ability to handle viscous liquids.  However, its cost must be compared to that of less expensive, conventional recovery technologies (e.g., distillation) and their associated residual treatment costs; e.g. thermal destruction, solidification, and land disposal.  The cost of the entire treatment train will ultimately dictate selection of the optimal recovery technique.

REFERENCES


1.    Kappenberger, P.F., P.G. Bhandarkar.  LUWA Ltd., Process Engineering
      Division.  Zurich, Switz.  Thin Film Technology in Environmental
      Protection.  Chemical Age of India V. 36(1).  January 1985.

2.    Allen, C.C., et al.  Research Triangle Institute.  Field Evaluation of
      Hazardous Waste Pretreatment as an Air Pollution Control Technique.
      U.S. EPA/ORD, Cincinnati, OH.  January 1986.

3.    Luwa Corporation, Luwa Thin-Film Evaporator Technology, Bulletin EV-24,
      1982.

4.    Kontro Company, Bulletin 7510.

5.    Cherry-Burrel, ANCO/Rotator Division, Turba-film Evaporator Bulletin.

6.    Pfaudler Co., Recover Wash Solvent with the Pfaudler Solvent Recovery
      System, Data Sheet 146, Supplement 1.

7.    Roeck, D., et al.  GCA Technology Division, Inc.  Sampling Data Collected
      at Milsolv Corporation, Milwaukee, WI. under Contract No. 68-03-3243 with
      the U.S. EPA Office of Solid Waste.  December 1986.

8.    Breton, M., et al.  Technical Resource Document - Solvent Bearing
      Wastes.  Prepared by GCA Technology Division for U.S. EPA, HWERL under
      Contract No. 68-03-3243, Work Assignment No. 2.  September 1986.

9.    Horsak, R.D., et al.  Pace Company Consultants and Engineers, Inc.
      Solvent Recovery in the United States:  1980 - 1990.  Houston, TX.
      Prepared for Harding Lawson Associates.  January 1983.

10.   Chemical Engineering Equipment Buyers' Guide.  McGraw Hill.  1986.

## 5.3   STEAM STRIPPING

### 5.3.1   Process Description

Steam stripping by steam injection, usually into a tray or packed distillation column, is used in both industrial chemical production and waste treatment to remove volatile organic chemicals from waste streams. This unit operation is most effectively applied to aqueous solutions for the removal of volatile components that are immiscible in water. It can also be used for stripping organic solutions when water forms low boiling point azeotropes and does not adversely affect overhead or bottoms quality. The presence of water must either be acceptable to or economically separable from the final product to achieve product purity specifications.

Steam stripping is commonly employed to separate halogenated and certain aromatic compounds from water. It is less effectively used to recover miscible organics such as ketones or alcohols. The process is preferable to conventional distillation processes for recovering high yields of contaminated wastes which would otherwise foul heat transfer surfaces. It is more economical and effective than air stripping for recovering wastes with high concentrations of volatiles and wastes with low volatility.[1]

Figure 5.3.1 illustrates a typical steam stripping process. Waste enters near the top of the column and then flows by gravity countercurrent to the steam. As the waste passes down through the column, volatile compounds within the waste are lost to the steam/organic vapor stream rising from the bottom of the column. The concentration of volatile compounds in the waste reaches a minimum at the bottom of the column. The overhead vapor is condensed as it exits the column and the condensate is then decanted to achieve solvent/water separation. Reflux may or may not be used, depending upon the desired composition of the overhead stream.

The common uses for steam distillation can be summarized as follows:[2]

1.  To separate relatively small amounts of a volatile impurity from a large amount of material.

2.  To separate appreciable quantities of low solubility, higher boiling point materials from nonvolatile wastes. This requires that the separable materials form low boiling point azeotropes with water.

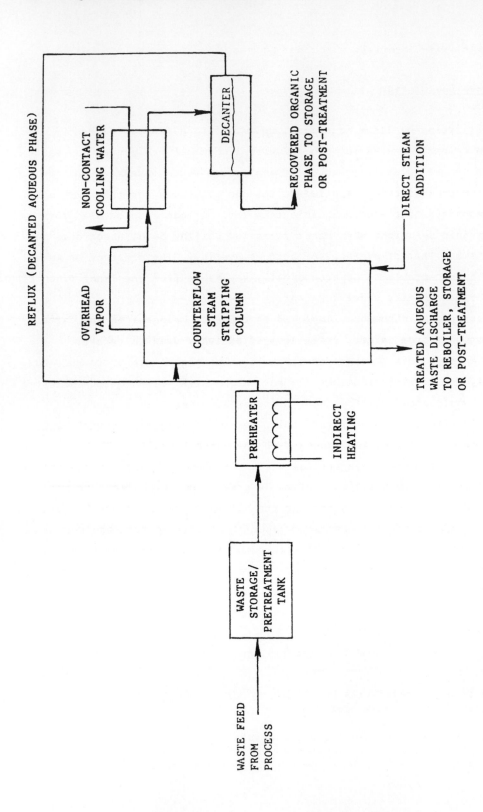

Figure 5.3.1.  Typical steam stripping process.

3.  To recover materials which are thermally unstable or react with other waste components at the boiling temperature.

4.  To recover materials which cannot be distilled by indirect heating even under low pressure, because of their high boiling temperatures.

5.  To recover materials in instances where direct-fired heaters cannot be used because of ignition or explosion hazards.

Theoretical Considerations--

The residual streams from steam stripping of aqueous wastes typically consist of decanted overhead products and treated waste stream bottoms. The stripped waste is sewered and undergoes additional treatment (e.g., carbon adsorption) as necessary to further reduce contaminant levels. Depending on specified purity, decanted overhead (usually organic) is either used directly or further purified through processes such as drying or fractionation. The overhead aqueous phase is typically returned to the stripping column if even slight solubility exists between water and the organic components.

Using steam in distillation permits a more complete separation of immiscible liquids at lower temperatures than non-steam distillation for the same conditions of total pressure or vacuum. The essential feature of an immiscible system is that each liquid phase exerts its own total vapor pressure, regardless of the quantity of the other liquid phase. At constant system pressure, the presence of steam reduces the total vapor pressure which is required to induce boiling, thereby lowering system temperature requirements. This permits separation of compounds which could not be accomplished through conventional distillation due to polymerization (e.g., cresols, vinyl-type monomers) or thermal decomposition (e.g., halogenated compounds) of waste constituents.

The Hausbrand diagram is very useful in steam distillation calculations.[2] As shown in Figure 5.3.2, it plots the total system pressure (760 mm Hg) minus the water vapor pressure versus temperature. The intersection of the water curve with the ordinary vapor pressure curves of the other materials gives the temperature at which steam distillation can take place (in this case at 1 atmosphere pressure).

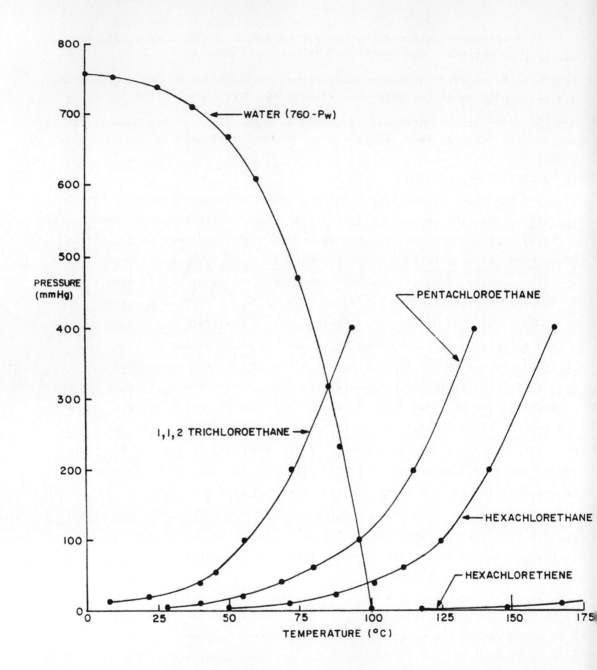

Figure 5.3.2.   Hausbrand diagram for various halogenated
organic liquids at 1 atmosphere.

    While the Hausbrand diagram makes possible a quick determination of the
temperature at which a steam distillation takes place, it also presents
graphically the molar content of the vapor.  In the case of the compound
1,1,2-trichloroethane, the intersection point comes at 86°C.  During steam
distillation, the vapor emitted consists of 1,1,2-trichloroethylene with a
partial pressure of about 320 mm and water vapor with a partial pressure of
440 mm Hg.  The corresponding weight fraction of the halogenated compound in
the vapor is 83.6 percent.  Since most of the halogenated organics have much
higher molecular weights than water, the distillate is much richer by weight
than would appear from the diagram.

    With the exception of certain acids, most organic compounds produce
minimum boiling point azeotropes with water.  This phenomenon is
characteristic of mixtures with dissimilar molecular species with activity
coefficients greater than unity.  Most halogenated organic compounds fall into
this category as manifested by their limited solubility in water.

    A minimum boiling point azeotrope forms at a temperature below that of
the boiling point of the pure compounds.  Unless this azeotrope is shifted to
more favorable equilibrium conditions through lowered operating pressure or
addition of a chemical complexor (entrainer), it will act to limit the
concentration which can be achieved in the overhead product.  With some
organic mixtures, water can act as an entrainer to preferentially distill
compounds by creating a low boiling point azeotrope.

    As compounds become dissimilar, they tend to approach liquid
immiscibility (e.g., chlorinated aromatics in water).  Their equilibrium vapor
concentration will be essentially constant over an increasing range in liquid
concentrations and only begin to deviate from this level at very high or very
low liquid concentrations.  Generally, the azeotrope will occur within the
liquid immiscibility composition range forming a heterogeneous overhead
product which can easily be separated into two phases.  Examples clarifying
this and other concepts are provided below.

    As noted above and shown in Figure 5.3.2, water and 1,1,2-trichloro-
ethane, two slightly immiscible liquids, boil at 86°C (1 atm) to form a
heterogeneous azeotrope consisting of 83.6 percent 1,1,2-trichloroethylene.
The normal boiling point of pure 1,1,2-trichloroethylene (113.7°C), is
substantially higher than the azeotrophic boiling point, thus heating costs

are reduced in the presence of steam.  The overhead product readily separates into two phases, the upper layer consisting of 99.55 percent water and the lower layer consisting of 99.5 percent 1,1,2-trichloroethylene (specific gravity = 1.443)[3].  Thus, steam distillation is readily applied to this compound, particularly if it must be removed from polymerizable, nonvolatile materials which can foul distillation equipment at the normal boiling point.

Steam stripping is also effective in instances where water acts as an entrainer to shift the vapor-liquid equilibrium toward more desirable conditions.  For example, mixtures can shift toward higher concentrations of the less ionic component in the overhead product in the presence of steam. Benzene-alcohol overhead products resulting from steam stripping are more highly concentrated with benzene relative to normal distillation and separate into two phases upon condensing (e.g., benzene and isopropanol/water), thereby further separating the components.[3]  Additional information regarding azeotrope theory and azeotropic and extractive distillation can be found in standard engineering texts.[4,5]

In the absence of azeotrope data for particular wastes, the principal indices used to predict steam stripping feasibility of halogenated organics from aqueous wastes are their boiling points and Henry's Law constants. Compounds with boiling points less than 150°C (i.e., volatile compounds), have good steam stripping potential, as do compounds with Henry's Law constants greater than $10^{-4}$ atm-m$^3$/mole.[6]  The Henry's Law constant expresses the equilibrium distribution of a compound between vapor and liquid for dilute solutions.  It is roughly proportional to the product of vapor pressure and the reciprocal of solubility, thus taking into consideration the miscibility of the compound in the liquid phase.  Therefore, increasing the value of the Henry's Law constant also correlates with increasing favorability of volatilization through the use of steam stripping.

Henry's Law constant is expressed as the ratio of mass per unit volume in air to mass per unit volume in water.  The expression is:

$$H = \frac{16.04 \text{ PM}}{TS}$$

Where P is the vapor pressure in mmHg, M is the gram molecular weight, T is the temperature in °K, and S is the solubility of the solute in milligrams per liter. The values obtained in this fashion are about 40 times greater than values of Henry's Law constant calculated in units of atm-m$^3$/mole. Values of Henry's Law constant are shown in Table 5.3.1 for a number of halogenated solvents and compounds addresses in this TRD. The compounds have been divided into three groups. A representative compound from each group was used to assess the cost of steam stripping in Section 5.3.3.[7]

It should be noted, however, that a study performed by the U.S. EPA/ OAQPS[1] suggested that the use of Henry's Law constants given in the literature for some chemicals could result in underestimating the required contact time and overestimating the removal efficiency of steam stripping. As part of the study, Henry's Law constants were calculated from headspace and liquid composition sampling data. These calculated constants were substantially less than their corresponding literature values, but did provide good correlation with test data. It is recommended that vapor-liquid equilibrium data be established through headspace analysis and activity coefficient models for more complex solutions.[1] Alternatively, relative volatilities in non-ideal situations can be modeled through the use of partition coefficients and critical constants.[8]

5.3.1.1  Pretreatment and Post-Treatment Requirements--

Certain waste characteristics affect the viability of steam stripping as a waste treatment technique. Restrictive waste characteristics include:[9]

- High solubility of the organic compound in water; usually more than 1,000 ppm;

- Organic compounds with high boiling points (more than  150°C);

- VOC concentrations in excess of 10 percent; above this concentration, distillation may be more cost effective; and

- Suspended solids concentrations in excess of 2 percent or the presence of materials that tend to polymerize at operating temperatures; these can cause fouling of packing material and eventual plugging of equipment.

TABLE 5.3.1.   HENRY'S LAW CONSTANTS (Hi)   $(mg/m^3/mg/m^3)$

| High Hi* $(3 \times 10^2 - 10^{-1})$ | Medium Hi $(10^{-2} - 10^{-3})$ | Low Hi $(10^{-4} - 10^{-8})$ |
|---|---|---|
| Alkyl chloride (14.0) | 1,2-dichloroethane (0.046) | Bis(2-chloroethyl)ether ($5.4 \times 10^{-4}$) |
| Carbon tetrachloride (1.0) | Epichlorhydrin (0.006) | Bis(2-chloroethoxy)methane ($1.17 \times 10^{-5}$) |
| Chlorobenzene (0.17) | Ethylene dibromide (0.026) | 3-chloropropionitrile ($6.4 \times 10^{-4}$) |
| 1,1,1-trichloroethane (0.15) | Hexachloroethane (0.046) | 2-chlorophenol ($4.29 \times 10^{-4}$) |
| Chloroethane (0.21) | 1,1,2-trichloroethane (0.032) | 2,4-dichlorophenol ($1.17 \times 10^{-4}$) |
| 1,1-dichloroethane (0.62) | 1,1,2,2-trichloroethane (0.032) | 2,4,6-trichlorophenol ($1.67 \times 10^{-4}$) |
| Chloroform (0.14) | Methylene chloride (0.085) | Pentachlorophenol ($1.17 \times 10^{-4}$) |
| Cyanogen chloride (0.11) | 1,2-dichloropropane (1.096) | p-chloro-m-cresol ($1.04 \times 10^{-4}$) |
| chloromethane (1..67) | 1,3-dichloropropane (0.055) | Pronamide ($7.8 \times 10^{-5}$) |
| methyl chloride | Dibromochloromethane (0.041) | DDD ($1.1 \times 10^{-4}$) |
| Pentachloroethane (0.67) | Tribromomethane (0.23) | Endrin ($1.6 \times 10^{-6}$) |
| Vinyl chloride (3.4) | Bis(chloromethyl)ether (0.00875) | Dieldrin ($3.4 \times 10^{-4}$) |
| 1,1-dichloroethene (7.92) | Bis(chloroisopropyl)ether ($4.58 \times 10^{-3}$) | Diallate ($1.6 \times 10^{-4}$) |
| 1,2-trans-dichloroethane (2.79) |  | Silvex ($1 \times 10^{-5}$) |
| Trichloroethene (0.379) | 4-chlorophenyl phenyl ether ($9.12 \times 10^{-3}$) |  |
| Tetrachloroethene (0.638) |  |  |
| Hexachloro-1,3-butadiene (1.07) | 4-bromophenyl phenyl ether ($4.17 \times 10^{-3}$) |  |
| Bromomethane (8.21) |  |  |
| Bromodichloromethane (0.100) | 1,2-dichlorobenzene (0.080) |  |
| Dichlorodifluoromethane (124.2) | 1,2,4-trichlorobenzene (0.096) |  |
| Trichlorofluoromethane (4.58) | Hexachlorobenzene (0.0015) |  |
| 1,3-dichlorobenzene (0.150) | 2-chloronapthalene (0.020) |  |
| 1,4-dichlorobenzene (0.125) | 2,6-dichlorophenol (0.026) |  |
| Heptachlor (0.11) | DDT (0.0011) |  |
| Toxaphene (1.5) | Chlordane (0.0245) |  |
|  | Aldrin (0.0017) |  |

Source:   Adapted from Reference 7.

Pretreatment requirements for wastes, therefore, consist of reducing high concentrations of volatiles, solids, and polymerizable organics.  Highly concentrated volatile wastes are more economically pretreated through conventional evaporation/distillation technologies.  The diluted bottoms product can then be treated via steam stripping.

Solids, metals, oil, and grease concentrations can be reduced through a variety of pretreatment techniques as discussed in Section 3.0.  These include precipitation, coagulation, flocculation, centrifugation, membrane separation processes, flotation and other chemical/physical separation processes.  For example, fouling of packing material with oxidized iron and manganese can be reduced through pretreatment via lime precipitation.  Membrane separation processes are effective in removing high molecular weight compounds.  Since these compounds are typically nonvolatile, and thus not amenable to steam stripping, pretreatment methods using membrane separation techniques compliment steam stripping removal efficiency while reducing column fouling.

Post-treatment is generally required of both the overhead and bottoms streams.  Data show that steam stripping may be capable of reducing organic concentrations in wastewater bottoms to levels which meet the land disposal ban treatment standards.  However, due to economic considerations, conventional wastewater treatment methods (e.g., adsorption, air stripping, biological or chemical treatment) are more commonly employed to remove residual organic levels from aqueous streams.

Concentrated organic bottoms such as those containing high boiling point or solid halogenated organics must be separated from the condensed steam through decanting, centrifugation, and other physical separation techniques. In some cases these bottoms may be recycled.  Overhead products undergo liquid-liquid separation, typically through decanting followed by dehydrating of the recovered organic.  Depending on its organic content levels, the decanted aqueous stream is reprocessed through the stripper or treated via other wastewater treatment processes.  Alternatively, it could be used along with a portion of the stripped wastewater bottoms to generate steam if the boilers are properly equipped to accommodate the presence of volatile components.[10]

5.3.1.2  Design Characteristics and Operating Parameters--

Stripping towers operate in a batch or continuous mode.  Generally, batch
stripping is of less commercial interest.  It is reserved for low volume
processing or for staged stripping of streams with multiple volatile
components which have different boiling points.  Continuous operation more
effectively separates components of comparable volatility, provides higher
purity of separated products, and uses less stripping medium for the same
degree of separation, particularly when stripping to low levels of organics.[2]

Three modes of flow are possible:  cocurrent, countercurrent, and
crossflow.  Cocurrent flow, being least efficient, is not usually used, while
crossflow operation is often preferred to counterflow since it provides
greater transfer efficiency over a wider operating range.

A tower can be operated isothermally or adiabatically.  Steam stripping
is typically performed isothermally; i.e., temperature is constant along the
length of the tower.  The feed is usually preheated to the boiling point
before entering the tower to minimize steam requirements and, consequently,
treated waste volume.

Reflux involves condensing a portion of the vapors from the top product
and returning it to the tower.  This can enhance separation by increasing the
concentration of the stripped organics in the vapor stream.  This occurs
because condensation of vapor in the column is required to heat the refluxed
liquid to its bubble point.  This condensation increases the concentration of
strippable components in the liquid stream which, in turn, will increase their
equilibrium vapor concentrations.  This effect is more important as the
solution components become more miscible in one another.

Similarly, for miscible fluids, introducing the feed at a lower tray in
addition to refluxing can increase the concentration of organics beyond that
obtainable by reflux alone.  Addition of reflux shifts the distribution in the
column from rectifying to stripping zones.  A column designed with variable
feed plate location can accommodate this shift, as well as changes in waste
feed, to permit column operation at maximum efficiency.

The optimal height of the rectifying zone depends on the waste being
treated.  If feed enters at the top of the column (i.e., no rectifying zone),
the limiting overhead concentration is given by the vapor equilibrium with the
feed.  As the rectifying zone is increased, the overhead concentration is
similarly increased and approached the azeotropic concentration limits.

Finally, stripping can be carried out in two types of towers. Tray towers provide staged contact between liquid and vapor streams. Alternatively, packed towers allow for continuous contact between the two phases. Packed towers are less expensive, have low liquid hold-up, low pressure drop, and are preferred for low pressure operation and treatment of corrosive, foaming, or viscous liquids.[7] However, tray columns have been more widely used in the past and consequently are more predictable in their performance. Tray columns are more flexible since they operate efficiently over a wide range of flow rates and can be readily adapted to process multiple feeds or sidestreams.[7] They are also more easily cleaned and are, therefore, preferred for processing wastes with high concentrations of metals, solids, or polymerizable materials.

Steam stripper design is ultimately dependent on the waste characteristics, throughput, and desired residual characteristics. Thus, tower height, diameter, packing material, and bed volume (or type and number of trays), materials of construction, and use of ancillary equipment (e.g., reflux, heat exchangers) are highly specific to the waste being treated. For example, a survey of commercial steam strippers currently in use to treat pollutants revealed tower diameter ranges of 1.0 to 9.5 feet, column height ranges of 10 to 180 feet, and throughputs ranging from 250 gpd to as high as 500,000 gpd.[7]

In typical applications of stripping volatiles from aqueous wastes, steam requirements range from 10 to 30 mole percent of the feed at system pressure of 1 atm and 100°C.[11] Steam consumption is directly related to the equilibrium vapor pressure of the material being stripped and its resistance to diffusion through the waste. The latter determines the extent to which equilibrium conditions are approached and becomes increasingly important as the concentration of volatile species is diminished. Equipment manufacturers provide steam consumption data for stripping organic streams which are appropriate if significant volatile quantities remain in the bottoms product. For example, one manufacturer reported 1,236 lb/hr of steam required to steam distill spent mineral spirits in a 100 gph capacity still. Processing xylene on the same unit would require 829 lb/hr whereas toluene steam requirements would be only 419 lb/hr.[10]

The reader is referred to design methodologies and cost estimation procedures which have been developed in the literature for more information on steam stripping design, optimization, and cost effectiveness.[1,2,4,5,7,11]

### 5.3.2  Demonstrated Performance

As in previous sections dealing with distillation and thin film
evaporation, very little data are available concerning the recovery of
nonsolvent halogenated compounds by steam stripping.  Most of the available
data concern halogenated solvents (see Reference 2).  However, the same
concepts apply to nonsolvent as apply to solvent halogenated compounds.
Azeotrophic behavior, including the use of the Hausbrand diagram, and relative
volatilization rates as indicated by Henry's Law constant can be used to
assess waste stream constituent behavior in steam stripping equipment.
Compounds in Table 5.3.1 which possess moderate to high Henry's Law constants
are good candidates for treatment by steam stripping.  However, steam
stripping could be considered for other halogenated organics if, for example,
decomposition would result from recovery by distillation.

Some performance data for halogenated organic compounds are reported in
References 9, and 12 through 16.  The removal data shown in Table 5.3.2., are
reported for low molecular weight halogenated organics, including some
solvents.  Additional solvent data can be found in Reference 9.  As noted in
the table, removal efficiencies can exceed 99 percent or more.  These data are
from two full-scale facilities in the pesticide industry.  Data found in all
references indicate that volatile halogenated organic compounds are readily
removed from aqueous waste streams due to their high activity coefficients.

### 5.3.3  Cost of Steam Stripping

Steam stripping costs are highly site-specific.  Large-scale units used
for stripping wastewater streams are custom designed for specific
applications.  As a result, equipment manufacturers are reluctant to supply
cost estimates without detailed waste characteristic and volume data for
specific applications.

Costs of these units are basically a function of throughput,
concentration, and relative volatility of the compounds to be stripped.  Flow
rate determines the diameter of the column (without reflux) and initial
concentration determines the required removal efficiency to meet effluent
treatment standards.  Relative volatility determines the ease with which

TABLE 5.3.2.    STEAM STRIPPING PERFORMANCE

|  | Influent (mg/L) | Effluent (mg/L) | Percent removal |
|---|---|---|---|
| **Stripper 1** | | | |
| Dichloromethane | 1,430 | <0.0153 | >99.99 |
| Carbon tetrachloride | <665 | <0.0549 | >99.99 |
| Chloroform | <8.81 | 1.15 | <86.9 |
| **Stripper 2** | | | |
| Dichloromethane | 4.73 | <0.0021 | >99.95 |
| Chloroform | <18.6 | <1.9 | 89.8 |
| 1,2-Dichloroethane | <36.2 | 4.36 | <88.0 |
| Carbon tetrachloride | <9.7 | <0.030 | 99.7 |
| **Stripper 3** | | | |
| Methylene chloride | 34 | <0.01 | >99.97 |
| Chloroform | 4,509 | <0.01 | >99.99 |
| 1,2-Dichloroethane | 9,030 | <0.01 | >99.99 |

Source:    Reference 14.

halogenated constituents can be stripped.  Depending on relative volatility, an optional tradeoff between column height, reflux, steam rate, operating pressure, and post-treatment costs can be established.

Treatment cost data presented in the literature were generally not useful for predicting total waste treatment costs for a range in waste characteristics.  RTI[13] presented cost data for four batches of solvent waste.  These data were for specific wastes treated in an offsite facility and, therefore, were not indicative of general onsite processing costs.  Water General[11] presented a detailed design and treatment cost modeling approach. This methodology does not include an evaluation of post-treatment costs or cost reductions achieved through waste constituent recovery.  For dilute wastewaters, these can effectively be ignored since they are relatively small and offset one another.  However, these costs can represent significant fractions of total waste processing costs for more concentrated wastes, and therefore, must then be taken into consideration.

Two cost analyses are presented below.  The first provides capital and operating cost equations for steam stripping of dilute wastewater streams.  It is based on a review of actual onsite steam stripping installations performed by JRB Associates.[7]  The second analysis, performed by GCA, is appropriate for developing cost estimates for more concentrated wastes.  This analysis takes into consideration three residual disposal options (wastewater treatment, use as a fuel, and incineration) and discusses the impact of various waste characteristics and cost centers on overall processing costs.

Steam Stripping Costs for Wastewater Streams--

JRB analyzed cost and design data for 15 industrial steam strippers used to recover secondary materials or organic priority pollutants from wastewaters that flowed into secondary biological treatment systems.[7]  Steam strippers used to recover or recycle primary products/raw materials were excluded from this analysis since they differ in design from units used to treat wastewater.  In addition, tray towers were chosen for use in the analysis instead of packed towers since tray tower data were more readily available.

Capital and operation and maintenance (O&M) costs were normalized to 1980 dollars using the appropriate Engineering News Record indices.  Where installation costs were not provided, they were assumed to be 50 percent of

capital costs.  Capital costs include:  stripping columns, feed tanks, feed
preheaters, condensers, decanters, organic phase pumps, bottom pumps, and
existing equipment modifications.  O&M costs include:  operation and
maintenance labor, maintenance materials, steam, and electricity.

An analysis was performed to determine a mathematical relationship
between capital and O&M costs and significant steam stripper design parameters
such as contaminant volatility, wastewater flows, column diameter, and column
height.  The results of this analysis showed that capital costs were best
related to the diameter (D, in inches) and height (H, in feet) of the column,
while O&M costs were best related to the diameter and wastewater flow (Q, in
million gallons/day) as follows:

$$\text{Capital cost (in million dollars)} = 0.246 - 2.88 \times 10^{-4} \, (D) + 1.546 \times 10^{-6} \, (D^2 H)$$

$$\text{O\&M cost (in million dollars)} = 3.68 \times 10^{-3} \, (D) + 0.809 \, (Q) - 0.023$$

Overall, predicted capital costs were within a factor of 3 of reported
costs, O&M cost estimates were within a factor of 5, and costs per gallon of
treated waste were within a factor of 3.7 of actual values.  With an annual
capital recovery factor of 0.177, capital costs accounted for an average of
26 percent of total cost per gallon of treated waste.  Excluding one facility
which had low capacity utilization (largest diameter tower but lowest flow
rate), costs per gallon of treated waste averaged 0.9 cents per gallon with a
range of 0.14 to 20.4 cents per gallon.  Costs for four packed towers,
excluded from the analysis, averaged 1.12 cents per gallon.  Since flow rates
to these units were only one-fourth of that in the average tray column, the
cost difference may be attributable to economies of scale.

The above data are applicable to steam stripping costs for continuous
flow columns treating dilute (i.e., less than 1 percent) organic contaminated
aqueous wastes.  They do not include waste pretreatment and bottoms
post-treatment costs or net cost benefits derived from material recovery.
Overhead products consist of a volatile organic-water mixture which requires
further treatment; e.g., distillation or additional stripping.  Additional
overhead processing costs may be offset by benefits resulting from recovery

(e.g., reduced raw material purchase requirements) and therefore these costs can be neglected.  Bottoms post-treatment will be required if organic compound concentrations continue to exceed disposal limits or if the waste is still considered to be hazardous due to the presence of other nonvolatile contaminants.  Post-treatment by activated carbon, biological treatment, or other methods add roughly 2 cents per gallon to waste treatment costs.

JRB also attempted to determine cost variability as a function of contaminant volatility.  JRB used the design methodology provided by Water General Corporation[11] to determine variability in column height and capital cost required to strip compounds with different Henry's Law constants.  Water General's methodology involves calculation of a stripping factor which is proportional to Henry's Law constant, the vapor/liquid flow ratio, and the reciprocal of tower operating pressure.

JRB assumed a steam-to-liquid feed ratio of 10 percent and atmospheric column operating pressure.  Costs were based on stripping to a maximum residual VOC concentration of 1 ppm.  Minimum column diameter was set at 1.0 feet and minimum height at 10 feet to reflect wastewater processing equipment currently in use.

Table 5.3.3 summarizes the resulting cost data based on the above assumptions.  As shown, cost per gallon of waste treated shows little variability between compounds with different volatility.  However, if JRB's column size constraints were removed, the data would show lower treatment costs for wastes with highly volatile constituents and low flow rates.  Also, operating costs would constitute a higher fraction of total costs since optimal operating conditions would, in some cases, be represented by higher steam rates instead of increases in column height.

Steam Stripping Costs for Nonaqueous Wastes--

GCA performed a cost analysis for steam stripping of nonaqueous wastes using three nominal flow rates (10, 50, and 500 gph) and three disposal methods; i.e. wastewater treatment, use as a fuel, incineration.[12]  The cost of stripping concentrated wastes (i.e., greater than 1 percent organic constituents) is highly dependent on disposal method.  These disposal methods were selected because they represent a wide range in stripped product characteristics (i.e., aqueous, concentrated organic with Btu value such as

TABLE 5.3.3.   STEAM STRIPPING COSTS FOR WASTEWATER STREAMS CONTAINING CONTAMINANTS OF VARYING HENRY'S LAW CONSTANT

| Henry's Law constant range | Example compound | Flow Rate (MGD) | Height (ft) | Diameter (ft) | Capitol Cost ($MM) | O&M Cost ($MM) | Unit cost (¢/gal)[a] |
|---|---|---|---|---|---|---|---|
| Greater than $10^{-1}$ | 1,1,1-Trichloroethane | 1.0 | 10.0 | 6.28 | 0.312 | 1.06 | 0.36 |
| | | 0.10 | 10.0 | 2.11 | 0.249 | 0.151 | 0.63 |
| | | 0.01 | 19.6 | 1.00 | 0.247 | 0.029 | 2.30 |
| $10^{-2}$ to $10^{-3}$ | Acrylonitrile | 1.0 | 14.2 | 6.28 | 0.349 | 1.06 | 0.36 |
| | | 0.10 | 18.8 | 2.11 | 0.257 | 0.151 | 0.63 |
| | | 0.01 | 30.9 | 1.00 | 0.249 | 0.029 | 2.30 |
| Less than $10^{-4}$ | Nitrobenzene | 1.0 | 36.1 | 6.28 | 0.540 | 1.06 | 0.37 |
| | | 0.10 | 42.9 | 2.11 | 0.281 | 0.151 | 0.64 |
| | | 0.01 | 65.2 | 1.00 | 0.257 | 0.029 | 2.40 |

[a]Assuming 312 operating days/year and an annual capital recovery factor of 0.177.

Source:  Adapted from JRB.  Reference No. 7.

oil, and highly contaminated organic material) and disposal costs.  Other
major cost variables considered included capital, installation, maintenance,
labor, overhead and utility costs and value of recovered organics.

Capital costs for process equipment, tanks, engineering, electricity and
instrumentation were taken from the literature.[17]  A contingency of
15 percent and an annualized cost of 17.7 percent of the total were assumed as
summarized in Table 5.3.4.  Maintenance costs for stream strippers were based
on an EPA estimate of 4.13 percent of annualized capital cost.[11]  Labor
costs were assumed to be $14.42/hour including overhead.[17]  Labor usage was
assumed to range linearly from 0.5 to 3.0 operators for the flow rates under
consideration, with a base case operating time of 2,080 hours/year.

Utility costs were assumed to average $0.04/gallon of recovered
volatiles.  This value is based on the cost of steam ($3.00/million Btu)[11],
electricity ($0.04/KWH), and cooling water ($0.25/1,000 gallons) necessary to
separate and condense volatiles from an organic-water mixture.  This value
will vary depending on the material to be recovered.  For example, at
1 percent volatile organic concentration, utility costs range from
approximately 2 cents (e.g., acetyl chloride) to 7 cents (e.g., hexachlorobutadiene) per gallon of recovered organic, depending primarily on the
compound's boiling point in the mixture.  Utility costs for more highly
concentrated waste depend on both the boiling temperature and heat of
vaporization of the constituent to be separated, and range from roughly 3 to
6 cents per gallon of recovered compound.[17]  Since utility costs generally
represent a small fraction of total treatment costs, an average value of
4 cents per gallon was assumed.

Three methods of bottoms disposal were used in this analysis.  Wastewater
treatment technologies such as adsorption and biological treatment were
assumed to cost 2 cents per gallon.  Organic bottoms which could be used as a
fuel substitute were assumed to cost $0.20/gallon and bottoms which required
incineration (e.g., liquid injection) were assumed to cost $2.00/gallon.

Finally, recovered organic compound (95 percent recovery) was assumed to
have a value of $2.00/gallon for the purposes of calculating overall unit cost
of waste treated.  Although many halogenated organics have higher purchase
prices, this value was used since many recovered materials will be in the form
of less valuable mixtures or require additional treatment to achieve purity
specifications.

TABLE 5.3.4.  COST COMPONENTS FOR ONSITE STEAM STRIPPING HALOGENATED ORGANIC COMPOUNDS RECOVERY:  (EXAMPLE CASE WITH 30 PERCENT HALOGENATED ORGANIC CONTENT AND BOTTOMS USED AS FUEL)

| Nominal feed rate (gph) | 10 | 50 | 500 |
|---|---|---|---|
| **Capital costs** | | | |
| Process equipment | 8,000 | 39,000 | 72,000 |
| Tanks | 18,500 | 18,500 | 74,000 |
| Subtotal (1) | 26,500 | 57,500 | 146,000 |
| Engineering, electrical, instrumentation (20% of (1)) | 5,300 | 11,500 | 29,200 |
| Subtotal (2) | 31,800 | 69,000 | 175,200 |
| Contingency (15% of (2)) | 4,800 | 10,400 | 26,300 |
| Subtotal (3) | 36,600 | 79,400 | 201,500 |
| Annualized capital cost (17.7% of Subtotal (3)) | 6,478 | 14,054 | 35,666 |
| **Operating and maintenance costs[a]** | | | |
| Maintenance (4.13% of annualized capital cost) | 268 | 580 | 1,473 |
| Labor ($30,000/man-year including overhead) | 15,000 | 22,800 | 90,000 |
| Utility costs (4¢/gallon of recovered solvent) | 237 | 1,186 | 11,856 |
| Compound recovery benefit cost ($2/gallon of recovered compound) | (11,850) | (59,300) | (592,800) |
| Disposal cost ($0.20/gallon) | 3,391 | 16,953 | 169,530 |
| Net O&M cost | 7,046 | (17,781) | (319,941) |
| Total capital and O&M cost | 13,524 | (3,727) | (284,275) |
| Cost/gallon of waste treated | 0.65 | (0.04) | (0.27) |
| Threshold cost of recovered compound ($/gallon) | 4.24 | 1.83 | 1.00 |

[a]Numbers in parenthesis represent revenues.

Table 5.3.4 provides an example cost analysis for treating waste containing volatile organics and using the residual bottoms product as a fuel.    Table 5.3.4 summarizes the results of the cost analysis for wastes ranging from 1 to 70 percent volatile organics for the three bottoms disposal scenarios.    Cost figures are presented on the basis of cost per gallon of waste treated and threshold cost per gallon of volatile organics recovered.

Table 5.3.5 presents steam stripping costs as a function of throughput, volatile organic compound content and disposal method.    The table demonstrates that stripping costs increase dramatically with increasing bottoms disposal cost, particularly for high flow rate units.    For example, at 10 gpm and 10 percent organic content, disposal costs for wastewater treatment account for only 4 percent of total treatment costs.    This fraction jumps to 16 and 66 percent for disposal by fuel and non-fuel incineration, respectively.    At a feed rate of 500 gpm, disposal costs became even more significant, accounting for 24, 61, and 93 percent of total costs as more expensive disposal methods are used.    These costs are lower for wastes with higher percentages of recoverable organics.    However, disposal costs remain significant for high-volume units with nonaqueous bottoms products, unless they too can be reused as a product stream.

As capacity and percent recoverable volatile organic increase, total treatment cost becomes increasingly sensitive to the value of recovered overhead.    For example, at 10 gpm with 10 percent volatile organics in the waste, the total value of recovered overhead is 16 percent of processing costs (assuming a $2.00 value per gallon of recovered overhead and using bottoms as fuel).    This percentage jumps to 113 percent as volatile organic content in the feed increases to 70 percent, and increases further to 554 percent of processing costs if capacity is then increased to 500 gpm.    Thus, unit value of recovered organics has an increasingly significant impact on processing economics as throughput and percent recoverable organics increase.

Utility costs show a similar relationship to throughput and recoverable organics content.    For wastes with less than 10 percent volatile organics, utility costs are only a few percent of total costs.    At high flow rate and high volatile organic content, however, these costs become significant.    For example, at 500 gpm and 70 percent volatiles in the feed, utilities account for 11 percent of total processing costs when bottoms are used as fuel (2.6 percent when bottoms are incinerated).

TABLE 5.3.5.   STEAM STRIPPING COST ESTIMATES AS A FUNCTION OF THROUGHPUT, VOLATILE ORGANIC COMPOUNDS CONTENT AND DISPOSAL METHOD

| Bottoms disposal method | Organic content in waste (%) | Cost ($)/gal of waste treated[a] | | | Cost ($)/gal of organic recovered[a] | | |
|---|---|---|---|---|---|---|---|
| | | 10 gph | 50 gph | 500 gph | 10 gph | 50 gph | 500 gph |
| Wastewater treatment | 1 | 1.05 | 0.36 | 0.12 | 108.00 | 38.58 | 14.59 |
| | 5 | 0.98 | 0.29 | 0.06 | 21.79 | 7.95 | 3.15 |
| | 10 | 0.89 | 0.21 | (0.03) | 11.01 | 4.08 | 1.68 |
| Use as fuel | 10 | 1.05 | 0.37 | 0.13 | 13.11 | 5.90 | 3.39 |
| | 30 | 0.65 | (0.04) | (0.27) | 4.24 | 1.83 | 1.00 |
| | 50 | 0.24 | (0.45) | (0.68) | 2.46 | 1.02 | 0.52 |
| | 70 | (0.17) | (0.85) | (1.09) | 1.70 | 0.67 | 0.31 |
| Incineration | 10 | 2.00[b] | 2.00[b] | 1.93 | b | b | 22.36 |
| | 30 | 2.00[b] | 1.43 | 1.19 | b | 6.98 | 6.15 |
| | 50 | 1.36 | 0.68 | 0.44 | 4.83 | 3.39 | 2.89 |
| | 70 | 0.61 | (0.07) | (0.31) | 2.88 | 1.85 | 1.49 |

[a] Numbers in parenthesis represent revenues.

[b] Cost of treatment via steam stripping exceeds incineration cost of raw waste ($2.00/gallon), therefore, incineration represents the lower cost alternative.

In contrast to disposal and utility costs, capital and labor represent the primary costs of processing low volume, dilute wastes. Capacity utilization is more critical for low volume units since capital costs are distributed over the total volume of waste processed. For example, increasing processing time from 8 to 16 hours per day reduces processing costs by 16 cents per gallon for 10 gpm units versus only 1 cent per gallon for 500 gpm strippers.

## 5.3.4   Overall Status of Process

### 5.3.4.1  Availability--

Steam stripping is a commonly applied waste treatment technology for separating low solubility halogenated organic compounds from water or low volatility organics (e.g., oil). The EPA has identified 27 industrial steam stripping wastewater treatment units; 11 units used by the pesticides industry, and 8 steam strippers (packed towers) used by a pharmaceutical manufacturer.[6] GCA's analysis of the commercial solvent recycling industry showed 25 percent of the reclaimers using steam distillation. This technology is directly applicable to waste streams containing halogenated organic compounds provided relative volatiles are sufficient to effect separation.

### 5.3.4.2  Application--

Steam stripping is commonly used as a pretreatment method, particularly when applied to concentrated solvent wastes. In many cases it can be used to reduce solvent concentrations to levels which permit direct discharge, although it is often more cost-effective to use other treatment methods for final bottoms processing. The extent to which steam stripping is applied to halogenated organic compound wastes is unknown, but it should be equally effective for many of these compounds. Steam stripping of wastewater streams is typically followed by biological treatment or adsorption systems for final effluent polishing. Stripped organic waste bottoms can be decanted and used as fuel provided that chlorine content has been sufficiently reduced in the stripper or diluted with other fuels. In the case of nonvolatile compounds, reuse of bottoms product as a feedstock may be possible.

5.3.4.3  Environmental Impact--

Post-treatment of both the overhead and bottoms streams is usually required to attain proposed effluent standards, although in certain instances the organics concentration in the bottoms stream may be below proposed standards.  Overhead products typically undergo liquid-liquid separation through decanting to achieve effluent specifications.  However, the organic compound may require drying and the aqueous phase may require further treatment to remove dissolved organics.  Air emissions from column vents can be significant[13] and should, at least, be monitored.

5.3.4.4  Advantages and Limitations--

Steam stripping is preferrable to other physical separation technologies in the following instances:

- Treating wastes which contain high solids or polymerizable materials which would otherwise foul heat transfer surfaces;

- Treating wastes which contain constituents that form low boiling point azeotropes with water, particularly those which require low processing temperatures due to thermal degradation; and

- Treating wastes to low residual organic compound content, particularly when the bottoms product would be rendered unpumpable in the absence of water.

Steam stripping is not well suited to treating wastes in which either the overhead or bottoms are difficult to separate from water.  It is better utilized for separating organics which decant readily and have low solubilities in water (e.g., halogenated organics) and less applicable to treating water soluble wastes such as alcohols.

# REFERENCES

1.   Allen, C. C., and B. L. Blaney.  Techniques for Treating Hazardous Waste to Remove Volatile Organic Constituents.  Research Triangle Institute performed for U.S. EPA HWERL.  EPA-600/D-85-127, PB85-218782/REB.  March 1985.

2.   Ellerbe, R. W.,  Steam Distillation/Stripping.  The Rust Engineering Company.  In:  Handbook of Separation Techniques for Chemical Engineers.  McGraw-Hill Book Company, New York, N.Y.  1979.

3.   Weast, R. C.,  Handbook of Chemistry and Physics.  65th Edition, CRC Press, Cleveland, OH.  1984-1985.

4.   Perry, R. H.,  Chemical Engineers' Handbook, 6th Edition, McGraw-Hill Book Company, New York, N.Y.  1984.

5.   Kirk-Othmer Encyclopedia of Chemical Technology.  John Wiley & Sons, New York, N.Y., 3rd Edition.  1978.

6.   U.S. EPA Office of Solid Waste.  Background Document for Solvents to Support 40 CFR Part 268 Land Disposal Restrictions.  Volume II.  Analysis of Treatment and Recycling Technologies for Solvents and Determination of Best Available Demonstrated Technology.  U.S. EPA Public Docket.  January 1986.

7.   JRB Associates and Science Applications International Corporation.  Costing Documentation and Notice of New Information Report. Draft Report prepared for U.S. EPA.  June 1985.

8.   U.S. EPA Office of Solid Waste.  Analysis of Organic Chemicals, Plastics and Synthetic Fibers (OCPSF) Industries Data Base.  U.S. EPA Public Docket.  January 1986.

9.   Arienti, M., et al.  GCA Technology Division, Inc.  Technical Assessment of Treatment Alternatives for Wastes Containing Halogenated Organics.  Draft Report prepared for U.S. EPA Office of Solid Waste under EPA Contract No. 68-01-6871.  October 1984.

10.  Michigan Department of Commerce.  Hazardous Waste Management in the Great Lakes:  Opportunities for Economic Development and Resource Recovery.  September 1982.

11.  Shukla, H. M., and R. E. Hicks.  Water General Corporation, Waltham, MA. Process Design Manual for Stripping of Organics.  Prepared for U.S. EPA IERL.  EPA-600/2-84-139.  August 1984.

12.  Breton, M., et al.  Technical Resource Document - Solvent Bearing Wastes.  Prepared by GCA Technology Division, Inc. for HWERL, CIN under Contract No. 68-03-3243, Work Assignment No. 2, September 1986.

13.  Allen, C. C., Research Triangle Institute, and Simpson, S., and G. Brant of Associated Technologies, Inc.  Field Evaluation of Hazardous Waste Pretreatment as an Air Pollution Control Technique.  Prepared for U.S. EPA HWERL under EPA Contract No. 68-02-3992.  January 1986.

14.  Jett, G. M.  Development Document for Expanded Best Practicable Control Technology, Best Conventional Control Technology, Best Available Control Technology in the Pesticides Chemicals Industry.  Effluents Guidelines Division, U.S. EPA.  EPA-440/1-82-079b.  November 1982.

15.  Coco, J. H., et al.  Gulf South Research Institute, New Orleans, LA. Development of Treatment and Control Technology for Refractory Petrochemical Wastes.  U.S. EPA, ADA, OK.  EPA-600/2-79-080.  April 1979.

16.  Stover, E. L., and D. F. Kincannon.  Contaminated Ground Water Treatability - A Case Study.  Journal of the American Water Works Association.  June 1983.

17.  Horsak, R. D., et al.  Pace Company Consultants and Engineers, Inc. Solvent Recovery in the United States:  1980-1990.  Houston, TX. Prepared for Harding Lawson Associates.  January 1983.

## 5.4    LIQUID - LIQUID EXTRACTION

Liquid - liquid extraction is the separation of constituents of a liquid solution by transfer to a second liquid, immiscible in the first liquid, but for which the constituents have a preferential affinity.  Although not a commonly used waste treatment technology, liquid-liquid extraction has potential for removal of many organic constituents from effluent waste streams.  Liquid extraction can be attractive in cases where the solutes in question are toxic or non-biodegradable, where the solutes are present at sufficiently high concentrations to provide economic recovery value, when steam stripping would be rendered less effective by low solute volatility or formation of azeotropes, or when high solute concentrations increase activated carbon adsorption costs to excessive levels.  Historically, its general application to wastewater treatment has been limited to removal of high concentrations (5 percent or greater) of phenol and phenolic compounds.  Steam stripping is not very effective on phenolic compounds because of their low Henry's Law constants and activated carbon adsorption is not feasible at these high concentrations.

### 5.4.1  Process Description

The liquid-liquid extraction process is shown in Figure 5.4.1.  The typical process includes the following basic steps:

1.    Extraction of organic pollutants from wastewater,

2.    Recovery of solute from the solvent phase or extract, and

3.    Removal of solvent from treated wastewater or raffinate.

The first extraction step brings two liquid phases (feed and solvent) into intimate contact to allow transfer of solute from the feed to the solvent.  Any method by which single or multistage mass-transfer processes can be conducted can conceivably be used to conduct liquid extractions.  For example, an extractor unit can be a mixer-settler device in which feed and solvent are mixed by agitation, then allowed to settle and separate into two

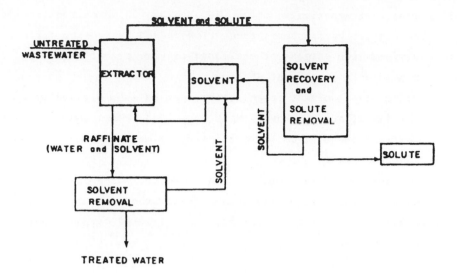

Figure 5.4.1.   Schematic of extraction process.

liquid streams.  Alternatively, it can be a column in which two liquids are brought into contact by counter-current flow caused by density differences. The process yields two streams, the cleaned stream, or raffinate, and the extract, or solute-laden solvent stream.

The second step, solvent regeneration, can be accomplished by a second extraction or distillation.  For example, a second extraction with caustic may be used to extract phenol from light oil (the primary solvent in dephenolizing coke plant wastewaters).

Distillation is a more common solvent regeneration process.  Potential difficulties with distillation may arise if azeotropes are formed, or if the relative volatilities of the solvent and the extracted compound are close enough to hinder separation.

The third step, removal of solvent in the treated wastewater or raffinate, is necessary when solvent concentrations are great enough to create solvent losses that would add significantly to the process cost or have a detrimental environmental impact.  Solvent removal can be accomplished by a number of technology options.  When treating large quantities of dilute wastes, an additional extraction step usually cannot economically compete with other technologies such as stripping, biological or adsorption post-treatments.

5.4.1.1  Pretreatment Requirements for Different Waste Forms and
        Characteristics--
    Pretreatment is necessary to remove material which will interfere with the mass transfer of the organic contaminant into the solvent extract. Reduced mass transfer efficiency results in the need for higher solvent/aqueous phase ratios to obtain desired levels of extraction.  Thus, any emulsion or organic phase droplets should normally be removed prior to treatment.  Solids, to the extent that they retain sorbed contaminants or hamper column performance, should also be removed.  In certain cases dissolved solids can also affect partitioning of the solute(s), and removal or addition of dissolved solids may be desirable to enhance the separation.  Similarly, changes in temperature may also modify partitioning behavior.  Phase distribution data, if not available in the literature, will generally have to be developed in the laboratory.  An estimation method based on vapor liquid data for binary systems can be used to estimate the distribution of an organic compound (at low concentrations) between water and an organic solvent.[1]

5.4.1.2  Operating Parameters--

In the simplest case of liquid extraction, the solvent is added to a liquid mixture, causing a second liquid phase to form.  It may be desirable to add a salt to an aqueous phase to enhance the activity of a component, causing it to transfer into a nonaqueous phase in which the salt is insoluble. Similarly, adjusting the pH of an aqueous phase containing organic acidic or basic solutes will depress their ionization potential and cause them to concentrate in the nonaqueous solvent phase.  It is also often helpful to change the temperature of the phases in contact to give the most favorable equilibrium at each step of the extraction.

Theoretically, any aqueous organic waste can be treated by extraction. However, determining potential feasibility requires a series of analyses to assess overall system applicability.  Much depends on how residual solvent is to be removed from the treated water stream, how the solvent is to be regenerated, and what restrictions exist for each unit operation.

In general, extraction is best suited for waste streams of consistent composition to assure satisfactory performance.  In cases where performance is less important, acceptable ranges in waste characteristics become broader, as in the case where extraction is to be used as a pretreatment.  For example, when several waste streams are to be combined for final treatment, a single waste stream with higher constituent concentration may be extracted to reduce the load on the final treatment process.

As noted in Perry's Chemical Engineers' Handbook, the removal mechanisms in extraction are primarily physical, since the solutes being transferred are ordinarily recovered without chemical change.  The physical equilibrium relationship on which such operations are based depend mainly on the chemical characteristics of the solutes and solvents.  The use of a solvent that chemically resembles one component of a mixture more than the other components will lead to concentration of that like component in the solvent phase.

The choice of solvent is a key factor in evaluating the utility of liquid extraction as a means of removing hazardous organic compounds from aqueous waste streams.  Perry, et al.,[1] lists characteristics which must be assessed in selecting a solvent.  These are:

- Selectivity-the ability of a solvent to extract the organic contaminant preferentially from the aqueous phase. It is a numerical measure that is equal to the ratio of the distribution constants of contaminant and water in the solvent. As such, it is analogous to relative volatility as used in distillation. Poor selectivity (ratios near unity) means large solvent feed ratios and a large number of extraction stages will be needed for good separation.

- Recoverability-the solvent must be recoverable from both extract and raffinate. Since distillation is the usual recovery method, relative volatilities of all components should be favorable and low latent heats for volatile solvents are desirable.

- Distribution Coefficient-the distribution coefficient of the contaminant should be large in order to achieve selectivity and reduce equipment size and costs.

- Contaminant Solubility-the solubility of the extracted contaminant in the solvent should be high in order to reduce solvent requirements.

- Solvent Solubility-the solubility of the solvent in the aqueous phase should be low. This will generally increase selectivity and the range of waste stream concentrations that can be handled, and reduce costs of solvent recovery or makeup.

- Density-a difference in density is essential since the flow rates and separation of the two phases is directly affected.

- Interfacial Tension-the interfacial tension should be large to assist in the coalescence of dispersed phase droplets.

- Other-other desirable solvent properties are low corrosivity, low viscosity for higher mass transfer rates, nonflammability, low toxicity, and low cost.

A guide to solvent selection may be provided by binary critical solution temperatures of solute components with prospective solvents. The solvent having the lower critical solution temperature with the solute compound will be more selective in an extraction from the aqueous phase.[1]

A significant quantity of data have been collected for the distribution of pollutants in water and various extractive solvents. These values, called equilibrium distribution coefficients ($K_D$), generally express the equilibrium concentration of the solute as the ratio of the weight percent in solvent relative to water:

$$K_D = \frac{X_{os}}{X_{oa}}$$

where:     $X_{os}$ the weight fraction of organic solute in the solvent
phase, and

$X_{oa}$ is the weight fraction of organic solute in the aqueous
phase, both at equilibrium.


$K_D$ values for the octanol/water system for the halogenated organic
compounds of interest are provided in Appendix A.   These data represent just a
small fraction of the data available in the literature for ternary systems
consisting of water and two organic compounds.   For example, a number of
references are provided in Perry[1] along with distribution coefficients for
over 200 selected water/organic solute/organic solvent systems which include
some of the organic compounds of concern.   Values of $K_D$ that are specific to
the compounds of concern are provided in a number of recent publications.
These data, of use in assessing the potential of liquid-liquid extractions as
a treatment technology are provided in Table 5.4.1 and in Table 5.4.2.   These
data were reported in the AIChE Symposium Series (1981) by S. T. Hwang,[2]
using References 3 through 6 as additional data sources.

Higher values of $K_D$ (or $K_V$) mean that less solvent is required to
extract a given amount of solute from the wastewater which usually translates
into a less expensive extraction processes.   The ratio of the distribution
coefficients of solvent systems for extraction of a specific compound from
water is a measure of the relative amounts of solvent that must be employed to
achieve a given level of extraction.   As shown in the table, distribution
coefficients increase with increasing chlorine content, paralleling the
solubility behavior of these compounds in water.   In general, chlorinated
compounds would appear to be well suited for recovery by extraction, however,
distribution coefficients are only one of the many solvent properties that
must be considered.[8]

The Chemical Engineers' Handbook, Kirk Othmer (Reference 7) and other
background materials present calculation and design methods that can be used
to assess the applicability of liquid-liquid extraction to specific waste
streams.   The techniques generally involve the use of equilibrium distribution
data to develop equilibrium and operating line curves.   These can be used to

## TABLE 5.4.1.   $K_V$ VALUES FOR AQUEOUS/ORGANIC SYSTEMS[a]

| Solute | Tricresyl phosphate | Undecane | MIBK | Tridecane | Benzene | Isobutylene | Isobutane | n-butyl acetate | Isobutyl acetate | Diisopropyl ether | Octanol |
|---|---|---|---|---|---|---|---|---|---|---|---|
| carbon tetrachloride (tetrachloromethane) | 480 | 380 | 930 | 300 | 2,200 | 1,200 | 810 | 870 | 900 | 990 | 436 |
| chlorobenzene | 250 | 260 | 1,200 | 200 | 2,900 | 685 | 445 | 1,200 | 1,060 | 460 | 692 |
| 1,2,4-trichlorobenzene | 17,000 | 12,000 | 20,000 | 10,000 | 66,000 | 15,000 | 10,000 | 19,900 | 19,000 | 14,400 | 18,200 |
| hexachlorobenzene | 93,000 | 290,000 | 750,000 | 200,000 | $1\times10^7$ | 68,000 | 37,000 | $7\times10^5$ | $6.9\times10^5$ | $5.2\times10^5$ | $1.5\times10^6$ |
| 1,2-dichloroethane | 31 | 10 | 38 | 8 | 100 | 41.6 | 24 | 37 | 36 | 32 | 36 |
| 1,1,1-trichloroethane | 81 | 120 | 220 | 110 | 270 | 190 | 150 | 210 | 210 | 150 | 92 |
| hexachloroethane | 11,000 | 3,500 | 9,600 | 2,500 | 62,000 | 8,600 | 5,300 | 9,400 | 9,000 | 7,600 | 8,800 |
| 1,1-dichloroethane | 51 | 56 | 140 | 52 | 100 | 94 | 70 | 140 | 130 | 68 | 138 |
| 1,1,2-trichloroethane | 74 | 70 | 220 | 64 | 200 | 110 | 80 | 210 | 200 | 86 | 224 |
| 1,1,2,2-tetrachloro-ethane | 75 | 27 | 85 | 20 | 590 | 61 | 37 | 85 | 80 | 60 | 86 |
| chloroethane | 28 | 31 | 62 | 30 | 43 | 50 | 40 | 60 | 60 | 37 | 35 |
| bis (chloroethyl) ether | | | | | | | | | | | |
| bis (2-chloroethyl) ether | 21 | 19 | 90 | 17 | 82 | 36 | 22 | 90 | 80 | 24 | 94 |
| 2-chloroethyl vinyl ether (mixed) | 15 | 35 | 34 | 31 | 60 | 57 | 45 | 30 | 33 | 44 | 20 |
| 2-chloronaphthalene | 33,000 | 2,500 | 28,000 | 2,200 | 29,000 | 5,000 | 2,500 | 29,000 | 25,000 | 8,000 | 24,000 |
| 2,4,6-trichlorophenol | 1,374 | 9 | 4,900 | 9 | 144 | 27 | 7 | 2,610 | 4,123 | 1,580 | 2,400 |
| parachlorometa cresol | 188 | 1.5 | 667 | 1.5 | 25 | 5 | 1.2 | 357 | 563 | 217 | 1,240 |
| chloroform (trichloromethane) | 80 | 46 | 117 | 44 | 88 | 76 | 57 | 116 | 111 | 56 | 93 |
| 2-chlorophenol | 90.5 | 0.9 | 432.25 | 0.9 | 15 | 3 | 0.7 | 255.8 | 272 | 105 | 141 |
| 1,2-dichlorobenzene | 2,700 | 1,900 | 3,800 | 1,600 | 11,000 | 2,300 | 1,600 | 3,800 | 3,500 | 2,300 | 2,400 |
| 1,3-dichlorobenzene | 2,700 | 2,800 | 4,600 | 2,400 | 13,700 | 3,500 | 2,500 | 4,580 | 4,350 | 3,400 | 2,400 |
| 1,4-dichlorobenzene | 2,800 | 830 | 2,500 | 600 | 12,600 | 2,000 | 1,260 | 2,400 | 2,300 | 2,000 | 2,455 |
| 1,1-dichloroethylene | 1,100 | 1,000 | 1,200 | 900 | 2,750 | 1,200 | 960 | 1,100 | 1,120 | 1,150 | 999 |
| 1,2-trans-dichloro-ethylene | 700 | 260 | 840 | 210 | 1,500 | 710 | 510 | 810 | 810 | 830 | 738 |
| 2,4-dichlorophenol | 304 | 3 | 1,080 | 3 | 46 | 9 | 3 | 580 | 911 | 351 | 562 |
| 1,2-dichloropropane | 120 | 125 | 330 | 114 | 275 | 200 | 150 | 320 | 340 | 150 | 305 |
| 1,3-dichloropropylene (1,3-dichloropropene) | 120 | 130 | 450 | 120 | 370 | 230 | 160 | 450 | 420 | 150 | 474 |
| bis(2-chloroisopropyl) ether | 162 | 430 | 430 | 360 | 850 | 980 | 680 | 350 | 410 | 630 | 176 |
| bis(2-chloroethoxy) ether | 7 | 9 | 8 | 8 | 15 | 26 | 17 | 6 | 8 | 14 | 3 |

(continued)

## TABLE 5.4.1 (continued)

| Solute | Solvent | | | | | | | | | | |
|---|---|---|---|---|---|---|---|---|---|---|---|
| | Tricresyl phosphate | Undecane | MIBK | Tridecane | Benzene | Isobutylene | Isobutane | n-butyl acetate | Isobutyl acetate | Diisopropyl ether | Octanol |
| methylene chloride (dichloromethane) | 13 | 15 | 50 | 12 | 106 | 36 | 26 | 50 | 46 | 24 | 72 |
| methyl chloride (chloromethane) | 9 | 10 | 18 | 9 | 14 | 14 | 11 | 18 | 17 | 11 | 30 |
| methyl bromide (bromomethane) | 432 | 460 | 960 | 450 | 750 | 680 | 550 | 960 | 930 | 530 | 1,000 |
| bromoform (tribromomethane) | 143 | 140 | 530 | 130 | 500 | 240 | 160 | 540 | 490 | 160 | 610 |
| dichlorobromomethane | 22 | 22 | 62 | 20 | 50 | 36 | 27 | 62 | 59 | 27 | 63 |
| trichlorofluoro-methane | 563 | 690 | 980 | 650 | 770 | 1,120 | 950 | 939 | 955 | 850 | 740 |
| dichlorodifluoro-methane | 2,400 | 7,400 | 6,700 | 6,000 | 5,300 | 38,000 | 33,000 | 5,000 | 6,330 | 15,000 | 4,500 |
| chlorodibromomethane | 30 | 28 | 140 | 26 | 120 | 53 | 32 | 140 | 130 | 34 | 173 |
| hexachlorobutadiene | 288 | 90 | 270 | 59 | 9,500 | 180 | 86 | 280 | 240 | 140 | 299 |
| hexachlorocyclopen-tadiene | 26,000 | 25,000 | 40,560 | 20,900 | 86,000 | 52,000 | 34,000 | 36,900 | 37,600 | 37,000 | $2.4 \times 10^4$ |

[a] $K_V = K_D \times \dfrac{\text{Specific Gravity of Solvent}}{\text{Specific Gravity of Water}} = \dfrac{\text{Grams Solute/1,000 ml Solvent}}{\text{Grams Solute/1,000 ml Water}}$

## TABLE·5.4.2.   Kv VALUES FOR AQUEOUS/ORGANIC SYSTEMS [a]

| Solute | Cumene | Mesityl oxide | MEK | Ethyl acetate | Ethyl ether | Ethyl benzene | n-hexanol | EDC | Toluene | Xylene | n-hexane |
|---|---|---|---|---|---|---|---|---|---|---|---|
| carbon tetrachloride (tetrachloromethane) | 1,460 | 770 | 690 | 668 | 1,200 | 1,266 | 236 | 2,329 | 1,570 | 2,069 | 866 |
| chlorobenzene | 1,080 | 1,700 | 3,480 | 2,360 | 1,000 | 1,450 | 1,180 | 3,700 | 1,660 | 2,500 | 610 |
| 1,2,4-trichlorobenzene | 35,000 | 19,900 | 19,600 | 17,000 | 20,000 | 43,000 | 9,090 | 73,000 | 51,000 | 75,700 | 18,500 |
| hexachlorobenzene | $2.6\times10^6$ | $6.3\times10^5$ | $4.7\times10^5$ | $4.2\times10^5$ | $7.5\times10^5$ | $4.2\times10^6$ | $1.2\times10^7$ | $1.9\times10^7$ | $5.7\times10^6$ | $6.8\times10^6$ | $8\times10^5$ |
| 1,2-dichloroethane | 72 | 34 | 30 | 28 | 40 | 55 | 14 | 142 | 70 | 92 | 24 |
| 1,1,1-trichloroethane | 196 | 212 | 221 | 208 | 239 | 22 | 96 | 212 | 248 | 388 | 177 |
| hexachloroethane | $2.44\times10^4$ | 8,183 | 6,600 | 6,100 | $1\times10^4$ | $2.8\times10^4$ | 1,184 | $9\times10^4$ | $3.7\times10^4$ | $4.7\times10^4$ | 8,600 |
| 1,1-dichloroethane | 66 | 163 | 200 | 186 | 127 | 93 | 130 | 55 | 101 | 163 | 73 |
| 1,1,2-trichloroethane | 108 | 256 | 309 | 278 | 174 | 165 | 202 | 115 | 184 | 290 | 95 |
| 1,1,2,2-tetra-chloroethane | 229 | 75 | 60 | 54 | 82 | 261 | 22 | 963 | 340 | 432 | 66 |
| bis(chloroethyl) ether | 33 | 67 | 79 | 75 | 61 | 40 | 50 | 25 | 43 | 70 | 39 |
| bis(2-chloroethyl) ether | 34 | 114 | 148 | 129 | 63 | 64 | 92 | 38 | 72 | 110 | 28 |
| 2-chloroethyl vinyl ether (mixed) | 49 | 26 | 23 | 23 | 54 | 50 | 6 | 45 | 56 | 86 | 50 |
| 2-chloronaphthalene | $1.6\times10^4$ | $3.7\times10^4$ | $4.9\times10^4$ | 41,000 | $1.9\times10^4$ | $2\times10^4$ | $3.8\times10^4$ | $1\times10^4$ | $2.3\times10^4$ | $3.5\times10^4$ | 4,000 |
| 2,4,6-trichlorophenol | 1,650 | 2,450 | 2,950 | 4,027 | 2,721 | 200 | 1,980 | 322 | 89 | 70 | 7 |
| parachlorometa cresol | 225 | 348 | 404 | 550 | 372 | 34 | 270 | 56 | 16 | 12 | 1.2 |
| chloroform (trichloromethane) | 56 | 134 | 162 | 150 | 103 | 79 | 109 | 48.3 | 86 | 138 | 60 |
| 2-chlorophenol | 110 | 200 | 200 | 265 | 179 | 20 | 130 | 32 | 9 | 7 | 0.7 |
| 1,2-dichlorobenzene | 5,970 | 4,000 | 4,057 | 3,560 | 3,400 | 7,327 | 2,248 | $1.2\times10^4$ | 5,670 | $1.3\times10^4$ | 2,880 |
| 1,3-dichlorobenzene | 7,800 | 4,560 | 4,500 | 4,047 | 4,660 | 9,295 | 2,240 | $1.5\times10^4$ | $1\times10^4$ | $1.6\times10^4$ | 4,280 |
| 1,4-dichlorobenzene | 5,800 | 2,170 | 1,788 | 1,656 | 2,570 | 6,000 | 673 | $1.8\times10^4$ | 7,766 | $9.9\times10^3$ | 1,968 |
| 3,3-dichlorobenzidine | 1 | 4 | 10 | 6 | 6.2 | 3 | 17 | 30 | 3 | 4 | 0.09 |
| 1,1-dichloroethylene | 1,900 | 990 | 870 | 840 | 1,360 | 2,127 | 359 | 2,770 | 2,400 | 3,800 | 1,330 |
| 1,2-trans-dichloroethylene | 1,300 | 738 | 665 | 640 | 982 | 903 | 294 | 1,768 | 1,100 | 1,488 | 562 |
| 2,4-dichlorophenol | 364 | 560 | 650 | 890 | 600 | 62 | 437 | 102 | 28 | 22 | 3 |
| 1,2-dichloropropane | 170 | 369 | 440 | 404 | 396 | 234 | 250 | 158.8 | 259 | 407 | 170.05 |
| 1,3-dichloropropylene (1,3-dichloropropene) | 168 | 555 | 703 | 632 | 333 | 315 | 492 | 180 | 246.6 | 548 | 178.8 |
| 4-chlorophenyl phenyl ether | 3,000 | 972 | 808 | 784 | 3,178 | 5,670 | 73 | 8,500 | 7,300 | 8,300 | 4,500 |

(continued)

# TABLE 5.4.2 (continued)

| Solute | | | | | | Solvent | | | | | |
|---|---|---|---|---|---|---|---|---|---|---|---|
| | Cumene | Mesityl oxide | MEK | Ethyl acetate | Ethyl ether | Ethyl benzene | n-hexanol | EDC | Toluene | Xylene | n-hexane |
| 4-bromophenyl phenyl ether | 2,090 | 349 | 243 | 211 | 4,437 | 4,630 | 46 | $2.7 \times 10^4$ | 6,580 | 7,400 | 562 |
| bis(2-chloroisoprophyl) ether | 656 | 287 | 256 | 253 | 881 | 662 | 31 | 518.7 | 751 | 1,066 | 763 |
| bis(2-chloroethoxy) methane | 12 | 5 | 5 | 5 | 21 | 12 | 0.4 | 8 | 14 | 19 | 18 |
| methylene chloride (dichloromethane) | 46 | 69 | 106 | 89 | 49 | 59 | 58 | 134 | 68 | 103 | 30 |
| methyl chloride (chloromethane) | 10 | 20 | 24 | 22 | 16 | 14 | 20 | 9 | 15 | 24 | 11 |
| methyl bromide (bromomethane) | 509 | 1,090 | 1,300 | 1,200 | 864 | 700 | 1,049 | 434 | 751 | 1,249 | 550 |
| bromoform (tribromomethane) | 208 | 682 | 880 | 778 | 366 | 418 | 696 | 238 | 462 | 734 | 188 |
| dichlorobromomethane | 30 | 73 | 89 | 81 | 52 | 44 | 61 | 28 | 48 | 77 | 29 |
| trichlorofluoro-methane | 701 | 925 | 1,000 | 975 | 1,200 | 718 | 454 | 503 | 773 | 1,249 | 894 |
| dichlorodifluoro-methane | 5,127 | 4,180 | 4,600 | 23,000 | $2.1 \times 10^4$ | 4,573 | 269 | 2,578 | 4,500 | 7,380 | 14,880 |
| chlorodibromomethane | 41 | 197 | 272 | 237 | 88 | 104 | 250 | 41 | 115 | 183 | 37 |
| hexachlorobutadiene | 1,450 | 232 | 158 | 135 | 227 | 3,000 | 40 | $2.3 \times 10^4$ | 4,340 | 4,976 | 277 |
| hexachlorocyclopentadiene | $5.5 \times 10^4$ | $3.6 \times 10^4$ | $4 \times 10^4$ | $3.6 \times 10^4$ | $5.8 \times 10^4$ | 60,000 | $1 \times 10^4$ | $6.2 \times 10^4$ | 69,000 | 975,000 | $4.7 \times 10^4$ |
| tetrachloroethylene | 5,500 | 8,700 | $1.3 \times 10^4$ | $1.1 \times 10^4$ | 5,500 | 7,300 | 5,200 | 17,000 | 8,200 | $1.2 \times 10^4$ | 3,400 |
| toluene | 1,300 | 606 | 527 | 506 | 919 | 1,290 | 170 | 2,674 | 1,690 | 2,080 | 708 |
| trichloroethylene | 375 | 1,074 | 1,320 | 1,220 | 771 | 595 | 840 | 344 | 650 | 1,030 | 436 |
| vinyl chloride | 39 | 58 | 63 | 62 | 65 | 42 | 35 | 30 | 45 | 74 | 49 |
| DDT | | | | | | | | | | | $9.1 \times 10^5$ |
| heptachlor | | | | | | | | | | | $1.1 \times 10^5$ |
| heptachlor epoxide | | | | | | | | | | | $4 \times 10^4$ |

$$= K_D \times \frac{\text{Specific Gravity of Solvent}}{\text{Specific Gravity of Water}} = \frac{\text{Grams Solute/1,000 ml Solvent}}{\text{Grams Solute/1,000 ml Water}}$$

provide graphical calculations of the number of theoretical stages required to achieve desired extraction levels.  The general method is analogous to the use of McCabe-Thiele diagrams to assess distillation performance.  Formulas, such as the Kremser equations, are also available that quantitatively express the effect of flow variations on exit concentration levels.  The use of such techniques (in conjunction with laboratory data) provide the basis for determining equipment size and post-treatment requirements and, therefore, the costs and applicability of the liquid extraction process.

5.4.1.3  Post-Treatment Requirements--

     The post-treatment requirements of a liquid-liquid extraction process are determined by many of the system component properties discussed above.  For example, solvent solubility in the aqueous phase will determine the need for further treatment to eliminate solvent discharge with the treated waste stream.  The relative volatilities of the solute and solvent will affect the ease of their separation following extraction.  Careful selection of solvent and proper design can minimize the cost and difficulty of such processes.  Those technologies most commonly used for raffinate post-treatment are established technologies such as steam or air stripping, carbon adsorption, and biological treatment.  Distillation will probably be used to separate solvent and solute.  These technologies are discussed in detail in other sections of this report.

5.4.2  Demonstrated Performance

     The use of liquid-liquid extraction for the treatment of aqueous organic waste streams has been limited.  Application of the technology has been primarily for the treatment of phenol contaminated waste streams from the petroleum and coal processing industries.  Liquid-liquid extraction has proven to be particularly well suited because of difficulties involved in removal of phenol from these waste streams by steam stripping and adsorption.

     Actual performance data for halogenated organic compounds are limited to bench and pilot runs conducted under EPA auspices to assess the extractability of priority pollutants from industrial waste streams.  Solvent extraction was explored in one EPA sponsored program as a method of treating wastewaters from

petroleum refineries and petrochemical plants.[9]  Results were obtained from
the use of both spray columns and rotating disc contactors (RDC).[9]  However,
data for halogenated organic compounds are limited to a few compounds;
i.e., ethylene dichloride, chloroacetaldehyde, and trichloroacetaldehyde.

Two earlier studies[10,11] conducted by the EPA summarized available
solvent extraction data as shown in Table 5.4.3.  These and other related data
from the same studies can also be found in Reference 12, the U.S. EPA
Treatability Manual.

A more relevant EPA sponsored study was conducted to determine the
feasibility of pesticide extraction from process waste streams.[13]  The study
examined partition coefficients for several pesticide (including DDT;
toxaphene; chlordane; 2,4-D; and bromacil) and solvent (n-butyl chloride,
monopropyl ether, hexane, pentane, and diethyl ether) combinations.  The
boiling points of the five solvents considered were in the 34°C-78°C range.
Efficiencies in excess of 99.9 percent for synthetic DDT solutions using
hexane as the extraction solvent were obtained in a bench scale rotary disk
contactor.  DDT concentrations ranged from 0.015 to 0.39 percent.  Subsequent
tests with process stream effluents also indicated that efficiencies in excess
of 99 percent were attainable.  As discussed below, costs were estimated and
compared favorably with carbon absorption.

## 5.4.3  Cost of Treatment

As noted in EPA's Treatability Manual, it is quite difficult to predict
costs of solvent extraction because of the wide variety of systems, feed
streams, and equipment that may be involved.  However, in Volume IV of the
Treatability Manual, the EPA does present some cost data based on a waste
phenol feed of 45,000 lbs/hr containing 1.5 percent phenol by weight and a
similar toluene solvent feed rate.  ($K_D$ = 2 for the phenol/toluene/water
distribution coefficient.)  The unit is a rotating disc contactor containing
an equivalent of about five theoretical stages to produce a wastewater
discharge containing 75 ppm phenol.  Using the equations provided in Kirk
Othmer (Volume 9, Liquid-Liquid extraction), approximately five additional
theoretical stages would be required to achieve a discharge level of 21 ppm
which is slightly above the design residual phenol level of the entering
solvent stream.

## TABLE 5.4.3.    RESULTS OF SOLVENT EXTRACTION STUDIES

| Chemical | Description of study | | | Results of study | Comments |
|---|---|---|---|---|---|
| | Study type[a] | Study type[b] | Influent conc. | | |
| Acrolein | R | U | | Extractable w/xylene. Solvent recovery by azeotropic distillation. | |
| Acrylonitrile | R | U | | Extractable w/ethyl ether. | |
| Chlorobenzene | R | U | 600 ppm | 3 ppm effluent conc. using chloroform solvent. | |
| bis-chloroethyl ether | R | U | | Extractable w/ethyl ether and benzene. | |
| Chloroethane | R | U | | Extractable w/alcohols and aromatics. | |
| 1,1-Dichloroethane | R | U | | Extractable w/alcohols, aromatics and ethers. | |
| Hexachloroethane | R | U | | Extractable w/aromatics, alcohols and ethers. | |
| 1,1,2,2-Tetrachloro-ethane | R | U | | Extractable w/aromatics, alcohols and ethers. | |
| 1,1,1-Trichloroethane | R | U | | Extractable w/alcohols and aromatics. | |
| 1,1,2-Trichloroethane | R | U | | Extractable w/aromatics, methanol and ethers. | |
| Dichloroethylene | L,B | I* | 49 ppm | Kerosene effluent conc. 2 ppm; $C_{10}$-$C_{12}$ effluent conc. 1+ ppm. | Solvent extraction w/kerosene & $C_{10}$-$C_{12}$ hydrocarbon at 7:1 solvent to wastewater ratio. |
| Ethylene Dichloride | P, C | I* | 23-1,804 ppm @ 2.76-3.76 L/min | A 5.5:1 water to solvent ratio gave 94-96% reduction. $C_{10}$-$C_{12}$ paraffin solvent at 5:1 to 16.5:1 water to solvent ratio showed 94-99% reduction. | Wastewater contained 14 other halocarbons including 30-350 ppm 1,1,2-trichloroethane and 5-197 ppm 1,1,2,2-tetrachloroethane. A 532 L/min extractor w/1,000 ppm influent estimated to have a capital cost of $315,000 and total annual cost of $143,000 including credit for recovered EDC. |

[a] Describes the scale of the referenced study:
   B - Batch Flow      P - Pilot Scale
   C - Continuous Flow    R - Literature Review
   L - Laboratory Scale

[b] Describes the type of wastewater used in the referenced study:
   I - Industrial Wastewater
   U - Unknown

Source: Reference 10.

*Reference 11.

The costs are presented in Table 5.4.4 with capital costs modified by ENR
index adjustment from 3119 to 4230 (May 1986) and minor changes made in the
costs of toluene and power to represent May 1986 values.  The annual costs of
$917,000/year represent costs in excess of $21 per 1,000 gallons of treated
wastewater, up from the value of $17 per 1,000 gallons provided in
Reference 12 (1980 dollars).  Assuming capital costs are related to size
(number of stages) through a 0.7 exponential factor as indicated in
Reference 12, the capital costs shown in Table 5.4.4 would increase from $1.5
to over $2.4 million to achieve a discharge level of 21 ppm phenol in the
effluent wastewater.  Total annual operating costs would be well in excess of
$30/1,000 gallons.

The above costs are far higher than the costs of $2/1,000 gallons
estimated in the Reference 13 study for a 300 million gallon per year plant.
Costs for carbon absorption were estimated at $3-$10/1,000 gallons.

Conceptual designs and economic analyses were also carried out for
several cases in the Reference 8 study.  These include extraction of
nitrobenzene with diisobutyl ketone (DIBK - a low boiling solvent), and
extraction of acrolein by methyl isobutyl ketone (MIBK), n-butyl acetate,
toluene, and 1,1,2,2-tetrachloroethane (all high boiling solvents).  The costs
(1982 dollars) range from $5-$13/1,000 gallons and appear to be in reasonable
agreement with the costs of Reference 12, given the differences in base year
and operating volumes and concentrations.

5.4.4  Overall Status of Process

5.4.4.1  Availability--

Although liquid-liquid extraction processes have not been widely applied
to the treatment of waste streams, they are extensively used within the
chemical process industry to affect separations and recoveries.  Table 5.4.5
lists a number of processing equipment units which can be used for
liquid-liquid extractions.  Advantages and disadvantages of each type are
listed in the table.  These are discussed in more detail in References 1, 7,
12 and other standard texts dealing with separation processes.  There are a
number of commercial suppliers of liquid-liquid extraction equipment and
accessory equipment such as that needed for the regeneration of solvent and
removal of solvent from the wastewater effluent.

TABLE 5.4.4.   ESTIMATED COSTS FOR A LIQUID-LIQUID EXTRACTION SYSTEM

| Operating Characteristics | Value |
|---|---|
| Water/phenol/feed | 45,000 lb/hr containing 1.5% phenol (by wt); temperature is 110°F. |
| Toluene feed | 45,000 lb/hr (containing 20 ppm phenol from steam stripper reecycle). |
| Discharge water | Contains 75 ppm phenol. |
| Extraction column | Rotating disc type; 6 ft diameter, 60 ft high; made from carbon steel; contains 50 compartments and equivalent of about 5 theoretical stages. (Equilibrium distribution coefficient of phenol between toluene and water is about 2.) |
| Loss of toluene/cycle | Approximately 0.1%/cycle. |
| Electrical requirements (column only) | One 10 hp electric motor. |
| Operation | 330 d/yr; 24 hr/d. |

Fixed capital costs based on an ENR index of 4,230 (May 1986) are estimated to be $1,500,000 (up from an estimate cost of $1,000,000 provided by **Reference** 12 in 1980).  Estimation of annual operating cost is presented below.

| Cost item | Annual quantity | Cost per unit quantity | Annual cost[a] $ | |
|---|---|---|---|---|
| Direct operating cost | | | | |
| Labor | | | | |
| Operating | 12,000 man hr | $16/hr | 192,000 | |
| Maintenance | | | 16,000 | |
| | | | | 208,000 |
| Chemicals – Toluene | 15,000 gal | $1.35 | 20,300 | |
| Materials | | | 16,700 | |
| Steam | 33 10⁶ lb | $5/1,000 lb | 165,000 | |
| Power | 150,000 kWh | $0.05/kWh | 7,500 | |
| Total | | | 417,500 | |
| Total indirect Operating Cost | | | 500,000 | |
| Total annual operating cost[a] | | | 917,500 | |

[a]Excludes annual credit for phenol recovery.

Source:  Reference 12 (modified to represent May 1986 dollars).

TABLE 5.4.5.  ADVANTAGES AND DISADVANTAGES OF EXTRACTION TYPES

| Class of equipment | Advantages | Disadvantages |
|---|---|---|
| Mixer-Settlers | Reliable scaleup<br>Good contacting<br>Handles wide flow ratio<br>Low headroom<br>Many stages available | Large holdup<br>High power costs<br>High capital costs<br>Large floor space<br>Interstage pumping may be required |
| Gravity Columns<br><br>Spray Column<br>Packed Column<br>Tray Column | Low capital cost<br>Low operating cost<br>Simple construction<br>Handles wide flow ratio<br>  (tray column)<br>Handles suspended solids<br>  (spray column) | Extensive backmixing (spray, packed column)<br>Limited throughput with small density difference<br>Cannot handle high flow rate (packed column)<br>High headroom<br>Low efficiency (spray column)<br>Difficult scaleup<br>Internals subject to fouling (packed column) |
| Mechanically<br>Agitated Columns<br><br>Agitated Column<br>Pulsed Column | Good dispersion<br>Reasonable cost<br>Many stages possible<br>Relatively easy scaleup<br>Handles systems of high<br>  interfacial tension | Limited throughput with small density difference<br>Cannot handle emulsifying systems<br>Cannot handle high flow ratio |
| Centrifugal<br>Extractors | Handles low density difference<br>  and high interfacial tension<br>  between phases<br>Low holdup<br>Low space requirements<br>Small inventory of solvent<br>Handles stable emulsions | High capital cost<br>High operating cost<br>High maintenance cost<br>Limited number of stages in single unit<br>Subject to fouling |

Source:  References 9-12.

5.4.4.2 Application--

Integration of equipment into an overall system for successful treatment of waste streams will require considerable analysis of the waste stream of interest and the candidate processes. Liquid-liquid extractions are most useful when separations involve materials that are not easily separated by distillation or other treatment processes. Generally, liquid-liquid extractions of water streams are conducted to remove materials which have high water solubility and therefore almost invariably a low Henry's Law constant. Air or stream stripping do not appear to be viable options for wastes of this type. Liquid-liquid extractions may be particularly applicable when the relative volatilities of solute/solvent compounds make separation by distillation difficult or when high waste concentrations make carbon adsorption uneconomical. Most of the halogenated organics addressed in this TRD are only sparingly soluble in water although some of the lower molecular weight compounds do exhibit appreciable solubility (see Appendix A). Extraction of the soluble compounds could be viable. Solvent extraction of emulsified material might also be considered provided mass transfer considerations are acceptable. This would have to be determined experimentally.

Solvent choice and design parameter options are many and varied. Although design and operation of a liquid-liquid extraction system to achieve acceptable effluent levels is theoretically possible, existing experimental and field data for halogenated and other organics indicate that most units, as presently designed and operated, fall short of this goal.

5.4.4.3 Environmental Impact--

Properly designed and operated, the liquid-liquid extraction process does not appear to pose significant environmental problems. Both process exit streams contain potential contaminants that must be addressed as part of the process. The solvent will contain solute (contaminant in the feed) that must be removed if the solvent is to function adequately in recycle. The treated waste stream (assuming all significant traces of contaminant have been trans- ferred to the solvent) could contain dissolved solvent which may or may not be significant and warrant additional treatment. Since these potential conditions are recognized and must be dealt with by system designers, the environmental impacts of a viable liquid-liquid extraction system should be minimal.

5.4.4.4   Advantages and Limitations--

Some potential advantages of liquid-liquid extraction processes are:

● Recovery of costly materials can be accomplished with little threat of thermal decomposition or chemical interaction.

● Recovery (separation) of materials which have similar relative volatilities or adsorption isotherms can be achieved.

Some potential limitations of liquid-liquid extraction are:

● Some residuals will generally be present in both the raffinate and extract streams, thus, some provision must be made for their removal and subsequent disposal.

● Economics may not be favorable.

● Deviations that limit the extent of removal may occur upon scale-up.

REFERENCES

1.    Perry, R. H., et al.  Editors; Chemical Engineers' Handbook, Sixth
      edition, McGraw Hill Book Company, NY. NY.  1984.

2.    Hwang, S. T., Treatability of Toxic Waste Water Pollutants by Solvent
      Extraction.  AIChE Symposium Series, No. 209 "Waste-1980", 1981.

3.    Earhart, J.P., Extraction of Chemical Pollutants from Industrial
      Wastewaters with Volatile Solvents, U.S. EPA, Ada, Oklahoma,
      EPA-600/2-76-220, December 1976

4.    Leo, A., C. Hansch, and D. Elkins, "Partition Coefficients and Their
      Uses," Chemical Reviews, 71, 525(1971).

5.    Earhart, J. P., K. W. Won, H. Y. Wong, J. M. Prausnitz, and C. J. King,
      "Recovery of Organic Pollutants via Solvent Extraction," Chem. Eng.
      Prog., May, 67 (1977)

6.    Murray, W. J., L. H. Hall, and L. B. Kier, "Molecular Connectivity III:
      Relationship to Partition Coefficients," J. of Phar. Sci., 64, 1978(1975).

7.    Kirk-Othmer.  Encyclopedia of Chemical Technology, Third Edition,
      Volume 9.  A Wiley-Intrascience Publication.  1978.

8.    King, C. J., D. K. Joshi, and J. J. Senetar, University of California,
      Berkeley, Department of Chemical Engineering.  Equilibrium Distribution
      Coefficients for Extraction of Organic Priority Pollutants From
      Water-II.  EPA-600/2-84-060b, February 1984.

9.    Earhart, J. P., et al.  University of California, Berkeley Department of
      Chemical Engineering.  Extraction of Chemical Pollutants from Industrial
      Wastewaters with Volatile Solvents.  EPA 600/2-76-220.  PB226241.
      December, 1976

10.   Dryden, F. E., J. H. Mayes, R. J. Planchet, and C. H. Woodard.  Priority
      Pollutant Treatability Review.  EPA Contract No. 68-03-2579, U.S.
      Environmental Protection Agency, Cincinnati, Ohio, 1978.

11.   Coco, J. H., et al.  Development of Treatment and Control Technology for
      Refractory Petrochemical Wastes.  EPA-600/2-79-080, U.S. Environmental
      Protection Agency, Ada, Oklahoma, 1979.  236 pp.

12.   U.S. EPA Treatability Manual, Volume III, EPA-600/2-82-001a.   September
      1981.

13.   Reynolds, S.L., "Extraction of Pesticides From Process Streams Using High
      Volatility Solvents".   IN:   International Conference on New Frontiers for
      Hazardous Waste Management.   EPA 600-9-85-025, September 1985.

## 5.5   CARBON ADSORPTION

Adsorption is a widely-used process for the removal of organic contaminants from gas or liquid waste streams. Activated carbon is the most commonly used adsorbent. Largely nonpolar, carbon is particularly effective for the removal of hydrophobic, high molecular weight organic compounds from aqueous streams.[1] Thus, it is a good adsorbent for many of the halogenated organic compounds considered in this document. Activated carbon adsorption must be considered a potentially viable treatment technology for many halogenated organic-bearing wastewater streams, either as a primary treatment for moderately high (up to 0.5 percent) concentrations of organic compounds or as a secondary polishing type treatment for much lower levels of contamination. The cost effectiveness of adsorption is dependent on flow rates and concentrations of the organic contaminants and on the adsorptive capacity of the carbon for the contaminants. Adsorption should be cost effective for concentrations of organic compounds up to about 1,000 mg/L, and could be cost effective for concentrations up to 5,000 mg/L. For concentrations above 5,000 mg/L, other unit processes are generally more cost effective,[2] unless nondestructive chemical regeneration can be used to recover the adsorbed materials.

Activated carbon is available in powder (PAC) or granular (GAC) form. GAC is more commonly used because its larger size is more amenable to handling in the equipment used to achieve contact and regeneration.[3] Both types of carbon adsorbent have large contact surface areas, far in excess of their nominal external surface areas. Surface areas, resulting from a network of internal pores 20 to 100 angstroms in diameter, are of the order of 500 to 1,500 square meters per gram. Porosities can be as large as 80 percent. The adsorption capacity of an activated carbon for a contaminant is a function of the surface area and the surface binding process and can approach 1 gram per gram of carbon.

Adsorbent binding forces result from the interaction of the contaminant surface molecules with the carbon surface atoms. The attractive forces are generally weaker and less specific than those of chemical bonds and, hence, the term physical adsorption is used to describe the binding mechanism. The effective attractive range is small and the adsorbed material is generally

present only as a monolayer upon the adsorbent surface.  The process is
considered analagous to condensation of gas molecules, or to crystallization
from a liquid.  The process is reversible, and molecules held at the surface
will subsequently return to the fluid stream.  The length of time elapsing
between adsorption and desorption is dependent upon the intensity of the
surface forces.  Adsorption is a direct result of this time lag.  Because
adsorption is a reversible process, the carbon surface can be regenerated
either thermally or chemically; e.g., by solvent extraction.  However, high
temperature thermal regeneration (which destroys the adsorbed organics) is
generally required to insure effective removal of adsorbed contaminants.

Water solubility and carbon affinity are two properties that, in general,
correlate with the adsorption of hazardous contaminants onto activated
carbon.  Generally, less soluble organic materials are more effectively
adsorbed.  Several factors are associated with decreased water solubility of
organics and, as a result, correlate with increased adsorption.  These
include:  high molecular weight, low polarity, low ionic character, low pH for
weak organic acids or high pH for weak organic bases, and aromatic
structures.  As a rule of thumb, molecules of higher molecular weights are
attracted more strongly to activated carbon than are molecules of lower
molecular weights.  Strongly ionized or highly polar compounds are more water
soluble and are usually poorly adsorbed.[4]  Compounds with solubilities of
less than 0.1 g/mL in water and molecular weights between 100 to 1,000 are
considered moderately to highly adsorbable.

Several other aspects of molecular structure also affect adsorbability.
In general, branch-chain compounds are more adsorbable than straight-chain
compounds.  Increasing hydrocarbon unsaturation also tends to decrease
solubility and increase carbon adsorption.  Thus, unsaturated organics such as
ethylenes tend to more readily adsorb on carbon than saturated compounds such
as ethanes.  Table 5.5.1 identifies the specific waste characteristics that
affect adsorption.  Table 5.5.2 summarizes the influence of substituent
chemical groups on adsorbability.

The adsorption of organic compounds by adsorbents is usually determined
in the laboratory through adsorption isotherm tests.  These tests measure, at
a given temperature, the amount of substance adsorbed and its concentration in
the surrounding solution at equilibrium.  Isotherms provide information on the

TABLE 5.5.1.   WASTE CHARACTERISTICS THAT AFFECT ADSORPTION
BY ACTIVATED CARBON

A.   General

1.   Polar, low-molecular weight compounds with high degrees of
solubility are poorly adsorbed.

2.   Conversely, nonpolar, high-molecular weight compounds with limited
solubility tend to be preferentially adsorbed.

B.   Molecular Structure

1.   Branched-chain compounds are more adsorbable than straight-chain
compounds.

2.   Type and location of substituent groups also affect the degree to
which a compound may be adsorbed from solution.   Table 5.5.2 gives
some general guidelines as to how substituent groups affect
adsorbability.

C.   Effect of pH

1.   The affect of pH on carbon sorbent equilibrium varies significantly
from compound to compound.   Adsorption isotherms for some compounds
are affected dramatically, whereas others show no significant change
as a function of pH.

2.   Dissolved organics generally adsorb most efficiently at that pH
which imparts the least polarity to the molecule.   For example, a
weak organic acid can be expected to adsorb best at a low pH value.

D.   Temperature Effects

1.   Adsorption reactions are generally exothermic; therefore lower
temperatures favor adsorption.   However, shifts in adsorbability
within the range of temperatures normally encountered in waste
stream applications are generally small.

E.   Physical Form

1.   Carbon adsorption is suitable for aqueous wastes, nonaqueous liquids
and gases.

2.   The oil and grease concentration should be less than 10 mg/L.

3.   Suspended solids concentrations higher than about 10-70 mg/L will
cause clogging of the bed.

Source:   Adapted from Reference 5.

TABLE 5.5.2.    INFLUENCE OF SUBSTITUENT GROUPS ON ADSORBABILITY

| Substituent | Nature of influence |
|---|---|
| Hydroxyl | Generally reduces adsorbability; extent of decrease depends on structure of host molecule. |
| Amino | Effect similar to that of hydroxyl but somewhat greater; many amino acids are not adsorbed to any appreciable extent. |
| Carbonyl | Effect varies according to host molecule; glyoxylic is more adsorbable than acetic but similar increase does not occur when introduced into higher fatty acids. |
| Double bonds | Variable effect as with carbonyl. |
| Halogens | Variable effect, but generally increased adsorbability. |
| Sulfonic | Usually decreases adsorbability. |
| Nitro | Often increases adsorbability. |
| Aromatic rings | Greatly increase adsorbability. |

Source:   Adapted from Reference 5.

relative affinity of an organic compound for the adsorbent and the adsorption
capacity.  Thus, isotherm tests can be useful in making qualitative
evaluations of different carbons for adsorption of specific components from a
given waste stream.

Isotherms data are frequently evaluated using the Freundlich Equation,
which describes the adsorbability characteristics of a constituent for a given
carbon.  This equation can be expressed as follows:

$$\frac{x}{m} = kC_f^{1/n} \tag{1}$$

where:   $x$   = mass of adsorbate, mg

$m$   = mass of dry adsorbent, g

$k$   = constant, adsorbability indicator

$C_f$   = solution concentration at equilibrium, mg/L

$1/n$ = constant, adsorption intensity

Values of $k$ and $1/n$ for a compound are found by a plot of experimentally
determined carbon adsorption data in which values of $x/m$ are plotted against
$C_f$ on log-log paper.

The adsorption data are useful in estimating the relative effectiveness
of an adsorbent for organic compounds.  However, care must be exercised in
assessing performance when the waste stream contains a large number of
competing contaminants.  It is possible to develop equilibrium equations that
apply to multi-component systems, as noted in standard texts on adsorption and
Perry's Chemical Engineers' Handbook.  However, most users will rely on
laboratory scale carbon adsorption/isotherm tests to assess performance and
design an appropriate system for a specific waste stream.

Carbon adsorption capacities are summarized in Table 5.5.3 for several
halogenated organic and other separate compounds.[6]  The capacity, $x/m$,
corresponds to the constant, $k$, expressed as mg compound per gram of carbon,
when the equilibrium concentration of the compound is 1.0 mg/L.  The compounds
have been ranked in decreasing order of their capacity ($k$ value) as determined
by isotherm tests using Filtrasorb 300 activated carbon.

TABLE 5.5.3.    SUMMARY OF CARBON ADSORPTION CAPACITIES

| Compound | Adsorption[a] capacity, mg/g | Compound | Adsorption[a] capacity, mg/g |
|---|---|---|---|
| Heptachlor | 1,220 | PCB-1221 | 242 |
| Heptachlor epoxide | 1,038 | DDE | 232 |
| Endosulfan sulfate | 686 | Benzidine dihydrochloride | 220 |
| Endrin | 666 | beta-BHC | 220 |
| Aldrin | 651 | alpha-Endosulfan | 194 |
| PCB-1232 | 630 | 4,4' Methylene-bis- | |
| beta-Endosulfan | 615 | (2-chloroaniline) | 190 |
| Dieldrin | 606 | 2,4-Dichlorophenol | 157 |
| Hexachlorobenzene | 450 | 1,2,4-Trichlorobenzene | 157 |
| Anthracene | 376 | 2,4,6-Trichlorophenol | 155 |
| DDT | 322 | Pentachlorophenol | 150 |
| alpha-BHC | 303 | Naphthalene | 132 |
| 3,3-Dichlorobenzidine | 300 | 1-Chloro-2-nitrobenzene | 130 |
| 2-Chloronaphthalene | 280 | 1,2-Dichlorobenzene | 129 |
| Hexachlorobutadiene | 258 | p-Chlorometacresol | 124 |
| gamma-BHC (lindane) | 156 | 1,4-Dichlorobenzene | 121 |
| Chlordane | 245 | 1,3-Dichlorobenzene | 118 |
| 4-Chlorophenyl phenyl ether | 111 | Hexachloroethane | 97 |
| Chlorobenzene | 91 | p-Xylene | 85 |
| 1,2-Dibromo-3-chloro-propane | 53 | Ethylbenzene | 53 |
| 2-Chlorophenol | 51 | Tetrachloroethene | 51 |
| 5 Bromouracil | 44 | Trichloroethene | 28 |
| Toluene | 26 | bis(2-Chloroisopropyl) ether | 24 |
| Phenol | 21 | Bromoform | 20 |
| Carbon tetrachloride | 11 | bis(2-Chloroethoxy) methane | 11 |
| 1,1,2,2-Tetrachloroethane | 11 | 1,2-Dichloropropene | 8.2 |
| Dichlorobromomethane | 7.9 | 1,2-Dichloropropane | 5.9 |
| Trichlorofluoromethane | 5.6 | 1,1-Dichloroethylene | 4.9 |
| Dibromochloromethane | 4.8 | 2-Chloroethyl vinyl ether | 3.9 |
| Chloroform | 2.6 | 1,1,1-Trichloroethane | 2.5 |
| 1,1-Dichloroethane | 1.8 | Methylene chloride | 1.3 |
| Benzene | 1.0 | | |

[a]Adsorption capabilities are calculated for an equilibrium
concentration of 1.0 mg/L at neutral pH.

Source:    Reference 6.

### 5.5.1   Process Description

A schematic of a carbon adsorption system utilizing a prefilter and a multiple hearth furnace regeneration system is shown in Figure 5.5.1.   In the regeneration process, adsorbed material is driven from the carbon surface by thermal forces.   However, other methods (e.g., extraction or steam stripping) can be used to drive off adsorbed material held largely by physical rather than chemical forces.   Regeneration is usually complete, although some loss of effective surface area over time (3 to 8 percent per cycle)[1] can result from build-up of hard to remove adsorbent, attrition, and other mechanisms. Collection or destruction of the desorbed material will also be necessary for these regeneration processes.

Carbon adsorption is applicable to single-phase aqueous solutions containing low concentrations of organic contaminants (up to 0.5 weight percent)[2] and inorganic contaminants (up to 0.1 weight percent).[4]   It is also applicable to some organic liquid solutions (e.g., those consisting of a poorly adsorbed solvent and a readily adsorbed solute), although it is less likely that the selectivity will approach that for adsorption from a water stream.

Carbon adsorption may be used as a pretreatment process for conventional biological treatment, but is more frequently used as a polishing step for biological treatment effluent to remove compounds that are resistant to biodegradation.   In this capacity, it is generally used for high volume waste streams which contain dilute organic constituents.

### 5.5.1.1   Pretreatment Requirements--

Pretreatment of the feed to carbon adsorption columns is often required to improve performance and/or prevent operational problems.   As discussed in Reference 7, there are four primary areas where pretreatment for different waste forms and characteristics may be required.   These include:

1.   Equilization of flow and concentrations of primary waste constituents.

2.   Filtration.

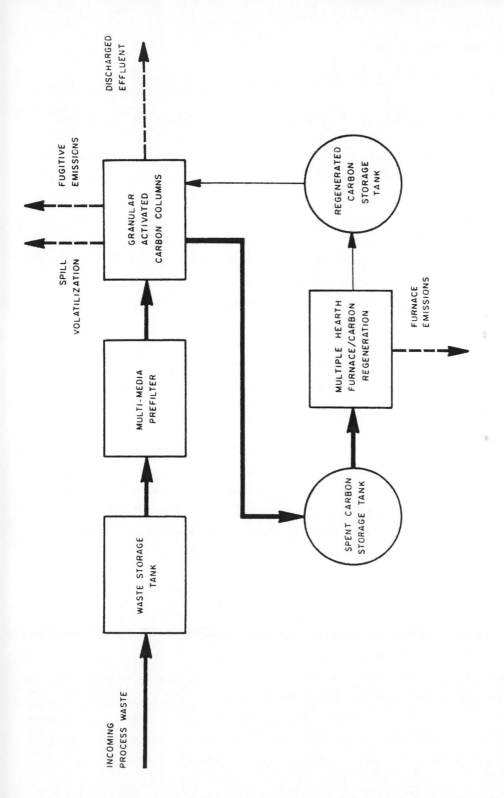

Figure 5.5.1.  Carbon adsorption flow diagram.

Source:  Reference 3.

3. Adjustment of pH.

4. Adjustment of temperature.


Equalization of flow and concentrations of primary waste constituents--It is generally assumed that both the flow to the GAC columns and the concentration of the primary waste constituent, namely the halogenated organic compound in the feed, are constant. Such is not generally the case, and since variations in either flow or concentration can have a detrimental impact on system performance, it is necessary to make provisions to equalize flow and minimize concentration surges.

Flow equalization is accomplished by employing a surge tank of sufficient capacity to accommodate flow variations. The result is a constant flow rate to the GAC columns. Concentration equalization can be handled by employing surge tanks in the same manner as flow equalization. However, provisions must be made for mixing tank contents prior to discharging to the GAC columns. Mixing prevents concentration surges which can lead to premature column leakage and breakthrough. Conversely, low concentration swings can result in premature regeneration of an underloaded GAC column.


Filtration--It is a general requirement for GAC processes that the feed of the column be low in suspended solids. In treating solvent and ignitable waste streams, it has been suggested that solids concentrations greater than 50 mg/L will interfere with column operation.[1] In addition to solids removal, many additional waste contaminants can interfere with carbon adsorption of solvent and ignitable waste streams. For example, if calcium or magnesium are present in concentrations greater than 500 mg/L, these constituents may precipitate out and plug or foul the column.[4] Oil and grease in excess of 10 mg/L can interfere in column operation.[1] Lead and mercury are also of concern because they may compete for adsorption sites and are difficult to remove from the carbon during the regeneration cycle.[8] The presence of many other compounds can influence adsorption as they compete for available adsorption sites on the carbon surface.

For efficient use of GAC for treating halogenated organic waste streams, removal of suspended solids and other waste contaminants noted above must be achieved by pretreatment with, for example, multi-media pressure filters.

Such filters are very compatible with fixed bed adsorption processes and can be readily integrated into a total design. Other possibilities include membrane filtration when a highly clarified feed is desired; ultrafiltration if high molecular weight (over 1,000) contaminants are present in the raw waste; and reverse osmosis to concentrate a feed containing numerous dissolved species, both organic and inorganic. Obviously other pretreatments (e.g. precipitation, clarification) will be needed to remove dissolved solids such as calcium and magnesium.

Adjustment of pH--GAC adsorption systems are sensitive to changes in pH. If the contaminants to be removed are either weakly acidic or weakly basic, then the pH of the feed will effect their adsorption. Weakly acidic organics are most readily adsorbed in the nonionized state and consequently a low pH (acid) favors adsorption. Weakly basic compounds such as aniline or dimethylamine are also most readily adsorbed in their nonionized state and, therefore, adsorption is favored by high pH (alkaline). The adsorption of neutral organic compounds is unaffected by pH.

The control of the feed pH should perhaps be considered a subcategory of the previously discussed concentration/equalization requirement. It can be readily controlled by applying pH measurement and feedback control for acid or base addition to the equalization system at the surge tank to achieve the desired pH feed to the GAC adsorption columns.

Adjustment of Temperature--Temperature adjustment is rarely required in GAC adsorption processes. High feed temperature could lead to increased VOC emissions to air in an open gravity feed system and is unfavorable for adsorption and retention of volatile constituents. If the possibility for temperature surges exists, temperature moderation through flow equalization should be considered.

5.5.1.2 Operating Parameters--

Process design activities must take into account a number of equipment design parameters to develop a system which is optimal for the characteristics of the waste or wastes to be treated. The design parameters will be considered in terms of both the adsorption system and the regeneration or desorption system.

Adsorption System--Isotherms, determined in a laboratory, measure the
affinity of activated carbon for the "target" adsorbates in the process
liquid.  This provides data for determining the amount of carbon which will be
required to treat the full scale process stream.  Carbon requirements will be
based on a limiting constituent for which attainment of effluent limitations
is the most difficult.  However, adsorption isotherms can vary widely for
different carbons, and isotherm data cannot be used interchangeably.

Table 5.5.4 gives properties of some commercially available granulated
activated carbons.  Properties of a typical powdered activated carbon are
shown in Table 5.5.5.  Adsorption properties of the two types of carbon are
generally comparable.  The principal difference is in the particle size; the
fine size of the PAC makes it unsuitable for use in the contacting and
regeneration equipment used for GAC applications.

A typical continuous adsorption system consists of multiple columns
filled with activated carbon and arranged in either parallel or series.  Total
carbon depth of the system must accommodate the "adsorption wavefront";
i.e., the carbon depth must be sufficient to purify a solution to required
specifications after equilibrium has been established.  Bed depths of 8 to
40 feet are common.  Minimum recommended height-to-diameter ratio of a column
is 2:1.  Ratios greater than 2:1 will improve removal efficiency, but result
in increased pressure drop for the same flow rate.  Optimum flow rate must be
determined in the laboratory for the specific design and carbon used.  For
most applications, 0.5 to 5 $gpm/ft^2$ of carbon is common.

Various configurations are available for GAC adsorption applications.
Based on influent characteristics, flow rate, size and type of carbon,
effluent criteria and economics, each design is unique in its mode of
operation.  Figure 5.5.2 illustrates several arrangements typically used for
GAC adsorption systems.

The adsorption beds of both series and parallel design can be operated in
either an upflow or downflow direction.  A downflow mode of operation must be
used where the GAC is relied upon to perform the dual role of adsorption and
filtration.  Although lower capital costs can be realized by eliminating
pretreatment filters, frequent backwashing of the adsorbers is required.
Application rates of 2 to 10 gallons per minute per square foot ($gpm/ft^2$)

TABLE 5.5.4.   PROPERTIES OF SEVERAL COMMERCIALLY AVAILABLE CARBONS

| | ICI<br>America<br>Hydrodarco<br>(lignite) | Calgon<br>Filtrasorb<br>300<br>(bituminous) | Westvaco<br>Nuchar<br>WV-L<br>(bituminous) | Witco<br>517<br>(12x30)<br>(bituminous) |
|---|---|---|---|---|
| **PHYSICAL PROPERTIES** | | | | |
| Surface area, $m^2/g$ (BET) | 600 - 650 | 950 - 1050 | 1000 | 1050 |
| Apparent density, $g/cm^3$ | 0.43 | 0.48 | 0.48 | 0.48 |
| Density, backwashed and drained, $lb/g^3$ | 22 | 26 | 26 | 30 |
| Real density, $g/cm^3$ | 2.0 | 2.1 | 2.1 | 2.1 |
| Particle density, $g/cm^3$ | 1.4 - 1.5 | 1.3 - 1.4 | 1.4 | 0.92 |
| Effective size, mm | 0.8 - 0.9 | 0.8 - 0.9 | 0.85 - 1.05 | 0.89 |
| Uniformity coefficient | 1.7 | 1.9 or less | 1.8 or less | 1.44 |
| Pore volume, $cm^3/g$ | 0.95 | 0.85 | 0.85 | 0.60 |
| Mean particle diameter, mm | 1.6 | 1.5 - 1.7 | 1.5 - 1.7 | 1.2 |
| | | | | |
| **SPECIFICATIONS** | | | | |
| Sieve size (U.S. std. series)[a] | | | | |
|   Larger than No. 8 (max. %) | 8 | 8 | 8 | -- |
|   Larger than No. 12 (max. %) | -- | -- | -- | 5 |
|   Smaller than No. 30 (max. %) | 5 | 5 | 5 | 5 |
|   Smaller than No. 40 (max. %) | -- | -- | -- | -- |
| Iodine No. | 650 | 900 | 950 | 1000 |
| Abrasion No. minimum | b | 70 | 70 | 85 |
| Ash (%) | b | 8 | 7.5 | 0.5 |
| Moisture as packed (max. %) | b | 2 | 2 | 1 |

[a]Other sizes of carbon are available on request from the manufacturers.

[b]No available data from the manufacturer.

-- Not applicable to this size carbon.

TYPICAL PROPERTIES OF 8 X 30-MESH CARBONS

| | Lignite<br>carbon | Bituminous<br>coal carbon |
|---|---|---|
| Total surface area, $m^2/g$ | 600 - 650 | 950 - 1,050 |
| Iodine number, min | 500 | 950 |
| Bulk density, $lb/ft^3$ backwashed and drained | 22 | 26 |
| Particle density wetted in water, $g/cm^3$ | 1.3 - 1.4 | 1.3 - 1.4 |
| Pore volume, $cm^3/g$ | 1.0 | 0.85 |
| Effective size, mm | 0.75 - 0.90 | 0.8 - 0.9 |
| Uniformity coefficient | 1.9 or less | 1.9 or less |
| Mean particle dia., mm | 1.5 | 1.6 |
| Pittsburgh abrasion number | 50 - 60 | 70 - 80 |
| Moisture as packed, max. | 9% | 2% |
| Molasses RE (Relative efficiency) | 100 - 120 | 40 - 60 |
| Ash | 12 - 18% | 5 - 8% |
| Mean-pore radius | 33 A | 14 A |

Source:   Reference 9.

TABLE 5.5.5.   TYPICAL PROPERTIES OF POWDERED ACTIVATED CARBON (PETROLEUM BASE)

| | |
|---|---|
| Surface Area $m^2/g$(BET) | 2,300 - 2,600 |
| Iodine No. | 2,700 - 3,300 |
| Methylene Blue Adsorption (mg/g) | 400 - 600 |
| Phenol No. | 10 - 12 |
| Total Organic Carbon Index (TOCI) | 400 - 800 |
| Pore Distribution (Radius Angstrom) | 15 - 60 |
| Average Pore Size (Radius Angstrom) | 20 - 30 |
| Cumulative Pore Volume ($cm^3/g$) | 0.1 - 0.4 |
| Bulk Density ($g/cm^3$) | 0.27 - 0.32 |
| Particle Size   Passes:   100 mesh (wt%) | 97 - 100 |
|                          200 mesh (wt%) | 93 - 98 |
|                          325 mesh (wt%) | 85 - 95 |
| Ash (wt%) | 1.5 |
| Water Solubles (wt%) | 1.0 |
| pH of Carbon | 8-9 |

Source:   Reference 5.

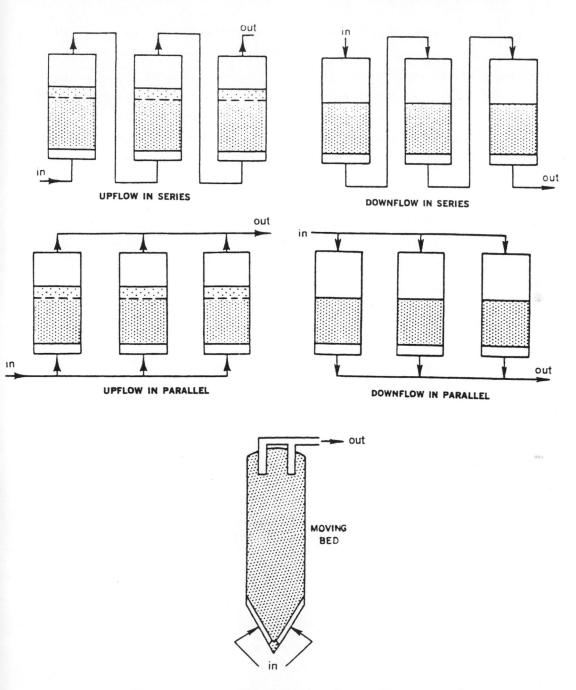

Figure 5.5.2.   Carbon bed configurations.

Source:   Reference 10.

are employed, and backwash rates of 12-20 gpm/ft$^2$ are required to achieve bed expansions of 20-50 percent. The use of supplemental air increases efficiency of the backwashing.

Prefiltration is normally required to prevent blinding upflow-expanded beds with solids. In this configuration, smaller particle sizes of GAC can be employed to increase adsorption rate and decrease adsorber size. Application rates can be increased even to the extent that the adsorbent may be in an expanded condition.

The design arrangements offer the following advantages and limitations as noted in Reference 10:

| Method | Comments |
|---|---|
| Adsorbers in Parallel | - For high volume applications<br>- Can handle higher than average suspended solids (<65-70 ppm) if downflow<br>- Relatively low capital costs<br>- Effluents from several columns blended, therefore, less suitable where effluent limitations are low |
| Adsorbers in Series | - Large volume systems<br>- Easy to monitor breakthrough at tap between units<br>- Effluent concentrations relatively low<br>- Can handle higher than average suspended solids (<65-70 ppm) if downflow<br>- Capital costs higher than for parallel systems |
| Moving Bed | - Countercurrent carbon use (most efficient use of carbon)<br>- Suspended solids must be low (<10 ppm)<br>- Best for smaller volume systems<br>- Capital and operating costs relatively high<br>- Can use such beds in parallel or series |
| Upflow-expanded | - Can handle high suspended solids (they are allowed to pass through)<br>- High flows in bed (~15 gpm/ft$^2$) |

The above systems are not generally used with powdered activated carbons. The PAC systems now used involve mixing the PAC with the waste stream to form a slurry which usually can be separated later by methods such

as filtration or sedimentation.    PAC is generally used simultaneously with
biological treatment to enhance organic removal by biological processes.[11]

Regeneration--The economic success of an adsorption system usually
depends on the regenerability of the adsorbent.    The exception is where there
are very long adsorption or loading cycles due to very low concentrations of
halogenated organic constituents in the inlet feed.    This type of system
usually operates on a "throw away" basis.    If very large quantities of
adsorbent are involved, then regeneration and reuse are required for
economical operation.    The regeneration techniques employed in industry are
thermal regeneration, steam regeneration, and acid or base regeneration.[7]
Solvent washing or biological treatment are other methods that are
occassionally used for regeneration.    Solvent recovery, if possible, can lead
to adsorbent recovery with attendant cost benefits.    Thermal regeneration is
the most commonly applied technique for GAC systems, since this is the only
method that can generally ensure effective regeneration.

Thermal regeneration involves high temperatures and a controlled gaseous
atmosphere.    Regeneration of spent carbon can be considered to take place in
three distinct phases.    First, wet carbon is dried at a temperature range of
approximately 100 to 150°C.    Water and some low boiling point organics will be
removed during this process but higher boiling point organics such as
1,4-dichlorobenzene (b.p. 173°C) will remain.    Next, the temperature is raised
to 250 to 750°C where more tightly bonded and higher boiling point organics
are removed by vaporization.    An inert gas atmosphere can be employed to
minimize oxidation.    Finally, the temperature is raised to 800 to 975°C where
residues and tars that may have accumulated are reacted and driven off the
carbon surface.    Steam is sometimes used to assist removal.    Even with careful
control, GAC losses are reported to be 3 to 8 percent/cycle due to both
oxidation and mechanical attrition.    Regeneration furnaces have been designed
to conduct all three steps of drying, vaporization under inert gases, and
regeneration separately in different zones.    Multiple hearth furnaces and
fluidized-bed furnaces are two types of thermal regenerators commonly found in
commercial use.

Steam regeneration can be used to displace the liquid in the adsorber bed, heat the adsorbent and, finally, strip the halogenated organics from the GAC.  However, not all halogenated organics, particularly some of the high molecular weight pesticides, are volatile enough to permit steam regeneration.  Average pressures of one to three atmospheres are utilized with steam flow rates of 0.5 to 4 lbs/min/ft$^3$.  The amount of steam required depends upon the size of the carbon bed.  The majority of steam used in regeneration is used to heat the carbon bed to the necessary temperature for vaporization to occur.  The heat capacities of the adsorbed constituents and their heats of vaporization do not represent a large fraction of the total steam requirement.  Thus units for steam usage are typically expressed as lb steam/lb carbon.

As discussed in the pretreatment section, the adsorption of weak organic acids and bases from aqueous solutions is dependent upon pH.  Therefore, if the adsorbed organic is acidic, regeneration with a basic solution is feasible.  Conversely, basic constituents can be regenerated with an acidic solution.  Acid or base regeneration is not as widely used as other regeneration techniques, but nonetheless, some organics such as cresols and ethylene diamine have been successfully recovered commercially by base and acid regeneration, respectively.  Solvent regeneration (with possible benefits resulting from sorbent recovery) may also be possible although removal of all sorbed materials is not likely.

5.5.1.3  Post-Treatment Requirements--

Air and water discharges from carbon adsorption systems employing carbon regeneration can be relatively innocuous.  Under proper design and operating conditions, the treated water will generally be suitable for discharge to surface waters.  Other aqueous streams such as backwash, carbon wash and transport waters are recycled or sent to a settling basin.  Emissions will result from thermal reactivation, but when afterburners and scrubbers are used, the controlled emissions are essentially non-polluting.  In some installations, particulates must be removed from the air stream (e.g., via a cyclone and baghouse) resulting in a solid waste.[1]

5.5.1.4  Treatment Combinations--

The high cost associated with the treatment of moderate to high total organic carbon (TOC) wastes and the ineffectiveness of carbon as an adsorbent for many low molecular weight water soluble organic compounds has impacted the use of carbon adsorption as a waste treatment technology.  Except when used alone as a polishing step for low levels of adsorbable materials in aqueous streams, carbon adsorption is usually employed in a "treatment train" with other treatment processes to achieve maximum efficiency at reduced cost.

An extensive discussion of treatment trains employing carbon adsorption can be found in References 12 and 13.

5.5.2  Demonstrated Performance

Information gathered from activated-carbon manufacturers and industry indicates that many granular-activated carbon systems are being used for the treatment of hazardous aqueous organic compound bearing wastes and wastewaters.  A 1982 EPA study[5] found that over 100 GAC systems were being used nationally to treat industrial wastewaters.  Another report[11] documented the use of PAC at seven and four facilities in the United States and Japan, respectively.

Despite the large number of units in use, data for full-scale applications are incomplete.  A major shortcoming of the available data base dealing with the removal of halogenated organics from aqueous waste streams by activated carbon, is the sparsity of performance data for higher concentration ( 0.1 percent) levels.  In addition, most of the data found in the literature do not consider the removal of individual compounds from concentrated waste streams, although data for BOD, COD, TOC, and other parameters are fairly common.  Data for individual compounds provided in EPA's background document for solvents (Reference 4) are, with few exceptions, for treatment of influent concentrations at the part per billion level.  All data presented, however, indicate that levels acceptable for direct discharge can be reached for essentially all solvents of concern.  Equal or better performance can be anticipated for most of the halogenated organic compounds.  Treatability ratings of some halogenated organics are shown in Table 5.5.6.  Ratings are high for most of these compounds, particularly the higher molecular weight,

TABLE 5.5.6.   TREATABILITY RATING OF SOME HALOGENATED ORGANICS
UTILIZING CARBON ADSORPTION

| Priority pollutant | Removal rating* |
|---|---|
| benzene | M |
| chlorobenzene | H |
| 1,2,4-trichlorobenzene | H |
| hexachlorobenzene | H |
| hexachloroethane | H |
| bis(chloromethyl)ether | - |
| bis(2-chloroethyl)ether | M |
| 2-chloroethyl vinyl ether | L |
| 2-chloronaphthalene | H |
| 2,4,6-trichlorophenol | H |
| parachlorometa cresol | H |
| 2-chlorophenol | H |
| 1,2-dichlorobenzene | H |
| 1,3-dichlorobenzene | H |
| 1,4-dichlorobenzene | H |
| 3,3'-dichlorobenzidine | H |
| 2,4-dichlorophenol | H |
| 4-chlorophenyl phenyl ether | H |
| bis(2-chloroisopropyl)ether | M |
| bis(2-chloroethoxy)methane | M |
| bromoform (tribromomethane) | H |
| dichlorobromomethane | M |
| chlorodibromomethane | M |
| hexachlorobutadiene | H |
| hexachlorocyclopentadiene | H |
| pentachlorophenol | H |
| vinyl chloride | L |
| PCB-1242 (Arochlor 1242) | H |
| PCB-1254 (Arochlor 1254) | H |
| PCB-1221 (Arochlor 1221) | H |
| PCB-1232 (Arochlor 1232) | H |
| PCB-1248 (Arochlor 1248) | H |
| PCB-1260 (Arochlor 1260) | H |
| PCB-1016 (Arochlor 1016) | H |

*Note:   Explanation of Removal Ratings.

Category H (high removal)
      adsorbs at levels  >100 mg/g carbon at C(f) = 10 mg/L
      adsorbs at levels  >100 mg/g carbon at C(f) <1.0 mg/L

Category M (moderate removal)
      adsorbs at levels  >100 mg/g carbon at C(f) = 10 mg/L
      adsorbs at levels  <100 mg/g carbon at C(f) <1.0 mg/L

Category L (low removal)
      adsorbs at levels  <100 mg/g carbon at C(f) = 10 mg/L
      adsorbs at levels  <10 mg/g carbon at C(f) <1.0 mg/L

C(f) - final concentratios of priority pollutants at equilibrium.

Source:   Reference 14.

aromatic compounds.  Thus, the utility of adsorption as a treatment process
hinges on the economics of the specific situation, which in turn depend
primarily on the costs of regeneration.

Data taken from Reference 5 have been summarized in Table 5.5.7.  These
data provide results of full scale GAC systems.  Because of the sparsity of
information concerning system design and operating conditions, including
carbon loading, no attempt has been made to include such information in the
table.  Additional data demonstrating effective (99+ efficiency) removal of
pesticides from aqueous waste streams is shown in Table 5.5.8.  However, these
data were obtained for loadings that were in the ppb range.  While indicative
of the effectiveness of carbon adsorption as a polishing step, the data do not
demonstrate effectiveness for the higher end (up to 5,000 mg/L) of the
reported cost effective range.

No data were found for systems using PAC, although PAC and GAC should
exhibit little, if any, difference in adsorption performance.  As noted
previously, the most significant difference between the two sorbents is in
their particle size.  The fine particle size of PAC is not suitable for use in
contacting equipment normally used for GAC systems.

## 5.5.3  Cost of Carbon Absorption

The cost of carbon adsorption treatment can be described in terms of
capital investment and operation and maintenance costs.  Capital costs consist
of direct and indirect expenses.  For the small scale system, direct capital
investment costs include the purchase of a waste storage tank, a prefilter,
carbon columns, a waste feed pump, piping and installation.  For the large
scale system, additional direct capital investment costs include storage tanks
for spent and regenerated carbon, a multiple hearth furnace and automatic
controls.[16]

A model has been developed by ICF, Inc. (Reference 17) for calculating
carbon adsorption costs.  Table 5.5.9 contains the equations used in this
model to calculate direct capital costs as a function of carbon consumption
rate and storage volume.  Indirect capital costs include the costs of
engineering, construction, contractor's fee, start-up expenses, spare parts
inventory, interest during construction, contingency and working capital.

TABLE 5.5.7.   COMPOUNDS REPORTED IN WASTESTREAMS BEING TREATED
BY FULL-SCALE, GRANULAR ACTIVATED CARBON UNITS

| Pollutant | Concentration (mg/L) | | Removal (%) |
|---|---|---|---|
| | Influent | Effluent | |
| Benzene | 590 | 210 | 64 |
| para-Chloronitrobenzene | 11.6 | 0.0093 | 99.9 |
| 2-Chlorophenol | 8.67 | 0.62 | 92.8 |
| 4-Chlorophenol | 5.64 | 0.010 | 99.8 |
| 2,4-D | 3,600 | 0.010 | 99.99 |
| 2,4-D | 58.4 | 0.037 | 99.9 |
| 2,6-Dichlorophenol | 3.47 | 0.26 | 92.5 |
| 2,4-Dichlorophenol | 42.6 | 0.64 | 98.5 |
| Dieldrin | 0.011 | 0.00001 | 99.91 |
| Kepone | 4 | 0.0001 | 99.99 |
| Pentachlorophenol | 10 | 0.0001 | 99.99 |
| Pentachlorophenol | 120 | 49 | 59 |
| Toxaphene | 0.036 | 0.001 | 97.2 |
| 2,4,6-Trichlorophenol | 35 | 0.01 | 99.97 |

Source:   Reference 5.

TABLE 5.5.8.   RESULTS OF ADSORPTION ISOTHERM TESTS ON TOXIC CHEMICALS

| Compound | pH | Initial concentration g/L | Carbon treated concentration g/L | Organic reduction % | Capacity[a] |
|---|---|---|---|---|---|
| Aldrin | 7.0 | 48 | 1.0 | 97.9 | 30 |
| Dieldrin | 7.0 | 19 | 0.05 | 99+ | 15 |
| Endrin | 7.0 | 62 | 0.05 | 99+ | 100 |
| DDT | 7.0 | 41 | 0.1 | 99+ | 11 |
| DDD | 7.0 | 56 | 0.1 | 99+ | 130 |
| DDE | 7.0 | 38 | 1.0 | 97.4 | 9.4 |
| Toxaphene | 7.0 | 155 | 1.0 | 99+ | 42 |
| Arochlor 1242 (PCB) | 7.0 | 45 | 0.5 | 98.9 | 25 |
| Arochlor 1254 (PCB) | 7.0 | 49 | 0.5 | 98.98 | 7.2 |

[a]mg of toxic chemicals adsorbed/g of carbon.

Source:   Reference 15.

TABLE 5.5.9.    DIRECT COSTS FOR CARBON ADSORPTION[a]

| Carbon consumption rate (lbs/day) | Direct capital costs ($) | Direct operation and maintenance cost[b] ($/yr) |
|---|---|---|
| Less than 400 | $1,256(c)^{.603} + 140(s)^{.54}$ | $29(c)^{.6} + 350(c)(cp) + 619(c)^{.168}(h) + 5(c)(p)$ |
| Greater than 400 | $14,231(c)^{.522} + 140(s)^{.54}$ | $58(c)^{.657} + 35(c)(cp) + 105(c)^{.455}(h) + 25,012^{.383}(c)(p) + 1.49\ 10^{6}(c)(f)$ |

where:  c  = carbon consumption rate in pounds per day

   s  = storage volume in gallons

   cp = carbon price in dollars per pound ($0.8/lb)

   h  = hourly wage rate in dollars per hour ($14.56/hr)

   p  = power price in dollars per kilowatt-hour ($0.05/KWh)

   f  = fuel price (natural gas) in dollars per Btu ($6x10^{-6}/Btu)

[a]Cost estimates were developed for three model treatment systems (three small scale and three large scale systems).  The cost estimates for these systems were then used to develop a cost equation in the form of a power curve.

[b]The power requirement is derived from the equipment specifications.

Source:  Reference 17.

These costs are expressed as percentages of either direct capital costs or the sum of direct and indirect capital costs as summarized in Table 5.5.10. Direct and indirect capital costs are assumed to be incurred in year zero.

Operation and maintenance costs also consist of direct and indirect costs. Direct operation and maintenance costs (in 1984 dollars) include the operating labor and electricity and carbon consumption. Table 5.5.9 also contains the equations used in the model to calculate direct operation and maintenance costs. As with the capital costs, the model considers operation and maintenance costs for carbon consumption rates less and greater than 400 lbs/day. For large-scale systems, the operation and maintenance costs also include the natural gas consumption necessary for the furnace. Indirect operation and maintenance costs include costs for insurance and overhead.

Based on the RCRA Risk-Cost Analysis Model, Table 5.5.11 shows carbon adsorption costs for 100, 400, 1,000 and 2,500 gal/hr processes.

## 5.5.4  Overall Status of Process

### 5.5.4.1  Availability--

Activated carbon adsorption is a widely used technology for treating waste streams containing organic compounds, including many hazardous halogenated compounds. Its ability to treat solvents and other organics has been demonstrated at bench, pilot, and full-scale levels by many firms. Manufacturers of activated carbon produce carbons to fit variable service needs. Companies that use these activated carbon systems, both GAC and PAC, are numerous as documented in several literature sources (see References 5, 13, and 18). Equipment designers and suppliers can be found in the Chemical Engineering Equipment Buyers' Guide published by McGraw-Hill, New York, NY. Many of these firms will provide assistance in developing a treatment system for specific waste streams.

### 5.5.4.2  Application--

Activated carbon adsorption systems are widely used in industry to process chemical product streams as well as waste streams. The technology has proven to be effective as a pretreatment for aqueous wastes prior to their introduction into biological treatment systems. Concurrent treatment of waste streams with PAC and biological treatment has also proven to be effective.

TABLE 5.5.10.   INDIRECT COSTS FOR CARBON ADSORPTION

| Item | Percent of direct capital costs | Percent of the sum of direct and indirect capital costs | Percent of total annual cost[a] |
|---|---|---|---|
| **Indirect Capital Costs** | | | |
| Engineering and Supervision | 12 | 0 | 0 |
| Construction and Field Expenses | 10 | 0 | 0 |
| Contractors Fee | 7 | 0 | 0 |
| Startup Expenses | 5 | 0 | 0 |
| Spare Parts Inventory | 2 | 0 | 0 |
| Interest During Construction | 10 | 0 | 0 |
| Contingency | 0 | 15 | 0 |
| Working Capital | 0 | 18 | 0 |
| **Indirect Operation and Maintenance Costs** | | | |
| Insurance, Taxes, General Administration | 0 | 5 | 0 |
| System Overhead | 0 | 5 | 10 |

[a]The total annual cost is defined as the sum of the total capital cost multiplied by the capital recovery factor and the total operation and maintenance costs.

Source:   Reference 17.

TABLE 5.5.11.   CARBON ADSORPTION COSTS[a]

| | Quantity processed (gal/hr) | | | |
|---|---|---|---|---|
| | 100 | 400 | 1,000 | 2,500 |
| **Capital Expenditures** | | | | |
| Capital Cost Including Installation[b] ($1,000) | 59 | 561 | 904 | 1,462 |
| **Annual Operation and Maintenance ($1,000)[c]** | | | | |
| Energy | 2 | 11 | 27 | 68 |
| Labor | 23 | 35 | 53 | 80 |
| Carbon | [e]7 | 27 | 67 | 168 |
| Other | 1 | 5 | 10 | 18 |
| Capital Recovery | 10 | 99 | 160 | 259 |
| Total Annual Cost | 42[e] | 177 | 317 | 593 |
| Cost/1,000 gal[d] | 210[e] | 221 | 159 | 119 |

[a]Costs are based on the RCRA Risk-Cost Analysis Model.[17]

[b]Capital costs for the 100 gal/hr system include waste storage tank, prefilter, carbon columns, waste feed pump, piping and installation; the other flow levels (400, 1,000, 2,500) include these units plus storage tanks for spent and regenerated carbon, a multiple hearth furnace and automatic controls.

[c]These costs are based on the following data:

    carbon price = $0.8/lb
    hourly wage rage = $14.56/hr
    power price = $0.05/kwh
    fuel price (natural gas) = $6 x $10^{-6}$/Btu
    capital recovery factor = 0.177

[d]Unit costs are based on 2000 hours of operation per year.

[e]Modified to reflect a direct relationship between carbon requirement and quantity processed.

*Note:   1984 dollars, prices are similar to 1986 values.

However, the most common application of carbon adsorption systems would appear to be as a polishing step for low concentration level effluents from other treatment technologies.  The use of carbon adsorption systems for treatment of wastes containing 0.5 percent or greater organic concentration levels is not considered to be cost effective.  Other technologies should be considered at these concentrations, unless regeneration can be used to achieve recovery of valuable adsorbed compounds.

Removal efficiencies which permit direct discharge can usually be met by GAC systems for most halogenated organics.  However, performance will depend upon the specifics of waste stream contamination, including the need for pretreatment, post-treatment, and other aspects of system operation.

5.5.4.3  Environmental Impacts--

Environmental impacts can result from emissions during the regeneration of carbon.  However, there will be no serious environmental impacts if the exit gases are treated by a control system; e.g., an afterburner and/or scrubber, and in some cases, a particulate filter.  Where the carbon is chemically regenerated (acid, base, or solvent), the regeneration stream will require future treatment; e.g., incineration or distillation to remove the organic contaminants.

The recovery or reuse of desorbed solutes from the adsorption process presents opportunities for both cost savings and reduction of environmental impacts.  Disposal of desorbed solutes as waste materials can be costly and also result in an environmental hazard.  Therefore, recycling of solute following desorption and recovery should be considered and practiced if possible.

5.5.4.4  Advantages and Limitations--

The principal advantages of carbon adsorption technology is its ability to achieve low effluent concentration levels for a large number of compounds, including many halogenated organics.  The technology appears particularly applicable to high molecular weight compounds such as the chlorinated pesticides.  It is also applicable to the treatment of many compounds which are normally toxic and resistant to biological treatment.  Material recovery may also be possible if regeneration methods other than thermal regeneration can be used.

Limitations are largely associated with high capital and operating costs, particulary when thermal reactivation must be used.   Thermal reactivation, the most effective means of regeneration, is cost effective only for relatively large installations (i.e., greater than 1000 lbs/day) and for wastes with relatively low (less than 1 percent) organic concentrations.   The technology is sensitive to other impurities such as suspended solids and oil and grease, thus, some degree of pretreatment is usually required to ensure effective performance.   The adsorption process is also not effective for many low molecular weight and highly water soluble organics.   However, most of the halogenated organics considered in this document do not fall into the low molecular weight, water soluble categories and are usually effectively adsorbed.

## REFERENCES

1.   Berkowitz, J.B. et al.  Physical, Chemical and Biological Treatment
     Techniques for Industrial Waste.  Noyes Data Corporation; Park Ridge, New
     Jersey.  1978.

2.   Rizzo, J.L.  Calgon Carbon Corporation.  Letter to Paul Frillici, GCA.
     June 11, 1986.

3.   ICF Inc. Survey of Selected Firms in the Commercial Hazardous Waste
     Management Industry:  1984 Update.  Final Report to U.S. EPA,
     Section II.  OSW Washington, DC.  1985.

4.   U.S. EPA Background Document for Solvents to Support 40 CFR Part 268,
     Land Disposal Restrictions, Volume II.  January 1986.

5.   II Enviroscience, Incorporated.  Survey of Industrial Applications of
     Aqueous-Phase Activated-Carbon Adsorption for Control of Pollutant
     Compounds from Manufacture of Organic Compounds.  Prepared for U.S. EPA
     IERL; EPA-600/2-83-034, PB-83-200-188.  April 1983.

6.   Dobbs, R.A., and J. Cohen.  Carbon Adsorption Isotherms for Toxic
     Organics.  EPA-600/8-80-023.  April 1980.

7.   Slejko, F.L.  Applied Adsorption Technology, Chemical Industry Series,
     Volume 19.  Marcel Dekker, Inc. NY, NY.  December 1985.

8.   Perrich, J.R., Editor.  Activated Carbon Adsorption for Wastewater
     Treatment.  CRC Press Inc., Boca Raton, Florida.  1982.

9.   U.S. EPA.  Activated Carbon Treatment of Industrial Wastewater-Selected
     Papers.  EPA-600/2-79-177.  Robert S. Kerr Environmental Research
     Laboratory.  August 1979.

10.  Lyman, W.J.  Carbon Adsorption, In:  Unit Operations for Treatment of
     Hazardous Industrial Wastes.  Pollution Technology Review No. 47, Noyes
     Data Corporation, Park Ridge, NJ.  1978.

11.  Meidl, J.A., Zimpco Inc.  PAC Process.  Engineering and Management.  June
     1982.

12.  Breton, M. et al.  Technical Resource Document - Treatment Technologies
     for Solvent-Containing Wastes.  Prepared for HWERL, Cincinnati under
     Contract No. 68-03-3243, Work Assignment No. 2.  August 1986.

13.  Touhill, Shuckrow & Associates, Inc.  Concentration Technologies for
     Hazardous Aqueous Waste Treatment.  Pittsburg, PA.  EPA-600/2-81-019.

14.  U.S. EPA.  Treatability Manual, Volume III.  EPA-600/2-82-001a, U.S. EPA
     ORD, Washington, DC.  1981.

15.  Hager, D.G., Calgon Corp.  "Wastewater Treatment by Activated Carbon."
     Chemical Engineering Progress.  October 1976.

16.  U.S. EPA.  Development Document for Effluent Limitation Guidelines and
     Standards for Petroleum Refining Point Source Category.
     EPA-440/1-82-014.  October 1982.

17.  ICF, Inc.  RCRA Risk-Cost Analysis Model, Phase III, U.S. EPA, OSW.
     March 1984.

18.  Radian Corporation.  Full-Scale Carbon Adsorption Applications Study.
     EPA-600/2-85-012.  May 1984.

5.6  RESIN ADSORPTION

Resin adsorption is an alternative treatment technology for the removal of organic contaminants from aqueous waste streams.  The underlying principle of operation is similar to that for carbon adsorption; organic molecules contacting the resin surface are held on the surface by physical forces and subsequently removed during the resin regeneration cycle.  Resin adsorbents can be made from a variety of monomeric compounds which differ in their polarity and thus, their affinity for different types of compounds.  The choice of resin type can lead to an adsorbent tailored specifically for effective removal of special classes of compounds.  For example, hydrophobic resins such as those prepared from styrene - divinyl benzene monomers, are most effective for nonpolar organics and bonding is largely the result of Van der Waal's forces; acrylic based resins on the other hand are more polar and dipole-dipole interactions may play the major role in the binding of polar molecules to the resin surface.  The general concept is that like molecules attract.  Polar resins will attract polar organics; nonpolar compounds will be attracted by the more hydrophobic or nonpolar resins.[1]

A significant aspect of resin adsorption is that the bonding forces are usually weaker than those encountered in granulated activated carbon (GAC) adsorption .  Regeneration can be accomplished by simple, nondestructive means such as solvent washing, thus providing the potential for solute recovery.  Thermal regeneration (generally not possible with resin adsorbents because of their temperature sensitivity) is usually required for carbon adsorbents, eliminating the possibility of solute recovery.  The resins differ in many other respects from activated carbon adsorbents.  In addition to differences in the ease and usual methods of regeneration associated with the chemical nature of the two adsorbents, there are significant differences in shape, size, porosity and surface area.  Resin adsorbents are generally spherical in shape rather than granular, and are smaller in size and lower in porosity and surface area than GAC adsorbents.  Surface areas for resins are generally in the range of 100-700 $m^2/g$, as opposed to 800-1,200 $m^2/g$ for activated carbon.[1]  Adsorptive capacities are thus less for the resin adsorbents, although the chemical nature and the pore structure of the resin can be tailored to enhance the selectivity of the resin and, therefore, its

adsorption capacity for specific organic components.  Other notable properties
of resin adsorbents include their nondusting characteristics, their low ash
content, and their resistance to bacterial growth.  The last characteristic is
primarily a result of the fine pore structure which inhibits bacterial
intrusion.  Pore diameter and other physical properties of resin adsorbents
are shown in Table 5.6.1.

Another significant difference between resin and carbon adsorbents is
their cost.  Resin adsorbents are much more expensive.  They generally will
not be competitive with carbon for the treatment of waste streams containing a
number of contaminants with no recovery value.  However, resin adsorption
should be considered if material recovery is practical, selectivity is
possible, and for cases where carbon regeneration is not effective.  Like
carbon adsorption systems, resin adsorption can produce an effluent with low
levels of contaminant concentrations, particularly in cases where contaminants
are well characterized and few in number.  Resin adsorption combined with
carbon adsorption may be effective for certain waste streams containing a
number of contaminants.[1]

## 5.6.1  Process Description

Resin adsorption systems are designed and operated in similar fashion to
GAC systems.  A principal difference will be in the regeneration step;
regeneration of the resin is usually performed in situ with aqueous solutions
or solvents.  Solute recovery from the regeneration liquor will also be
required, with distillation the most likely method.

### 5.6.1.1  Pretreatment Requirements--

Polymeric adsorbents require pretreatment of feed streams to remove
suspended solids, oils and greases, and to adjust pH and temperatures, as
appropriate.  Suspended solids in the influent should be less than 50 mg/L
and, in the case of oil and grease, less than 10 mg/L to prevent clogging of
the resin bed.[1]  The control of pH may be necessary to prevent resin attack
and to enhance adsorbability.  Low temperature will also generally enhance
adsorption.  Resin adsorbents, although generally resistant to chemical attack
because of their cross-linked structure, should not be brought into contact

TABLE 5.6.1.   PHYSICAL PROPERTIES OF ADSORBENTS

| Manufacturer | Adsorbent | Chemical nature | Pore volume ($cm^3/g$) | Surface area ($m^2/g$) | Pore diameter average (A) | Surface polarity |
|---|---|---|---|---|---|---|
| Rohm and Haas | Amberlite XAD-2 | Polystyrene | 0.68 | 300 | 100 | Low |
| | Amberlite XAD-4 | Polystyrene | 0.96 | 725 | 50 | Low |
| | Amberlite XAD-7 | Acrylic Ester | 0.97 | 450 | 85 | Intermediate |
| | Amberlite XAD-8 | Acrylic Ester | 0.82 | 160 | 150 | Intermediate |
| | Ambersorb XE-347 | Polymer Carbon | 0.41 | 350 | 200, 15[a] | Low |
| | Ambersorb XE-348 | Polymer Carbon | 0.58 | 500 | 200, 15[a] | Intermediate |
| Mitsubishi | Diaion HP-10 | Polystyrene | 0.64 | 500 | b | Low |
| | Diaion HP-20 | Polystyrene | 1.16 | 720 | 70 | Low |
| | Diaion HP-30 | Polystyrene | 0.87 | 570 | b | Low |

[a]Average pore diameter of the macropores and micropores, respectively.

[b]Average pore diameter not available.

with compounds such as chemical oxidants and functional reagents which may degrade the resin or poison adsorption sites. High levels of dissolved solids, particularly inorganic salts, do not compete with organics for adsorption sites, and their presence may in some instances increase the adsorption of organics.

Pretreatment options are similar to those proposed previously for carbon adsorption systems. For example, filtration or coagulation/sedimentation type separations can be used for suspended solids, and flotation/extraction procedures can be used for removal of oils and greases. Each pretreatment option will result in a residual which may or may not require additional processing prior to disposal.

There are no definite limitations on the upper or lower contaminant concentration levels that can be treated. An upper limit of 8 percent (for phenol) is suggested in Reference 1, however, this is to maintain cycle time and regeneration frequency within reasonable limits. As with carbon adsorption, the efficiency of resin adsorption (weight of adsorbed material per weight of adsorbent) is greater at high concentrations.

## 5.6.1.2  Operating Parameters--

The design of a resin adsorption system requires the development of basic information such as feed stream flow rate, contaminant concentration, and adsorbent type and capacity. Other information such as flow rate variations, suspended solid level, pH, and temperature will be required to ensure that adequate pretreatment precautions and operating practices are followed.

The choice of adsorbent type can be guided by the concept that attractive forces will be greatest for similar molecules. The solubility concept is also useful in identifying regeneration solvents. The similarity of the adsorbate in the regeneration solvent is quite important. The solvent not only must be capable of overcoming the attractive forces of adsorbate/adsorbent but must also remove the adsorbate in the smallest possible volume.

Although the relative strengths of the attractive forces between solute, solvent, and resin can be predicted through the use of solubility parameters,[4] there is no practical method for determining the actual capacity of an adsorbent for contaminants, particularly those existing in complex waste streams. It is, therefore, necessary to carry out experimental studies to determine working capacities for candidate adsorbents.  Costs may also be prohibitive, and activated carbon may often be a more attractive adsorbent, particularly where solute recovery is not desirable or practical.

Assuming a resin adsorbent can be found that can achieve required treatment levels, additional tests will be required to identify and select a regeneration process.  The selection of a regeneration solvent can be guided by use of solubility parameters.  However, other factors such as cost of solvent regeneration and adsorbate recovery must be considered.  Distillation appears to be the most likely solvent and solute recovery technology assuming a solvent/solute match can be found that is amenable to such a separation process.

Design of a resin adsorption process operation would include the following steps as a general procedure:  1) determine wastewater effluent purity desired, 2) select adsorbent and determine adsorption capacity, 3) select regeneration process based on bench or pilot scale tests, 4) size adsorbent bed, 5) check loading run length and determine if it is compatible with the regeneration time cycle, 6) repeat 4 and 5 until loading and regeneration cycles are compatible, 7) determine bed dimensions by hydraulic considerations, 8) design and size pumps, storage tanks, pretreatment equipment and auxiliary equipment.[5]

As noted in Reference 1, a system for treating low volume waste streams will commonly consist of two beds.  One bed will be on stream while the second is being regenerated as shown in Figure 5.6.1.

The adsorption bed is usually fed downflow at flow rates in the range of 0.25 to 2 gpm per cubic foot of resin; this is equivalent to 2-16 bed volumes/hr, and thus contact times are in the range of 3-30 minutes.  Linear flow rates are in the range of 1-10 gpm/ft$^2$.  Adsorption is stopped when the bed is fully loaded and/or the concentration in the effluent rises above a certain level.  A time of 30 minutes may not be adequate for attainment of minimum concentration levels.  EPA has suggested that limited contact times may play an important role in reducing column loadings in the field to values less than those predicted from isotherm testing.[6]  Reference was made to a study which attributed carbon contact times of greater than 230 minutes to applications which requires high degree of pollutant removal.  Although rapid

Resin Adsorption

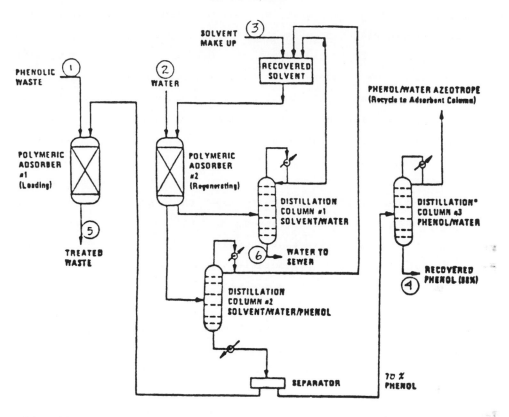

Column #3 utilized when highly pure phenol is required.

Material Balance, lb/hr

| | ① | ② | ③ | ④ | ⑤ + ⑥ |
|---|---|---|---|---|---|
| Phenol | 264 | | | 264 | < 10 ppm |
| Water | 21,736 | 1480 | | 9 | 23,207 |
| Acetone | | | 4 | | 4 |
| Total | 22,000 | 1480 | 4 | 273 | 23,211 |

Figure 5.6.1.   Phenol removal and recovery system – solvent
regeneration of Amberlite adsorbent.

Source:   Reference 1.

adsorption kinetics are attributed to resin adsorbents, caution should be exercised in assessing the contacting time requirements and design and operating features needed to meet acceptable concentration levels.

Regeneration of the resin bed is performed in situ with basic, acidic, and salt solutions or recoverable nonaqueous solvents being most commonly used. Basic solutions may be used for the removal of weakly acidic solutes and acidic solutions for the removal of weakly basic solutes; hot water or steam could be used for volatile solutes; and methanol and acetone are often used for the removal of nonionic organic solutes. A prerinse and/or a postrinse with water will be required in some cases to remove certain contaminants such as salts. As a rule, about three bed volumes of regenerant will be required for resin regeneration; as little as one-and-a-half bed volumes may suffice in certain applications.[1]

The use of steam as the regenerating agent should be considered; steam regeneration for volatile organics may provide some cost benefits in that it can reduce the need for subsequent treatment to separate the waste solvent from the dissolved organics. However, the condensed steam may also require additional treatment prior to discharge to also eliminate dissolved organics.

When using steam regeneration for polymeric adsorbents, one must consider the upper temperature limit of the resin in choosing the steam pressure. The styrene based polymeric adsorbents are usually stable to 200°C; acrylic based resins up to 150°C. Since the adsorbed solvent and other organic constituents can cause the adsorbent resin matrix to swell and weaken, removal of these constituents by steaming could result in disruption and breakup of the resin matrix. Therefore, adsorbent stability is of concern when using steam regeneration and should be studied using multi-cycling tests to confirm the integrity of the adsorbent before proceeding with design of the regeneration system.

Steam requirements are normally significantly lower for the polymeric adsorbents than those for granular activated carbon to achieve a certain desorption level of a given constituent. The reason for this is that the attractive forces binding the organic constituent to the adsorbent are much lower for the polymeric adsorbent.

5.6.1.3  Post-Treatment Requirement--

Assuming effluent goals are realized, the post-treatment requirements are restricted to treatment of the regeneration effluent.  Other possible waste streams requiring further processing could include the washing effluents (if required for the prerinse and/or postrinse of the resin), the regeneration solvent, and the condensed regeneration steam.  Requirements will depend upon the process scheme used.

5.6.1.4  Treatment Combinations--

Resin adsorption will normally be given consideration in applications for which carbon adsorption would be considered as a potentially viable treatment alternative.  However, it will not generally be economically competitive with carbon adsorption.  In certain situations a combination of resin and carbon adsorption could be used to advantage.  For example it may be attractive as a polishing step to remove specific contaminants (particularly if the contaminants have recovery value) passing a carbon adsorption bed, e.g., polar, low molecular weight compounds.

5.6.2  Demonstrated Performance

Resin adsorption technology is not as established as activated carbon adsorption is for full scale treatment of waste streams containing halogenated organic contaminants.  Studies have been conducted to determine the performance of resins as adsorbents for several types of organic chemical compounds.  The results of one such study for pesticides is shown in Table 5.6.2.  Although high efficiencies were obtained, the initial pesticide concentrations were in the ppm range and did not approach the higher levels (8 percent) suggested in Reference 1 as appropriate for the technology.  Further information concerning the performance of resin adsorbents for removal of halogenated and other organic solvents is provided in Reference 8.

TABLE 5.6.2.   REMOVAL OF POLYNUCLEAR AROMATICS, CHLORINATED PESTICIDES, AND POLYCHLORINATED BIPHENYLS FROM TWO TYPES OF SPIKED MIAMI TAP WATER

| | % Removed (66,800 BV)[a] | | | |
| | Ambersorb XE-340 | | FS-400 | |
| | Water A[b] | Water B[c] | Water A[b] | Water B[c] |
|---|---|---|---|---|
| Dibromochlorophenol | 81.5 | 99.4 | 94.3 | 100 |
| Hexachlorobenzene | 98.6 | 99.9 | | |
| BHC (1,2,3,4,5,6-hexa-chlorocyclohexane) | 98.7 | 100 | | |
| BHC (1,2,3,4,5,6-hexa-chlorocyclohexane) | 98.9 | 99.8 | | |
| Aldrin | 92.5 | 95.3 | | |
| Heptachlor | 97.6 | 99.8 | | |

[a]Bed depth, 2.5 ft; column diameter, 1 in.; flow rate, 1.2 gpm/ft$^3$ (decreasing to 0.6 near end of test); BV, 386 ml; EBCT, 6.2 min (increasing to 12 min near end of test); duration, 320 days.

[b]18 ppm $Cl_2$ needed for breakpoint chlorination, 7.2 ppm TOC in finished water.

[c]5 ppm $Cl_2$ needed for breakpoint chlorination, 5 ppm TOC in finished water.

Source:   Reference 7.

## 5.6.3  Cost of Resin Adsorption

Resin adsorbents are quite expensive (Table 5.6.3).  The cost exceeds that of granular activated carbon (GAC) ($0.80 to $1.00 per pound).  However, the economics of using resins or polymeric adsorbents may in certain cases be more favorable than those for granular activated carbon.

Thermal regeneration costs for GAC adsorption systems are quite high and carbon losses are of the order of 3 to 8 percent per regeneration.  Even though macroreticular (resin) adsorbents cost more per pound, they are relatively cheaper to regenerate and regeneration does not result in any appreciable adsorbent loss.  Thus, smaller beds and more frequent regenerations may be economically viable with resin adsorbents.

Design criteria for a one million gallon per day treatment plant are shown in Table 5.6.4.  Assuming influent concentrations of 300-1,000 ppb, the operation is designed to remove greater than 90 percent of the incoming contaminant.  A comparable GAC system is analyzed simultaneously for comparison.  The capital and operating costs for each system are given in Table 5.6.5.  It can be seen that both the capital investment and the operating costs are lower when the more expensive (by volume) adsorbent is used.  This comes about primarily because fewer and smaller contactors are utilized and expensive thermal regeneration furnaces are not required.

The resin system looks very promising because of the many assumptions made concerning design and performance, e.g., high capacity, rapid kinetics, and a 5 year resin lifetime.  The assumptions have not yet been demonstrated. Moreover, the design is for a waste influent loading (1 ppm) that is extremely low for an industrial waste stream.  Costs, already high relative to many other technologies, will increase drastically as influent loadings (and system size) increase.

However, the example does indicate that resin adsorption may be more economical than carbon adsorption.  Similar reasoning has been applied in Reference 1 where costs have been estimated for resin adsorption applied to three different waste streams.  Costs ranged from $38.60 per 1,000 gallons for a phenol recovery system (at 5 percent phenol in waste) to $0.83 per 1,000 gallons for a chlorinated pesticide removal system.  In the latter case, the cost of a GAC treatment system was estimated at $1.33 per 1,000 gallons.

TABLE 5.6.3.    COST OF ADSORBENTS[a]

| Adsorbent | Chemical nature | Cost $/ft$^{3b}$ |
|-----------|-----------------|------------------|
| Amberlite XAD-2 | Polystyrene | 282.95 |
| Amberlite XAD-4 | Polystyrene | 355.05 |
| Amberlite XAD-7 | Acrylic ester | 223.25 |
| Amberlite XAD-8 | Acrylic ester | 337.25 |

[a]Personal communication with Rohm and Haas Company, Fluid Process Chemicals Department, Philadelphia, PA, April 3, 1986.

[b]At a bulk density of 37 lbs/ft$^3$, costs are roughly $6 to $10 per pound.

TABLE 5.6.4.    DESIGN CRITERIA--TRIHALOMETHANE REMOVAL

| | Adsorbent | |
| Parameter | Ambersorb XE-340 | Granular activated carbon |
| --- | --- | --- |
| Density | 37 lb/ft$^3$ | 25 lb/ft$^3$ |
| Nominal Flowrate | 6.0 gpm/ft$^3$ 1.25-min EBCT[a] | 1.0 gpm/ft$^3$ 7.48-min EBCT[a] |
| Contactors | 58 ft$^3$ each 2 on-stream 1 regeneration/standby | 348 ft$^3$ each 2 on-stream 1 regeneration 1 standby |
| On-stream Time | 3.3 days | 20 days |
| Regeneration Type Time/Contactor | in-place steam 8 hr | thermal reactivation 11 days |
| Absorbent Lifetime | 5 yr (fouling limited) | 8 months @ 8% loss/cycle |

Design Basis:    1.0 mgd average flow, 1.43 mgd peak flow.

[a]Empty bed contact time.

Source:    References 9 and 10.

TABLE 5.6.5.  COST COMPARISON--GAC VS. RESIN[a]

| | 1.0 mgd Plant | |
|---|---|---|

| Capital Cost | Resin | Granular Activated Carbon |
|---|---|---|
| Contactor, Pumps, Regeneration Facilities Plus 25% for Engineering Contingencies | $350,000 | $950,000 |
| Adsorbent Cost | | |
| | $ 45,000 @ $7.00/lb | $ 20,000 @ $0.55/lb |
| Total | $395,000 | $970,000 |

| Operating Costs | $/yr | ¢/1,000 gal | $/yr | ¢/1,000 gal |
|---|---|---|---|---|
| Adsorber Power | 7,100 | 1.945 | 3,550 | 0.973 |
| Regeneration Fuel | 3,000 | 0.822 | 4,203 | 1.152 |
| Solvent Regeneration | 6,188 | 1.695 | --- | --- |
| Adsorbent Makeup | 9,000 | 2.466 | 18,000 | 4.932 |
| | (5 yr) | | (8% loss/cycle) | |
| Subtotal | $25,288 | 6.928 | $25,753 | 7.057 |
| Capital Related Costs (exclude adsorbent): | | | | |
| Depreciation        9% Maintenance        3% Property Overhead 2% | 49,000 | 13.420 | 133,000 | 36.438 |
| Quality Control | 9,000 | 2.460 | 9,000 | 2.460 |
| Total | $83,288 | 22.8¢/ 1,000 gal | $167,753 | 46.0¢/ 1,000 gal |

[a]No specific GAC or resin product.  Values taken at average costs.

Source:  References 9 and 10.

The cost data are outdated (from the 1970's); costs in 1986 dollars would be about 50 percent greater, based on changes in the chemical engineering plant cost index.

The high costs of resin adsorption for the treatment of moderate to high concentration contaminant levels can only be justified in situations where cost benefit is realized from product recovery.  In the case of the phenol recovery system used in the example above, credit from the sale of phenol exceeded total annual operating costs, therefore justifying use of the process on an economics basis.

### 5.6.4  Overall Status

#### 5.6.4.1  Availability--

Resin adsorption technology parallels that for carbon adsorption. Equipment requirements are similar and available from a number of manufacturers serving the chemical process industries.  However, there appears to be some question about the commercial availability of many of the resin adsorbents for which data are reported in the literature.  Ambersorb XE-340, for example, manufactured by Rohm and Haas and the subject of numerous technical studies, is not available in commercial quantities.  The availability of some other resin adsorbents may also be questionable.

#### 5.6.4.2  Application--

Because of their expense, resins are not commonly used full-scale to remove organics from wastewaters.[6]  There is also little publicly available information on current or proposed industrial applications.  Information of a general nature does report that resins are being used for color removal from dyestuff and paper mill waste streams, for phenol removal, and for polishing of high purity waters.

The following applicants have been identified as being particularly attractive for resin adsorption technology.[1]

- Treatment of highly colored wastes where color is associated with organic compounds

- Material recovery where solvents of commercial value are present in high enough concentration to warrant material recovery since it is relatively easy to recover solutes from resin adsorbents

- Where selective adsorption is an advantage and resins can be tailored to meet selectivity needs

- Where low leakage rates are required; resins exhibit low leakage apparently as a result of rapid adsorption kinetics

- Where carbon regenerations is not practical, e.g., in cases when thermal regeneration is not safe

- Where the waste stream contains high levels of inorganic dissolved solids which drastically lowers carbon activity; resins activity can usually be retained ,although prerinses may be required

5.6.4.3  Environmental Impacts--

The only major environmental impacts resulting from resin adsorption systems are associated with the disposal of the regeneration solution and the extracted solutes when they can not be recycled.  Distillation to recover solvent and incineration of the separated solute are likely treatment/disposal options.  Air emissions would have to be considered as a result of these treatment processes.

5.6.4.4  Advantages and Limitations--

As noted, resin adsorption appears to offer advantages in certain situations; e.g., for treatment of highly colored wastes, for material recovery, where low leakage is required, and in instances where carbon adsorption is not practical.  The advantages of resin adsorption are a result of their potential for selectivity, rapid adsorption kinetics, and ease of chemical regeneration.

Major limitations of resin adsorbents result from:  1) the generally lower surface area and usually lower adsorption capacities than those found in activated carbon; 2) possible susceptibility to fouling due to poisoning by materials that are not removed by the regenerant; and 3) their relatively high cost.  The high cost of the resin may be balanced by its ease of regeneration and their predicted long lifetimes in situations where carbon must be thermally regenerated and carbon losses become appreciable (up to 10 percent).

## REFERENCES

1.  Lyman. W.J., Resin Adsorption in: Unit Operations for Treatment of Hazardous Wastes, Pollution Technologies Review No. 47.  Noyes Data Corporation, Park Ridge, New Jersey, 1978.

2.  Rohm and Haas Company, Fluid Process Chemicals Department, Amber-Hi-Lites, Winter 1980 (Technical Bulletin).

3.  Neely, J.W. and E.G. Isacoff, Carbonaceous Adsorbent For the Treatment of Ground and Surface Waters, Marcel Dekker, Inc, New York, N.Y.,  1982.

4.  Mark, H., et al.  Encyclopedia of Polymer Science and Technology, Cohesive-Energy Density.  Vol. 3, p. 833.  John Wiley & Sons, Inc., 1970.

5.  Slejko, F.L., Applied Adsorption Technology, Chemical Industry Series, Marcel Dekker, Inc, New York, N.Y. 1985.

6.  U.S. EPA, Background Documents for Solvents to Support 40 CFR Part 268 Land Disposal Restrictions, Volume II, January 1986.

7.  Symons, J.M., J.K. Carswell, J. DeMarco, and O.T. Love, Jr., Removal of Organic Contaminants from Drinking Water Using Techniques Other Than GAC Alone, A Progress Report, U.S. EPA, Cincinnati, 1979.

8.  Breton, M. A., et al.  Technical Resource Document - Treatment Technologies for Solvent - Containing Wastes.  Prepared for U.S. EPA, HWERL, Cincinnati under Contract No. 68-03-3243, Work Assignment No. 2. August 1986.

9.  U.S. EPA.  Synthetic Resin Adsorbents in Treatment of Industrial Waste Streams, EPA 600/2-84-105, May 1982.

10. McGuire, M.J. and Sublet, I.A., Activated Carbon Adsorption of Organics from the Aqueous Phase, Volume 2; Economic Analysis Employing Ambersorb XE-340 Carbonaceous Adsorbent in Trace Organic Removal from Drinking Water, Ann Arbor Science 1980.

# 6. Chemical Treatment Processes

The chemical treatment methods discussed in this section include some processes which could equally well be classified as thermal processes (i.e., wet air and supercritical water oxidation) since the general result of these high temperature processes is the conversion of the organic contaminants to fundamental products of oxidation such as carbon dioxide and water. Other technologies, like the other oxidation processes do not achieve total destruction and must be considered as pretreatment steps for a second treatment technology, usually a biotreatment process. The processes addressed in this section are:

6.1   Wet Air Oxidation

6.2   Supercritical Water Oxidation

6.3   Other Chemical Oxidation Processes

6.4   Dechlorination Processes

Discussions of these chemical treatment processes are provided using the same format as was used for the discussions of physical treatment processes in the previous section. Parallel discussions for the above processes can also be found in the TRD for solvent-containing wastes.

6.1  WET AIR OXIDATION

Wet air oxidation (WAO) is the oxidation of dissolved or suspended
contaminants in aqueous waste streams at elevated temperatures and pressures.
It is generally considered applicable for the treatment of certain organic-
containing media that are too toxic to treat biologically and yet too dilute
to incinerate economically.[1,2] The leading manufacturer of commercially
available WAO equipment reports that WAO takes place at temperatures of 175 to
320°C (347 to 608°F) and pressures of 2,169 to 20,708 kPa (300 to
3,000 psig).[1] Although the process is operated at subcritical conditions
(i.e., below 374°C and 218 atmospheres), the high temperatures and the high
solubility of oxygen in the aqueous phase greatly enhances the reaction rates
over those experienced at lower temperatures and pressures.  In practice, the
three variables of pressure, temperature and time are controlled to achieve
the desired reductions in contaminant levels.

In addition to serving as the source of oxygen for the process, the
aqueous phase also moderates the reaction rates by providing a medium for heat
transfer and heat dissipation through vaporization.  Generally, pressures are
maintained above the vapor pressure of water to limit water evaporation rates,
thus limiting the heat requirement for the process.  The reactions proceed
without the need for auxiliary fuel at feed chemical oxygen demand (COD)
concentrations of 20 to 30 grams per liter.[3] The extent of contaminant
destruction will depend upon the wastes to be oxidized and the reaction
conditions.  Typically, 80 percent of the organic contaminants will be
oxidized to $CO_2$ and $H_2O$.  Residual organics will generally be low
molecular weight, biodegradable compounds such as acetic acid and formic acid.

However, halogenated aromatic compounds, e.g. chlorobenzenes and many
pesticides, are resistant to wet air oxidation.  Information concerning the
extent of reduction achievable and the nature of the residuals for these
compounds is largely unknown.  Wet air oxidation should definitely be
considered a pretreatment alternative for waste streams containing these
difficult to oxidize halogenated compounds.  A secondary treatment
(e.g., biological treatment), will generally be needed to achieve acceptable
destruction levels.

### 6.1.1   Process Description

A schematic of a continuous WAO system is shown in Figure 6.1.1.[4]   The Zimmerman WAO System,[5] as shown in the figure, has been developed by Zimpro, Inc., Rothschild, Wisconsin.   It represents an established technology for the treatment of municipal sludges and certain industrial wastes.   Full scale treatment of halogenated organic compound wastes has not yet been demonstrated.   However a 10 gpm pilot unit has been used to treat pesticides, solvent still bottoms and general organic wastes at a commercial waste treatment facility in California.[3,6-8]   As will be noted later, the effectiveness of WAO as an alternative to land disposal for certain halogenated organic containing waste streams will depend upon a number of factors including the molecular structure and concentration of the contaminants and the processing conditions.[1,3,6-12]

In the WAO process, the waste stream containing oxidizable contaminants is pumped to a vertical bubble tower reactor using a positive displacement, high pressure pump.   The feed stream is preheated by heat exchange with the hot, treated effluent stream.   Air (or pure oxygen) is injected following the high pressure pump.   Steam is added as required to increase the temperature within the reactor to a level necessary to support the oxidation reactions in the unit.   As oxidation proceeds, heat of combustion is liberated.   At feed COD concentrations of roughly 2 percent the heat of combustion will generally be sufficient to bring about a temperature rise and some vaporization of volatile components.   Depending upon the temperature of the effluent following heat exchange with the feed stream, energy recovery may be possible or final cooling may be required.   Following energy removal, the oxidized effluent, consisting mainly of water, carbon dioxide, and nitrogen, is reduced in pressure through a specially designed automatic control valve.   The effluent liquor is either suitable for final discharge (contaminant reduction achieves acceptable standards) or is now readily biodegradable and can be sent to a biotreatment unit for further reduction of contamination levels.   Similarly, noncondensible gases can either be released to the atmosphere or passed through a secondary control device (e.g., carbon adsorption unit) if additional treatment is required to reduce air contaminant emissions to acceptable levels.[8]

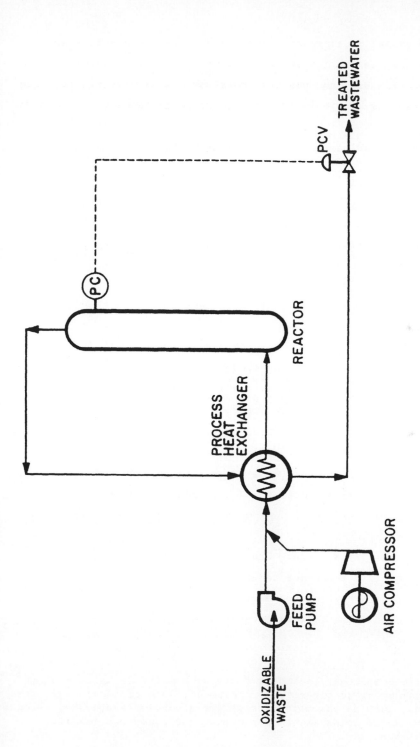

Figure 6.1.1. Wet air oxidation general flow diagram.

Source: Reference 4.

The pressure vessel is sized to accommodate a fixed waste flow and residence time. Based on the characteristics of the waste, a combination of time, temperature, pressure, and possibly catalyst can be utilized to bring about the destruction of many halogenated organic contaminants.

### 6.1.1.1 Pretreatment Requirements for Different Waste Forms and Characteristics--

Very little discussion is found in the literature concerning the physical form of wastes treatable by WAO. However, WAO equipment and designs have been used successfully to treat a number of municipal and industrial sludges. According to a Zimpro representative, wastes containing up to 15 percent COD (roughly equivalent to 7 to 8 percent organics) are now being treated successfully in commercial equipment.[13]

Treatment of solid bearing wastes is dependent upon selection of suitable pump designs and control devices. WAO units used for activated carbon regeneration now operate at the 5 to 6 percent solids range.[13] Treatment of higher solid levels is not precluded by fundamental process or design limitations. Column design must also be consistent with the need to avoid settling within the column under operating flow conditions. Thus, pretreatment to remove high density solids (e.g., metals by precipitation) and accomplish size reduction (e.g. filtration, gravity settling) would be required for some slurries. It should be noted that the WAO unit operated by Casmalia Resources in California does not accept slurries or sludges for treatment. This may be a result of design factors precluding their introduction into the system.[14]

Several bench scale studies have been conducted to determine the susceptibility of specific compounds to wet air oxidations. Results of these studies and other studies have been summarized in the literature.[1,8,10,15] The results indicate that the following types of compounds can be destroyed in wet air oxidation units.

- Aliphatic compounds, including those with multiple halogen atoms. Depending upon the severity of treatment, some residual oxygenated compounds such as low molecular weight alcohols, aldehydes, ketones, and carboxylic acids might be present, but these are readily biotreatable.

- Aromatic hydrocarbons, such as toluene and pyrene are easily oxidized.

- Halogenated aromatics can be oxidized provided there is at least one nonhalogen functional group present on the ring; the group should be an electron donating constituent such as an hydroxyl, amino, or methyl group.

- Halogenated aromatics, such as 1,2-dichlorobenzene, PCBs, and TCDDs, are resistant to oxidation under conventional conditions although these compounds are destroyed to a greater extent as conditions are made more severe or catalysts are employed. However, Casmalia Resources does not accept chlorinated aromatics.[14]

- Casmalia Resources also does not accept for WAO treatment wastes containing highly volatile organics like Freon which would enter the unit in the gas phase, and tin, which is corrosive to heat exchanger surfaces.[14]

Batch process results obtained in the laboratory are applicable to continuous process design for pure compounds and complex sludges, i.e., specific compound destruction is similar and predictable for pure compounds and those compounds contained in complex industrial wastes.[1,16]

6.1.1.2  Operating Parameters--

Although operation of a WAO system is possible, by definition, under all subcritical conditions; i.e., below 374°C and 218 atm (3,220 psig), commercially available equipment is designed to operate at temperatures ranging from 175 to 320°C and at pressures of 300 to 3,000 psig.[1]

Of all variables affecting WAO, temperature has the greatest effect on reaction rates. In most cases, about 150°C (300°F) is the lower limit for appreciable reaction. About 250°C (482°F) is needed for 80 percent reduction of COD, and at least 300°C (572°F) is needed for 95 percent reduction of COD within practical reaction times. Destruction rates for specific constituents may be greater or less than that shown for COD reductions.[2]

Initial reaction rates and rates during the first 30 minutes are relatively fast. After about 60 minutes, rates become so slow that generally little increase in percent oxidation is gained in extended reaction.[2]

An increase in reaction temperature will lead to increased oxidation but generally will require an increase in system pressure to maintain the liquid phase and promote wet oxidation. A drawback to increasing the temperature and pressure of the reaction is the greater stress placed on the equipment and its components, e.g., the increased potential for corrosion problems.  Corrosion is controlled by the use of corrosion resistant materials such as titanium.

As noted by Zimmerman, et al., the object of WAO is to intimately mix the right portion of air with the feed, so that under the required pressure, combustion will occur at a speed and temperature which will effectively reduce the organic waste to desired levels.  Pressures should be maintained at a level that will provide an oxygen rich liquid phase so that oxidation is maintained.[5]  Charts and curves are provided in Reference 5 to aid in the determination of waste heating value, stoichiometric oxygen requirement, and the distribution of water between the liquid and vapor phases at given temperatures and pressures.

Previous experience with the design of wet oxidation systems has shown that batch results are applicable to continuous process design when the oxygen transfer efficiency is 90 percent (11 percent excess air) or less.  A model was developed to gain insight into the key system parameters using a common industrial waste stream and fixed temperature, residence time, and COD reduction.  The model was also used to estimate costs for the system.[16]  Its value, as a predictive tool, along with that of supplementary kinetic studies[17] of batch wet oxidation, is limited by the sparsity of experimental data concerning reaction products and their phase distributions at the elevated temperatures and pressures encountered during WAO.

6.1.1.3  Post-Treatment Requirements--

The use of WAO to meet acceptable treatment levels halogenated organic for wastes has not yet been demonstrated.[8]  As will be noted later, WAO has been used under certain conditions to achieve destruction levels that are essentially complete.  However, for the most part, this level of performance has been achieved for specific compounds oxidized in batch reactors under conditions that are more rigorous than those normally used in commercial systems.

Destruction levels will vary for different compounds in complex waste mixtures and there is evidence that certain of the low molecular weight WAO breakdown products (e.g., methanol, acetone, acetaldehyde, formic acid, etc.) are resistant to further oxidation.  Thus, under typical WAO operating conditions it is likely that both contaminant residuals (unreacted halogenated organics) and low molecular weight process by-product residuals may be present. While it is entirely possible that imposition of more stringent operating conditions will serve to reduce these residuals to acceptable levels, the manufacturers and users of commercial WAO systems stress that the major applications involve the pretreatment of waste, usually for subsequent biological treatment.

Even under conditions that are favorable for wet oxidation, it is also likely that certain contaminants or byproducts, particularly some of the more volatile components, will partition between the vapor phase and the liquid phase. The partitioning will be a function of operating conditions and the contaminant partial pressure. The Henry's Law constant at the temperature of operation will fix the distribution; however, Henry's Law constant is not generally known under most conditions of WAO system operation. Although a method of estimation has been proposed by researchers at Michigan Technological University,[17] empirical tests will be necessary to establish vapor and liquid phase residuals and some post-treatment of both streams may be necessary. Existing post-treatment methods for the liquid generally involves bacteriological treatment. Although the results of post-treatment schemes for vapors from the WAO system have not been found in the literature, a two-stage water scrubber/activated carbon adsorption system has been used to treat WAO vapor emissions.[3] Presumably carbon adsorption or scrubbing systems could be routinely employed if necessary.

## 6.1.1.4  Treatment System Combinations--

Most of the commercial WAO systems in operation today are employed as pretreatment devices to enhance the biotreatability of municipal and industrial wastes. Wet air oxidation is also used as a means of regenerating spent activated carbon used as an adsorbent. In the latter case the WAO regenerates the activated carbon through oxidation of the organics adsorbed on the carbon surfaces.[18]

The application of WAO to industrial organic wastes has generally been limited to treating specific, homogeneous waste streams, including soda pulping liquors at pulp mills and n-nitrosodimethylamine and acrylonitrile wastes. However, WAO has been used since 1983 to treat varied waste streams

at the Casmalia Class I disposal site, located near Santa Maria, California.
Phenolics, solvent still bottoms, and other organic wastes have all been
treated at Casmalia, in certain instances in conjunction with a powdered
activated carbon treatment system and a two-stage scrubber-carbon adsorption
system for vapor treatment.[19]

Treatment of specific waste streams to meet acceptable halogenated
organic effluent levels by a WAO system is not precluded, as evidenced by some
of the performance data shown below for removal of specific contaminants.
However, in most instances reaction conditions would have to be tailored to
the waste stream and pollutant.  Generally an increase in the
pressure/temperature conditions normally employed by the users of WAO systems
would be required.  Equipment problems associated with the more stringent
operating conditions would have to be considered.

### 6.1.2   Demonstrated Performance of WAO Systems

As noted by EPA,[8] full scale use of WAO technology is well demonstrated
for the treatment of municipal sludge but full scale treatment of halogenated
organic wastes is not demonstrated.  However, data showing the WAO destruction
of specific organic compounds including some halogenated organics of concern
to EPA, have been provided in the literature.  These data are largely the
result of bench scale testing, but do include results of pilot-scale and
full-scale performance tests.  The data indicate that WAO can be effective in
treating specific organic contaminants, including many industrial wastewaters
containing halogenated organics.  However, chlorinated organics appear to be
the most difficult compounds to destroy.  Residuals in both the gas and liquid
phase would also have to be considered on a case by case basis if WAO
technology is to be used for the treatment of specific halogenated organic
compound containing waste streams.

### 6.1.2.1   Bench-Scale Studies--

Bench scale studies of the destruction of specific organic substances by
wet oxidation have been conducted at Zimpro, Inc.[1,8,15]  Some of these data
are shown in Table 6.1.1.  The tables include destruction data for halogenated
organic compounds and for halogenated solvents in order to illustrate the

TABLE 6.1.1.    BENCH-SCALE WET AIR OXIDATION OF PURE COMPOUNDS

| Compound | Wet oxidation conditions °C/minutes | Starting concentration (mg/L) | Final concentration (mg/L) | Percent destroyed |
|---|---|---|---|---|
| Arochlor 1254 | 320/120 | 20,000 | 7,400 | 63.0 |
| Carbon Tetrachloride | 275/60 | 4,330 | 12 | 99.7 |
| Chlorobenzene | [a]275/60 | 5,535 | 1,550 | 72.0 |
| Chloroform | 275/60 | 4,450 | 3 | 99.9 |
| 1-Chloronaphthalene | [a]275/60 | 5,970 | 5 | 99.91 |
| 2-Chlorophenol | 275/60 | 12,400 | 625 | 95.0 |
| 2-Chlorophenol | 320/60 | 12,400 | 17 | 99.9 |
| 2,4-Dichloroaniline | [a]275/60 | 259 | 0.5 | 99.8 |
| 1,2-Dichlorobenzene | [a]320/60 | 6,530 | 2,017 | 69.1 |
| 1,2-Dichloroethane | 275/60 | 6,280 | 13 | 99.8 |
| Hexachlorocyclopentadiene | 300/60 | 10,000 | 15 | 99.9 |
| Kepone | [a]280/60 | 1,000 | 690 | 31.0 |
| Pentachlorophenol | 275/60 | 5,000 | 902 | 82.0 |
| Pentachlorophenol | 320/60 | 5,000 | 6 | 99.9 |
| 2,4,6-Trichloroaniline | [a]320/120 | 10,000 | 2.5 | 99.9 |

[a]Catalyzed.

effect of operating variables, catalysts, and chemical structure on the effectiveness of wet air oxidation.  Further detail on treatment of halogenated solvents can be found in the TRD for solvents.[20]

As shown in Table 6.1.1, most compounds were destroyed to an appreciable extent at 320°C.  As noted above, the halogenated organic devoid of other functional groups (i.e. chlorobenzene, dichlorobenzene, Arochlor PCB, and kepone) were the most resistant to oxidations.  The oxidation resistant compounds showed a marked increase in destruction efficiency with temperature through the 275°C to 320°C range.  Presumably destruction efficiencies would be somewhat higher at even more elevated temperatures.

Although no attempt was made to measure vapor phase residuals, Reference 15 does present data for liquid phase residuals.  Formic acid and acetic acid were identified in these residue in amounts representing as much as 20 weight percent of the original charge of the specific test compound.  However, the two low molecular weight acids formed are readily biodegradable by conventional treatment methods.  Thus, it was concluded that wet oxidation of the waste constituents followed by biotreatment would yield an effluent suitable for discharge to a publicly owned treatment plant.

### 6.1.2.2  Pilot-Scale Studies--

The results of several pilot scale studies have been reported in the literature.[11]  The flow rates of systems used in these studies ranged from 2.5 to 28.9 gallons per hour (0.23 to 2.6 cubic meters per day).

Only one series of pilot-scale tests were conducted with a wastewater containing a halogenated organic pesticide/herbicide (2,5-dichloro-6-nitrobenzoic acid).  Removal was 90.5 percent.  In another test series, unexpectedly high destruction efficiencies of 98.2 percent were obtained for 1,2-dichlorobenzene, a compound that was not readily oxidized in the bench-scale tests.  The higher than anticipated efficiency was attributed to synergetic effects due to interaction with break down products (free radicals) of other contaminants.  Enhanced oxidation has been noted in other tests of multicontaminant industrial waste streams.[13]

6.1.2.3  Full-Scale Studies--

Several full-scale studies have been conducted at the Casmalia Resources facility in Santa Barbara County, California using a skid mounted WAO systems, capable of 10 gallon per minute flow rate for waste materials with a COD of 40 g/liter.  Tests have been conducted on wastewater containing phenolics, organic sulfur, cyanides, nonhalogenated pesticides, solvent still bottoms, and general organics.  In the case of the general organic wastewater, COD was reduced 96.7 percent to a level of 2.5 g/liter.[4]  During this test the wet oxidation unit was operated at 277°C (531°F), 1550 psig, and a residence time of 120 minutes.[2]  For the solvent still bottoms, the unit was operated at an average reactor temperature of 268°C (514°F), a reactor pressure of 1,550 psig, and a nominal residence time of 118 minutes.  COD, BOD, and TOC reductions of 95.3, 93.8, and 96.1 percent, respectively, were measured.[2]  However, no specific organic compound destruction efficiencies were reported for solvents or halogenated organics.

6.1.2.4  Studies of Treatment Systems Using WAO--

The use of pilot-scale and full-scale treatment systems combining PACT[TM] (powdered activated carbon addition to the reaction basin of an activated sludge process) with wet air oxidation regeneration have been reported in the literature.[18,21]  Reference 21 reports destruction efficiencies of greater than 99 percent for several priority pollutants present in a domestic and organic chemical wastewater not treatable by conventional biological treatment systems (see Table 6.1.2).

Reference 18 presents results obtained during treatment of RCRA wastewater and CERCLA ground water at the Bofors-Nobel facility in Muskegon, Michigan.  Cleanup at the site was conducted in accordance with the system schematic shown in Figure 6.1.2.  Two WAO units are used, one dedicated solely to detoxification, the other used primarily as a carbon regeneration unit with occasional use as an additional detoxification unit.  Although no data are provided for specific organic solvent components of the waste, an average efficiency of 99.8 percent is stated for toxics in the feed.

TABLE 6.1.2.    PRIORITY POLLUTANT REMOVALS USING A PACT[TM]/WET AIR
REGENERATION SYSTEM FOR DOMESTIC AND ORGANIC
CHEMICALS WASTEWATER[a]

| Parameter | Influent ($\mu$g/L) | Effluent ($\mu$g/L) | Removal efficiency (%) |
|---|---|---|---|
| Benzene | 907 | 3.0 | 99.6 |
| Chlorobenzene | 597 | 3.7 | 99.3 |
| 1,2,4-Trichlorobenzene | 62 | ND | ~ 100 |
| 1,1,1-Trichloroethane | 7 | ND | ~ 100 |
| 2,4,6-Trichlorophenol | 81 | ND | ~ 100 |
| Chloroform[b] | 87 | 25 | 71 |
| 2-Chlorophenol | 98 | ND | ~ 100 |
| 1,2-Dichlorobenzene | 113 | ND | ~ 100 |
| 1,3-Dichlorobenzene | 67 | ND | ~ 100 |
| 3,4-Dichlorophenol | 116 | 1 | 99 |
| Dichlorobromomethane | 2.3 | ND | ~ 100 |
| Pentachlorophenol | 35 | ND | ~ 100 |
| Toluene | 1,195 | 2 | 99.8 |

[a]4.0 hour aeration time; regeneration temperature = 230°C.

[b]Drinking water background exceeds 50 $\mu$g/L chloroform.

Source:  Reference 21.

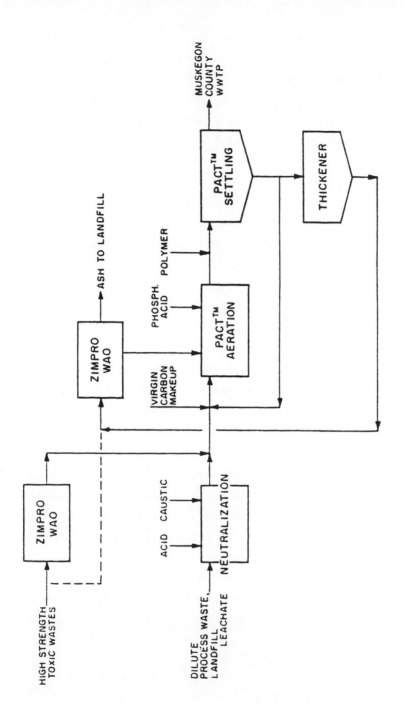

Figure 6.1.2.  4.5 MGD wastewater treatment facility.

Source:  Reference 18.

6.1.3  Cost of Treatment

6.1.3.1  Wet Air Oxidation Costs--

Treatment costs for wet air oxidation systems will be affected by a number of parameters including the amount of oxidation occurring, the hydraulic flow, the design operating conditions necessary to meet the treatment objectives, and the materials of construction. These factors account for the band of capital costs shown in Figure 6.1.3. The figure was taken from Reference 2 and updated to reflect changes in the 1982 to 1986 Chemical Engineering (CE) plant cost index. The costs do not include any costs associated with pretreatment of the feed or post-treatment of the vapor phase component of the treated liquor. However, post-treatment costs were included in another capital cost estimate of $2.45 million (adjusted to 1986 using the CE plant cost index) for a 20 gpm plant.[4] This estimate is within the capital cost band shown in Figure 6.1.3.

Operating costs for the wet oxidation unit are shown in Figure 6.1.4. These data were also derived from data given in Reference 2 with adjustment made for the costs of labor and cooling water. As noted in Reference 2, power accounts for the largest element of cost. This power cost is primarily the result of air compressor operation. Additional power for supplying energy for the oxidation of very dilute wastewaters would be at most 500 Btu/gallon. The associated costs for this energy would be less than one (1) cent/gallon.

Total costs, capital plus operating, on a per unit of feed basis, requires assumptions on life cycle, depreciation, taxes, and current interest rates for the capital cost. One avenue for financing that has been used commercially, common lease terms, are 5 years and 20 percent value at end of term.* Table 6.1.3 illustrates the effect on total costs per unit of feed.

At Casmalia Resources, the prices (April 1985) for treatment of wastes are computed based on the oxygen demand of the material. Prices range from a minimum of $120 per ton to a maximum of $700 per ton versus $15 per ton for the land disposal of low risk wastes.[14]

---

*Assume charges of $17/$1,000 per month based on total installed cost.

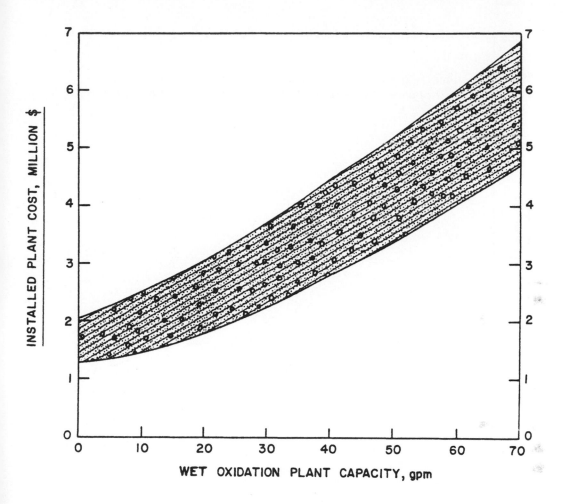

Figure 6.1.3.   Installed plant costs versus capacity.

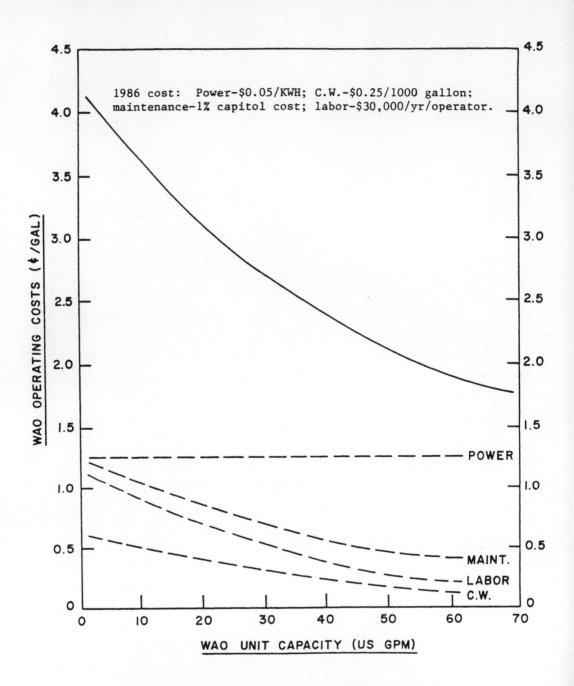

1986 cost:   Power-$0.05/KWH; C.W.-$0.25/1000 gallon;
maintenance-1% capitol cost; labor-$30,000/yr/operator.

Figure 6.1.4.   Unit operating costs versus unit flow rate.

TABLE 6.1.3.    WAO COSTS VERSUS FLOW

| Hydraulic flow (gpm) | Cost elements per gallon, cents | | |
|---|---|---|---|
| | operating | capital | total |
| 2.0 | 23 | 31 | 54 |
| 10 | 6 | 7 | 13 |
| 20 | 3 | 5 | 8 |
| 40 | 2-3 | 4-5 | 6-8 |

6.1.3.2  Comparison of WAO Costs with Other Alternative Treatment Costs--

A cost comparison between a 20 gpm WAO system and a comparable incinerator was presented in Reference 4 for a wastewater containing 7 percent COD.  It was concluded that, although the installed capital cost for WAO was 50 percent higher than that for incineration, operating costs were appreciably less ($132,000 annual operating cost for WAO versus $463,500 for incineration in 1979 dollars) despite a charge for scrubbing of the WAO off gases and an operating surcharge for BOD discharges to an average municipal wastewater treatment plant.  It was concluded that total operating costs including amortization favor WAO when the fuel value of the waste organics is low (less than approximately 50 g/liter Chemical Oxygen Demand).[4]

Other sources of cost data, including comparative costs, are References 3 and 16.  Reference 3 notes that WAO is generally less expensive than incineration when the COD concentration ranges between 10 to 150 g/liter. Rough cost estimates of from about 10 to 50 cents per gallon were proposed depending upon type of waste, concentration, and amount to be treated.  For comparison, landfilling costs of 12 to 25 cents per gallon for drummed wastes were provided.  Reference 16 provides cost data for a WAO system designed to treat a 7 percent COD waste at a 10 gallon per minute treatment rate.  Net operating costs of $90,780 per year (December 1980) were estimated, a value roughly equivalent to 3 cents per gallon, assuming a zero rate of return on investment.  This relatively low operating cost was compared to a landfilling cost of roughly $1 per gallon for barrelled waste and $0.55 to $0.75 per

gallon for bulk waste.  Although WAO costs were roughly two orders of
magnitude greater than typical costs for secondary biological municipal
wastewater treatment, the cost of $0.07 per pound of COD removed was suggested
as comparable to the typical municipal charge to industry of $0.05 to $0.10
per pound of COD removed.

Another source of cost data, Reference 22, provides data showing that
costs are a strong function of the contaminant type, its concentration, and
the amount of waste to be treated.  Costs ranged from $0.12 per pound of
pentachlorophenol to $1.04 per pound of hexachlorobutadiene treated.

### 6.1.4  Overall Status of WAO Process

#### 6.1.4.1  Availability and Application of WAO Systems--

The WAO process is available commercially, and reportedly well over
150 units are now operating in the field treating municipal and various
industrial sludges.[13]  The process is used predominately as a pretreatment
step to enhance biodegradability.  Only a few units are now being used to
treat industrial wastes.  These include the 10 gallon per minute unit at
Casmalia Resources in California and other units operating at Bofors-Nobel in
Muskegon, Michigan and Northern Petrochemical in Morris, Illinois.

The oxidation of specific contaminants in waste streams by the wet
oxidation process is not highly predictable.  Equipment manufacturers rely
largely on the result of bench-scale results to tailor the design of
full-scale WAO continuous units for specific wastes.  Full-scale data confirm
the results of WAO performance data obtained in bench and pilot-scale
studies.[1]  The use of WAO for halogenated aromatic compound bearing wastes
may pose particularly difficult problems and its use as a pretreatment should
be considered with caution.

#### 6.1.4.2  Energy and Environmental Impacts--

As noted, the process is thermally self-sustaining when the amount of
oxygen uptake is in the 15-20 g/liter range.  Below this range, some energy
input will be required to initiate and sustain reaction.  However, the energy
requirement will be appreciably less than that required for incineration.

The environmental impacts of WAO will hinge upon the residuals remaining after treatment. Wet scrubbing and carbon adsorption cleanup systems have been used to treat the HCl formed as a product of chlorinated organic oxidation and to remove volatile organics from the waste off gases. Residuals in the liquid phase may also require post treatment if, for example, 100 percent conversion to $CO_2$ and $H_2O$ is not realized when treating halogenated contaminants. The available data do suggest that some form of post treatment of both liquid and vapor phases will be required to meet EPA treatment standards.

### 6.1.4.3  Advantages and Limitations--

There are several advantages associated with the use of WAO as noted by the developer and stated in Reference 2.

1.   The process is thermally self-sustaining when the amount of oxygen uptake is in the 15-20 grams/liter range.

2.   The process is well suited for wastes that are too dilute to incinerate economically, yet too toxic to treat biologically.

3.   Condensed phase processing requires less equipment volume than gas phase processing.

4.   The products of WAO stay in the liquid phase. Offgases from a WAO system are free of $NO_x$, $SO_2$, and particulate. Water scrubbing and, if need be, carbon adsorption or fume incineration are used to reduce hydrocarbon emissions or odors.

5.   WAO also has application for inorganic compounds combined with organics. The oxidation cleans up the mixture for further removal of the inorganics. WAO can detoxify most of the EPA priority pollutants. Toxic removal parameters are in the order of 99+ percent using short-term, acute, static toxicity measurements.

Limitations of the WAO process relate to the sensitivity of destruction efficiency associated with the chemical nature of the contaminant, the possible influence of metals and other contaminants on performance, the unfavorable economics associated with low and high concentration levels, and the presence of residuals in both the vapor and liquid phases which may require additional treatment. Costly materials of construction and design features may also be required for halogenated wastes which will form corrosive reaction products or require extreme temperature/pressure conditions to achieve destruction to acceptable levels. In particular, chlorinated aromatic compounds are more resistant to degradation and can result in the production of HCl byproduct.

REFERENCES

1.    Dietrich, M.J., T.L. Randall, and P.J. Canney.  Wet Air Oxidation of
      Hazardous Organics in Wastewater, Environmental Progress, Vol. 4, No. 3,
      August 1985.

2.    Freeman, H.  Innovative Thermal Hazardous Treatment Processes, U.S. EPA,
      Hazardous Waste Engineering Research Laboratory, Cincinnati, Ohio, 1985.

3.    California Air Resources Board.  Air Pollution Impacts of Hazardous Waste
      Incineration:  A California Perspective, December 1983.

4.    Wilhelmi, A.R., and P.V. Knopp.  Wet Air Oxidation - An Alternative to
      Incineration, Chemical Engineering Progress, August 1979.

5.    Zimmerman, F.J., and D.G. Diddams, The Zimmerman Process and its
      Applications in the Pulp and Paper Industry, TAPPI Vol. 43, No. 8,
      August 1960.

6.    Copa, William, James Heimbuch, and Phillip Schaeffer.  Full Scale
      Demonstration of Wet Air Oxidation as a Hazardous Waste Treatment
      Technology.  In:  Incineration and Treatment of Hazardous Waste,
      Proceedings of the Ninth Annual Research Symposium, U.S. EPA
      600/9-84-015, July 1984.

7.    Copa, William, Marvin J. Dietrich, Patrick J. Cannery, and
      Tipton L. Randall.  Demonstration of Wet Air Oxidation of Hazardous
      Waste.  In Proceedings of Tenth Annual Research Symposium, U.S. EPA
      600/9-84-022, September 1984.

8.    U.S. Environmental Protection Agency, Background Document for Solvents to
      Support 40 CFR Part 268, Land Disposal Restrictions, Volume II,
      January 1986.

9.    Reible, Danny D., and David M. Wetzel.  Louisiana State University, A
      Literature Survey of Three Selected Hazardous Waste Destruction
      Techniques In Proceedings of Ninth Annual Symposium on Land Disposal of
      Hazardous Waste.  May 2-4, 1983.

10.   Randall, T.R.  Wet Oxidation of Toxic and Hazardous Compounds.
      Zimpro, Inc. Technical Bulletin 1-610, 1981.

11. Canney, P.J., and P.T. Schaeffer.  Detoxification of Hazardous Industrial Wastewaters by Wet Air Oxidation.  Presented at 1983 National AIChE Meeting, Houston, TX, March 27-31, 1983.

12. Baillod, C. Robert, and Bonnie M. Faith.  Wet Oxidation and Ozonation of Specific Organic Pollutants, U.S. EPA 600/S2-83-060, October 1983.

13. Telephone Conversation with A. Wilhelmi on April 3, 1986.

14. Metcalf & Eddy, Inc.  Hazardous Waste Treatment Storage and Disposal Facility - Site Evaluation Report, Casmalia Resources, Casmalia, California, Publication NS J-1074, April 8, 1985.

15. Randall, Tipton L., and Paul V. Knopp.  Detoxification of Specific Organic Substances by Wet Air Oxidation, Journal WPCF, Vol. 52, No. 8, August 1980.

16. Baillod, C.R., R.A. Lamporter, and B.A. Barna.  Wet Oxidation for Industrial Waste Treatment, Chemical Engineering Progress, March 1985.

17. Baillod, C.R., B.M. Faith, and D. Masi.  Fate of Specific Pollutants During Wet Oxidation and Ozonation, Environmental Progress, August 1982.

18. Meidl, J.A., and A.R. Wilhelmi, PACT[TM]/Wet Oxidation:  Economical Solutions to Solving Toxic Waste Treatment Problems.  Paper presented at Indiana Water Pollution Control Association Annual Meeting, August 20, 1985.

19. California Department of Health Services, Alternative Technology for Recycling and Treatment of Hazardous Wastes, Second Biennial Report, July 1984.

20. Breton, M. A., et al.  Technical Resource Directive - Treatment Technologies for Solvent Containing Wastes.  Prepared for U.S. EPA, HWERL, Cincinnati Under Contract No. 68-03-3243, Work Assignment No. 2, August 1986.

21. Randall, T.L.  Wet Oxidation of PACT[R] Process Carbon Loaded with Toxic Compounds.  Paper presented at 38th Industrial Waste Conference, Purdue University, West Lafayette, Indiana, May 10-12, 1983.

22. Miller, R.A., and M.D. Swietoniewski.  IT Enviroscience The Destruction of Various Organic Substances by a Catalyzed Wet Oxidation Process, Work Done Under U.S. EPA Contract No. 68-03-2568, 1982.

6.2   SUPERCRITICAL WATER OXIDATION

Supercritical water oxidation is a technology that has been proposed for the destruction of organic contaminants in wastewaters.  It is basically an oxidation process conducted in a water medium at temperatures and pressures that are supercritical for water; i.e., above 374°C (705°F) and 218 atmospheres.  In the supercritical region, water exhibits properties that are far different from liquid water under normal conditions; oxygen and organic compounds become totally miscible with the supercritical water (SCW) and inorganic compounds, such as salts, become very sparingly soluble.  When these materials are combined in the SCW process, organics are oxidized and any inorganic salts present in the feed or formed during the oxidation are precipitated from the SCW.

The oxidation reactions proceed rapidly and completely.  Reaction times are less than 1 minute, as compared to reaction times of about 60 minutes used in the subcritical wet air oxidation (WAO) process.  Moreover, the reaction is essentially complete.  Carbon and hydrogen atoms within the organic contaminants are reacted to form $CO_2$ and $H_2O$ (residuals such as the low molecular weight organic acids and alcohols found in the treated WAO effluent are not found in the SCW process effluent).  Heteroatoms (e.g., chlorine and sulfur) are oxidized to their corresponding acidic anion groupings.  These anions, and those occurring naturally in the feed, can be neutralized by cation addition to the feed, and the total inorganic content of the waste, save that soluble in the SCW, can be precipitated and recovered by mechanical separators operating at SCW conditions.

6.2.1   Process Description

In the supercritical region, water exhibits properties that are far different from liquid water at normal ambient conditions.  The density, dielectric constant, hydrogen bonding, and certain other physical properties change significantly with the result that SCW behaves very much like a moderately polar organic liquid.[1]  Thus, solvents such as n-heptane and benzene, for example, become miscible with SCW in all proportions.  On the other hand, the solubility of salts such as sodium chloride (NaCl) is as low

as 100 ppm and that of calcium chloride ($CaCl_2$) as low as 10 ppm.  These
solubilities are far different from those found under ambient conditions where
the solubilities of NaCl and $CaCl_2$ are about 37 weight percent and up to
70 percent, respectively (Josephson, 1982).

The solubility characteristics of SCW are strongly dependent upon
density.[2]  A temperature-density diagram is shown in Figure 6.2.1.  The
critical point which is located on the dome of the vapor-liquid saturation
curve is at 374°C and 0.3 gram/cubic centimeter.  The supercritical region is
that above 374° and the 218 atmosphere isobar.  Near the critical point
(e.g., between 300° and 450°) the density varies greatly with relatively small
changes in temperature at constant pressure.

Insight into the structure of the fluid in this region has been obtained
from measurements of the static dielectric constant, values of which are shown
in Figure 6.2.1.[3,4]  The dielectric constants of some common solvents are
given for comparison in Table 6.2.1.

TABLE 6.2.1.  DIELECTRIC CONSTANTS OF SOME COMMON SOLVENTS

| | |
|---|---|
| Carbon dioxide | 1.60 |
| n-Hexane | 1.89 |
| Benzene | 2.28 |
| Ethyl ether | 4.34 |
| Ethyl acetate | 6.02 |
| Benzyl alcohol | 13.1 |
| Ammonia | 16.9 |
| Isopropanol | 18.3 |
| Acetone | 20.7 |
| Ethanol | 24.3 |
| Methanol | 32.6 |
| Ethylene glycol | 37. |
| Formic acid | 58. |

Source:  Reference 5.

The dielectric constant is a measure of the degree of molecular
association.  While dielectric constant is not the sole determinant of
solubility, the solvent power of water for organics is consistent with
variations in the dielectric constant.  According to Figure 6.2.1, as
temperature rises along the saturated liquid-vapor curve the dielectric

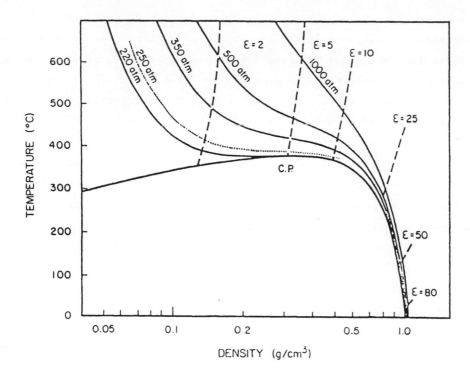

Figure 6.2.1.   Temperature-density diagram.

Source:   Reference 2.

constant (normally at about 80 due largely to strong hydrogen bonding) decreases rapidly despite only small changes in density. The large decreases in the dielectric reflect the strong dependence of hydrogen bonding forces on distance, with small decreases in density leading to large decreases in dielectric constant. At 130°C (d = 0.9 g/cm$^3$), the dielectric constant is about 50, which is near that of formic acid; at 260°C (d = 0.8 g/cm$^3$) the dielectric constant is 25 similar to that of ethanol. At the critical point the dielectric constant is 5, and little, if any, residual hydrogen bonding is present. The major contribution to the dielectric constant is due to dipole-dipole interactions, which gradually decrease with density.[5]

Depending upon the pressure and temperature, the dielectric constant can be varied to achieve values similar to those of moderately polar to nonpolar organic solvents. Solubility behavior parallels the changes in dielectric and at some points supercritical conditions are reached and the components are miscible in all proportions.

The solubilities of inorganic salts in water exhibit different behavior from that shown by the organic compounds. At 250 atmospheres, the solubilities of salts reach a maximum at 350–400°C. Beyond the maximum, the solubilities drop very rapidly with increasing temperature. For example, NaCl solubility is above 40 weight percent at 300°C and 100 ppm at 450°C; CaCl$_2$ has a maximum solubility of 70 percent at subcritical temperatures which drop to 10 ppm at 500°C.[2]

The properties of water, as a function of temperature, are summarized in Figure 6.2.2. The figure shows that water goes through a complete reversal in solubility behavior between 300–500°C. Above 450°C, inorganic salts are practically insoluble, and organic substances are completely miscible.[2]

Given the complete miscibility of oxygen and organic contaminants in the supercritical fluid and the high temperature of operation, oxidation reactions proceed rapidly and completely. In the MODAR process described below, organics, air and water wastes are brought together at 250 atmospheres and at temperatures above 400°C. The heat of oxidation is released within the fluid and results generally in a rise in temperature to 600–650°C.

The products of supercritical water reforming are subjected to oxidation while under these homogenous (i.e., single phase) supercritical conditions. The residence time required for oxidation is very short, which greatly reduces the volume of the oxidizer vessel.

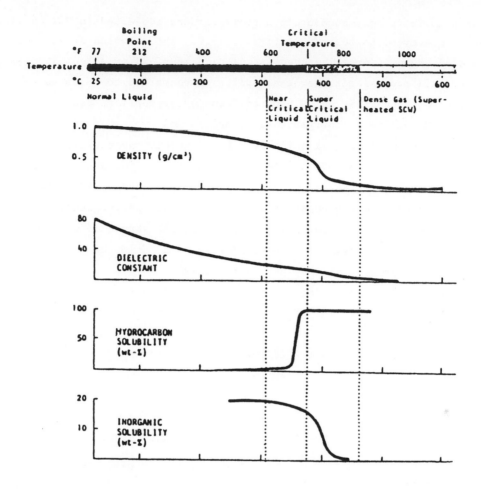

Figure 6.2.2.   Properties of water at 250 atm.

Source:   Reference 2.

When toxic or hazardous organic chemicals are subjected to SCW oxidation, carbon is converted to $CO_2$ and hydrogen to $H_2O$. The chlorine atoms from chlorinated organics are liberated as chloride ions. Similarly, nitrogen compounds will produce nitrogen gas, sulfur is converted to sulfates, phosphorus to phosphates, etc. Upon addition of appropriate cations (e.g., Na+, Mg++, Ca++), inorganic salts are formed.

The heat of oxidation is sufficient to bring the supercritical stream to temperatures in excess of 550°C. At these conditions, inorganic salts have extremely low solubilities in water. Inorganic salts are precipitated out and readily separated from the supercritical fluid phase. After removal of inorganics, the resulting fluid is a highly purified stream of water at high temperature and high pressure. The fluid is used as a source of high-temperature process heat by generating steam.

A schematic flow sheet for the MODAR process as applied to liquid wastes is presented in Figure 6.2.3. This figure and subsequent discussion was provided by MODAR, Inc.[6]

The process consists of the following steps:

A. Feed

1.  Organic waste materials in an aqueous medium are pumped from atmospheric pressure to the pressure in the reaction vessel.

2.  Oxygen, stored as a liquid, is pumped to the pressure of the reaction vessel and then vaporized.

3.  Feed to the process is controlled to an upper limit heating value of 1800 Btu/lb by adding dilution water or blending higher heating value waste material with lower heating value waste material prior to feeding to the reactor.

4.  When the aqueous waste has a heating value below 1,800 Btu/lb, fuel may be added in order to utilize a cold feed to the oxidizer.

5.  Optionally for wastes with heating value below 1,800 Btu/lb, a combination of preheat by exchange with process effluent and fuel additon, or preheat alone may be used.

6.  When organic wastes contain heteroatoms which produce mineral acids, and it is desired to neutralize these acids and form appropriate salts, caustic is injected as part of the feed stream.

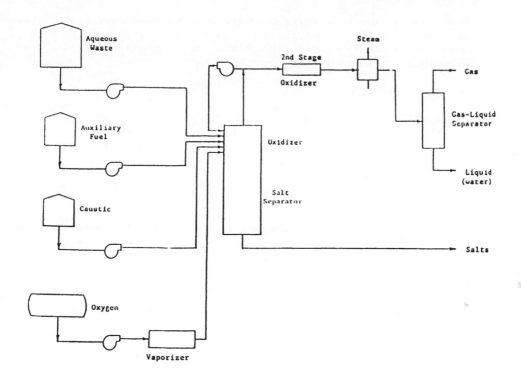

Figure 6.2.3.    Schematic flow sheet of MODAR process.

Source:    Reference 6.

7.    A recycle stream of a portion of the supercritical process effluent
      is mixed with the feed streams to raise the combined fluids to a
      high enough temperature to ensure that the oxidation reaction goes
      rapidly to completion.

B.    Reaction and Salt Separation

1.    Because the water is supercritical, the oxidant is completely
      miscible with the solution; i.e., the mixture is a single,
      homogenous phase.  Organics are oxidized in a controlled but rapid
      reaction.  Since the oxidizer operates adiabatically, the heat
      released by the readily oxidized components is sufficient to raise
      the fluid phase to temperatures at which all organics are oxidized
      rapidly.

2.    Since the salts have very low solubility in SCW they separate from
      the other homogenous fluids and fall to the bottom of the separation
      vessel where they are removed.

3.    The gaseous products of reaction along with the supercritical water
      leave the reactor at the top.  A portion of the supercritical fluid
      is recycled to the SCW oxidizer by a high temperature, high pressure
      pump.  This operation provides for sufficient heating of the feed to
      bring the oxidizer influent to optimum reactor conditions.

4.    The remaining reactor effluent (other than that recycled) consisting
      of superheated SCW and carbon dioxide is cooled in order to
      discharge $CO_2$ and water at atmospheric conditions.

C.    Cooling and Heat Recovery

1.    Most of the heat contained in the effluent is used to generate steam
      for use outside the MODAR Process.

2.    The heat remaining in the effluent stream is used for lower level
      heating requirements and is also dissipated.

D.    Pressure Letdown

1.    The cooled effluent from the process separates into a liquid water
      phase and a gaseous phase containing primarily carbon dioxide along
      with oxygen which is in excess of the stoichiometric requirements.

2.    The separation is carried out in multiple stages in order to
      minimize erosion of valves as well as to optimize equilibria.

3.    Salts are removed from the separator as a cool brine through
      multiple letdown stages and are either dried (and water recovered)
      or discharged as a brine depending upon client requirements.

### 6.2.1.1  Pretreatment Requirements--

Very little information exists in the literature to assess pretreatment requirements for the process and its feed streams.  The process reportedly can handle slurries, thus, filtration or some other solids removal process may not be required or even desirable if the contaminant is partitioned in the feed between the aqueous phase and the suspended solids.  Similarly, the need to remove inorganic constituents may not exist since these constituents will precipitate under the supercritical conditions of operation and presumably will be removed by the mechanical separator shown in Figure 6.2.3.  Adverse effects such as interference with pump operations, abrasion of internal parts, and fouling of internal surfaces resulting from existing or formed solids are possible problem areas but were not considered such by MODAR.[6]

### 6.2.1.2  Operating Parameters--

The operating conditions are specified by MODAR as follows:

- Form of Feed Materials:  Aqueous slurry or solution of organics.

- Temperature Range:  400°-650°C (750°-1200°F)

- Pressure Range:  220-250 atm

- Residence Time Range:  Less than one minute

- Energy Type and Requirements:  Thermal, to reaction conditions, with provisions for useful recovery of latent heat of oxidation.

These conditions are capable of achieving destruction efficiencies in excess of 99.999 percent.  The technology should be applicable to all halogenated organics considered in this TRD.  The principal question related to the applicability of the technology is associated with cost, including the durability of the system under the harsh supercritical conditions.

### 6.2.1.3  Post Treatment Requirements--

Because the oxidation reactions go essentially to completion and provision can be made for neutralization and removal of inorganic products and feed stock components the post treatment requirements should be minimal.  Off gases from the subcritical treated effluent should be largely $CO_2$ and $H_2O$ and liquid effluent residuals will consist mainly of dissolved salts at the 10 to 100 ppm levels.  These salts will consist, at least in part, of salts resulting from the addition of caustic to neutralize the halogen acids formed during the oxidation of halogenated compounds in the waste.

Along with $N_2$, $N_2O$ may also be a possible off gas component from the SCW oxidation of nitrogen containing organics.  A possible $N_2O$ component would not be considered an air contaminant since there is no evidence involving it in the series of complex chemical reactions producing photochemical smog.[7]

Apart from the modest impacts anticipated as a result of $N_2O$ emissions and the dissolved inorganic salt loading of the liquid effluent the only other residual stream requiring possible attention is the largely solid inorganic stream from the separator.  EP toxicity could be a characteristic of possible concern for some wastes.

### 6.2.1.4  Treatment Combinations-

SCW oxidation systems can be considered for aqueous waste streams containing one or more weight percent of organic constituents.  Below 1 percent, other treatment technologies appear to have a cost advantage.  The highest practical organic content again will depend upon costs; specifically the cost of SCW oxidation versus incineration for wastes in the 10 to 20 weight percent and higher range.  Largely unproven, the SCW oxidation system will, if cost effective, function as a finishing technology discharging effluents that can be expected to meet acceptable levels of discharge.

### 6.2.2  Demonstrated Performance

The destruction of organic contaminants is a function of reactor temperature and residence time.  MODAR reports that a reactor temperature in the range of 600 to 650°C (1120° to 1200°F) and a 5 second residence time are sufficient to achieve destruction efficiencies of 99.999 percent.  Higher temperatures could be used to reduce the residence time.  However, at a 5 second reaction time, the reactor cost is a small fraction of total capital cost and, thus, there is not much incentive to reduce reactor volume by operating above 650°C.[2]

Theoretically, increasing residence time will also result in increased destruction efficiency. The oxidation kinetics appear to be first order in organic concentration. Assuming perfect mixing and first order kinetics at all concentrations, doubling the residence time could result in a doubling of the destruction efficiency. Thus, a 99.999 percent efficiency could become 99.99999999 (ten nines).

MODAR has conducted more than 200 laboratory (bench) and pilot plant tests in order to study the technical feasibility of SCW oxidation for a variety of organic contaminants. In most cases MODAR does not attempt to measure destruction and removal efficiency to the greatest possible precision. Test objectives are rather to measure the levels of organic carbon in the liquid effluent, and in most cases, residual levels are below detection limits of the analytical equipment. Consequently, destruction removal efficiency, which may be claimed in many of MODAR's tests, are limited to between 99.9 percent and 99.99+ percent by precision of the analytical equipment (See Reference 8). When the objective is to demonstrate the maximum degree of waste destruction, richer feeds and more sensitive analytical equipment are used. Tests of this sort (e.g., on dioxins) show destruction and removal efficiencies of more than 99.9999 percent[6]. Equal or greater destruction efficiencies could be expected for most if not all halogenated organic compounds of concern.

### 6.2.3  Cost of Treatment

The most significant operating cost factor is the cost of oxygen consumed. Although compressed air can be used as the source of oxygen, the cost of power as well as the high capital cost of appropriate compressors has led MODAR to use liquefied oxygen as the primary oxygen source. Oxygen demand and heat content of an organic waste are usually directly related, and therefore the heating value of the waste and waste throughput can be used to make a preliminary estimate of waste treatment costs.

Table 6.2.2 presents waste treatment costs based on an aqueous waste with a 10 percent by weight benzene-equivalent and a heat content of 1,800 Btu/lb. This is the optimal heat content of a cold feed for this process to attain a reactor exit temperature of 600 to 650°C. Other factors on which the costs in

Table 6.2.2 are based are:  the system is installed at the site of the waste
generator; the units are owned and operated by the waste disposer; and the
units are not equipped with power recovery turbines.

TABLE 6.2.2.  MODAR TREATMENT COSTS FOR ORGANIC CONTAMINATED AQUEOUS WASTES

| Waste capacity | | Processing cost[a] | |
|---|---|---|---|
| Gal/day | Ton/day | $/gal | $/ton |
| 5,000 | 20 | $0.75 - $2.00 | $180 - $480 |
| 10,000 | 40 | $0.50 - $0.90 | $120 - $216 |
| 20,000 | 80 | $0.36 - $0.62 | $ 86 - $149 |
| 30,000 | 120 | $0.32 - $0.58 | $ 77 - $139 |

[a]Based upon an aqueous waste with 1,800 Btu/lb heating value (equivalent to
a 10 percent  organic waste).  Does not include energy recovery value of
approximately $0.05 per gallon.

Source:  Reference 6.

If the waste has a fuel value of greater than 1,800 Btu/lb, the cost will
be higher per unit of waste processed.  In treating a waste with a higher
organic content, it is recommended that the waste is diluted to a 10 percent
benzene-equivalent.  Therefore, the increase in cost will be in proportion to
the increase in organic content.

If the waste has a heat content of between 5 and 10 percent benzene-
equivalent, fuel can be added to the waste to bring the heat content up to
10 percent benzene-equivalent without appreciable cost increases.  If,
however, the waste is very dilute (2 to 3 percent benzene-equivalent), it is
more economical to use a combination of fuel with regenerative heat exchange.

6.2.4   Overall Status of Process

6.2.4.1  Availability--

A pilot plant with capacity to oxidize 30 gal/day of benzene equivalent
has been in operation at MODAR's laboratory as well as at a field site since
late 1984.  As a result of these activities, the MODAR SCW oxidation process

has been declared commercial and design of the first plant is underway.  The plant will be installed late in 1987 and will treat 10,000 to 30,000 gallons of aqueous waste per day.

6.2.4.2  Application--

SCW oxidation would appear to be applicable to aqueous wastes containing 1 to 20 weight percent organics.  As noted in previous discussions above, complete destruction of all halogenated organics can be anticipated on the basis of evidence presented by the developer.  The high efficiency of destruction can be related to the unique and stringent conditions associated with SCW oxidation which unites oxygen and organic contaminants under relatively high temperatures and pressures.

Restrictive waste characteristics have not been identified in the literature as a problem.  The effect of heteroatoms (halogens) and their reaction products can be anticipated and steps taken to essentially eliminate any deleterious impacts.  However, the applicability of solid content wastes to SCW oxidation systems may be problematical.  The effectiveness of removal of precipitated inorganic salts by the mechanical separators proposed for the MODAR system may also be a problem.  In the absence of particle size and flow and design data it is difficult to predict mechanical separator performance, although separation should be enhanced under the low density SCW conditions. If particles are present, abrasion problems could occur both within the oxidation system and in any subsequent system designed to recover energy from the treated stream.

Supercritical fluid technology is also being considered for a number of applications other than that concerned with the destruction of organic wastes, e.g., supercritical fluid extractions, including the extraction of adsorbed components from granular activated carbon.  Fluids such as $CO_2$, ethane, and ethylene can be used at critical temperature and pressure conditions which are much less severe than those of SCW.[1]  However, no data were found which relates the performance of such systems to the extraction of halogenated organics from wastes.

6.2.4.3  Environmental Impacts--

Liquid, solid, and gaseous emissions are generated from the SCW oxidation process. Gaseous emissions consist primarily of carbon dioxide with smaller amounts of oxygen and nitrogen gas. Effluent gas cleaning is not required. $N_2O$ is the most abundant nitrogen oxide in the atmosphere. It does not appear to interact with the nitrogen dioxide photolytic cycle. Any $N_2O$ which might be in the gaseous effluent is not classified as an atmospheric pollutant.

Solid emissions consist of the precipitated inorganic salts. When halogenated compounds are processed, halogen salts will be formed, and similarly sulfur is converted to sulfates, and phosphorous to phosphates.[2]

Liquid effluents consist of a purified water stream. Although no data are available for halogenated organic contaminants, six nines destruction has been measured for dioxins. On the basis of these data it is anticipated that essentially all halogenated compounds will be found only at the ppb level.

6.2.4.4  Advantages and Limitations--

The developer states that the MODAR process for supercritical water oxidation of organics is an improvement in:

- enhanced solubility of gases including oxygen and air in water, which eliminates two-phase flow;

- rapid oxidation of organics, which approaches adiabatic conditions as well as high outlet temperatures, and very short residence times;

- complete oxidation of organics, which eliminates the need for auxiliary offgas processing;

- removal of inorganic constituents, which precipitate out of the reactor effluent at temperatures above 450°C (840°F); and

- recovery of the heat of combustion in the form of supercritical water, which can be a source of high-temperature process heat.[9]

The above advantages are generally relative to the wet air oxidation process which could be considered as an alternative technology to SCW oxidation. The limitations of the process have yet to be determined through commercial operation. Potential limitations relate to cost and equipment limitations due to the stringent temperature and pressure requirements.

REFERENCES

1.    Josephson, J.  Supercritical Fluids.  Environmental Science and
      Technology.  Volume 16, No. 10.  October 1982.

2.    Thomason, T. B. and M. Modell.  Supercritical Water Destruction of
      Aqueous Wastes.  Hazardous Waste.  Volume 1, No. 4.  1984.

3.    Quist, A. S. and W. L. Marshall.  Estimation of the Dielectric Constant
      of Water to 800°, J. Phys. Chem., 69, 3165.  1965.

4.    Uematsu, M. and E. U. Franck.  J. Phys. Chem. Reference Data, 9(4),
      1291-1306.  1980.

5.    Franck, E. U.  Properties of Water in High Temperature, High Pressure
      Electrochemistry in Aqueous Solutions (NACE-4).  p. 109.  1976.

6.    Sieber, F. MODAR Inc.  Review of Draft Section, Supercritical Water
      Oxidation.  May 16 1986.

7.    National Academy of Sciences, Medical and Biological Effects of
      Environmental Pollution:  Nitrogen Oxides.  1977.

8.    Modell, M., G. Gaudet, M. Simson, G. T. Hong, and K. Biemann.
      Supercritical Water Testing Reveals New Process Holds Promise, Solid
      Wastes Management.  August 1982.

9.    Freeman, H.  Innovative Thermal Hazardous Waste Treatment Processes.
      U.S. EPA, HWERL Cincinnati, Ohio.  1985.

## 6.3   ULTRAVIOLET/OZONE OXIDATION

Chemical oxidation processes in addition to the wet air and supercritical water oxidation processes discussed previously, are potential options for the treatment of hazardous organic wastes.  Oxidants such as ozone, hydrogen peroxide, and potassium permanganate are among the strongest oxidants known (see Table 6.3.1) and are used industrially to treat specific waste streams containing phenols, cyanides, organic sulfur compounds, and other rapidly oxidized organics.  The use of these oxidants alone and in combination with ultraviolet light, for the treatment of hazardous waste streams has been described in the solvent TRD[3] and the dioxin TRD[4] and other publications.[5-14]  As noted in Reference 3 and other references, ozone and other commercial oxidants are not generally effective oxidants for halogenated organics.  Nevertheless, the potential for their use has been studied and oxidation processes have achieved some success in treating aqueous waste streams containing chlorinated pesticides.  These processes must be considered developing technologies.  Only the most prominent process, UV/ozone oxidation, is discussed here; other potentially useful processes are described in the previously cited references.  It should be noted that at the present time most chlorinated aliphatic compounds must be considered nonreactive.  The applicability of UV/ozone treatment for waste streams containing these compounds will require careful experimental documentation.

### 6.3.1   Process Description

Ozone, as an oxidant, is sufficiently strong to break many carbon-carbon bonds and even to cleave aromatic ring systems.  Oxidation of organic species to carbon dioxide, water etc., is not improbable if ozone dosage and contact times are sufficiently high, although many compounds are highly resistant to ozone degradation.  These compounds, which include oxalic and acetic acids, ketones, and chlorinated aliphatic organics, are not affected significantly by treatment conditions (1 to 10 mg/liter concentration levels and 5 to 10 minute contact times) normally used for treating drinking waters or for disinfecting wastewaters.[4]

TABLE 6.3.1.   RELATIVE OXIDATION POWER OF
OXIDIZING SPECIES

| Species | Oxidation potential, volts | Relative oxidation power[a] |
|---|---|---|
| Fluorine | 3.06 | 2.25 |
| Hydroxyl radical | 2.80 | 2.05 |
| Atomic oxygen | 2.42 | 1.78 |
| Ozone | 2.07 | 1.52 |
| Hydrogen peroxide | 1.77 | 1.30 |
| Perhydroxyl radicals | 1.70 | 1.25 |
| Permanganate | 1.70 | 1.25 |
| Hypochlorous acid | 1.49 | 1.10 |
| Chlorine | 1.36 | 1.00 |

[a]Based on chlorine as reference (= 1.00)

Source:   References 1 and 2.

Ozone has been used for years in Europe to purify, deodorize, and disinfect drinking water.  More recently, it has been used in the waste treatment area to oxidize phenolic and cyanide wastewaters.  Cost considerations and mass transfer factors limit the use of ozonation to applications involving 1 percent or lower contaminant concentration levels. Since oxidation by ozone occurs nonselectively, it is also generally used only for aqueous wastes which contain a high proportion of hazardous constituents versus nonhazardous oxidizable compounds, thus focusing ozone usage on contaminants of concern.  Ozonation may be particularly useful as a final treatment for waste streams which are dilute in oxidizable contaminants, but which did not quite meet effluent standards.

Ozone is generated onsite by the use of corona discharge technology. Electrons within the corona discharge split the oxygen-oxygen double bonds upon impact with oxygen molecules.  The two oxygen atoms formed from the molecule react with other oxygen molecules to form the gas ozone, at equilibrium concentration levels of roughly 2 percent in air and 3 percent in oxygen (maximum values of 4 and 8 percent, respectively).  Ozone must be produced onsite (ozone decomposes in a matter of hours to simple, molecular oxygen).[4]  Primarily, because of this, and solubility limitations (300 mg/L is considered a high dose level),[15] ozonation is restricted to treatment of streams with low quantities of oxidizable materials.  Using a rule of thumb, two parts of ozone are required per pound of contaminant.  A large commercial ozone generator producing 500 lb/day of ozone could treat 1 million gallons/day of wastewater containing 30 ppm of oxidizable matter, or equivalently, 3,000 gallons/day of wastewater containing 1 percent of oxidizable matter.[2]

While direct ozonation of industrial wastewater is possible and is practiced commercially, other technologies have been combined with ozonation to enhance the efficiency and rate of the oxidation reactions particularly for difficult to oxidize compounds such as halogenated organics.  These technologies, which supply additional energy to the reactants, involve the use of ultraviolet light or ultrasonics.  The use of ultraviolet light has received the most attention and will be the subject of further discussion.

Ultraviolet (UV) radiation is electromagnetic radiation having a
wavelength shorter than visible light, but longer than x-ray radiation.  The
energy content of light increases as the wave length decreases.  For wave
lengths in the UV region the energy is sufficient to break chemical bonds and
bring about rearrangement or dislocation of molecular structures.  The energy
corresponding to the absorption of a quantum (photon) of light is 95 kilo
calories per gram-mole for UV light with a wave length of 3,900 angstroms and
is 142 kilo calories per gram-mole for a wave length of 2,000 angstroms.

Table 6.3.2 lists the dissociation energies for many common chemical
bonds, along with the wavelength corresponding to the energy at which UV
photons will cause dissociation.  As can be seen from the data in Table 6.3.2,
bond dissociation energies range from a low of 47 cal/gmole for the peroxide
bond to a high of 226 kcal/gmole for the nitrogen triple bond.  Of particular
interest in the case of dioxins is the C-Cl bond, with a dissociation energy
of 81 kcal/gmole, corresponding to an optimum UV wavelength of 353 nm.  For
reference purposes, this can be compared to the violet end of the visible
spectrum with a wavelength of about 420 nm.  Thus, the UV radiation of
interest is in the electromagnetic spectrum close to visible light.

6.3.1.1  Pretreatment Requirements--

Due to the nonselective nature of the ozonation reactions it is important
that the concentration levels of nonhazardous, but oxidizable, contaminants in
the feed stream be reduced as much as possible prior to treatment.  The strong
electrophilic nature of ozone imparts to it the ability to react with a wide
variety of organic functional groups, including aliphatic and aromatic
carbon-carbon double and triple bonds, alcohols, organometallic functional
groups, and some carbon-chloride bonds.  It is important to recognize that
many functional groups can be present which compete for the oxidant and can
add significantly to the cost of treatment.

The waste to be treated should also be relatively free of suspended
solids, since a high concentration of suspended solids can foul the columns
often used to bring about contact between ozone and the aqueous phase
contaminants.  When ozonation is combined with UV radiation, a high
concentration of suspended solids also can impede the passage of UV radiation
and reduction reaction rates.

TABLE 6.3.2.   DISSOCIATION ENERGIES FOR SOME CHEMICAL BONDS

| Bond | Dissociation energy (kcal/gmol) | Wavelength to break bond (nm) |
|---|---|---|
| C-C | 82.6 | 346.1 |
| C=C | 145.8 | 196.1 |
| C≡C | 199.6 | 143.2 |
| C-Cl | 81.0 | 353.0 |
| C-F | 116.0 | 246.5 |
| C-H | 98.7 | 289.7 |
| C-N | 72.8 | 392.7 |
| C=N | 147.0 | 194.5 |
| C≡N | 212.6 | 134.5 |
| C-O | 85.0 | 334.5 |
| C=O (aldehydes) | 176.0 | 162.4 |
| C=O (ketones) | 179.0 | 159.7 |
| C-S | 65.0 | 439.9 |
| C=S | 166.0 | 172.2 |
| Hydrogen | | |
| H-H | 104.2 | 274.4 |
| Nitrogen | | |
| N-N | 52.0 | 540.8 |
| N=N | 60.0 | 476.5 |
| N≡N | 226.0 | 126.6 |
| N-H (NH) | 85.0 | 336.4 |
| N-H (NH$_3$) | 102.0 | 280.3 |
| N-O | 48.0 | 595.6 |
| N=O | 162.0 | 176.5 |
| Oxygen | | |
| O-O (O$_2$) | 119.1 | 240.1 |
| -O-O- | 47.0 | 608.3 |
| O-H (water) | 117.5 | 243.3 |
| Sulfur | | |
| S-H | 83. | 344.5 |
| S-N | 115. | 248.6 |
| S-O | 119. | 240.3 |

Source:   Legan, R.W.   1982.

6.3.1.2  Operating Parameters--

To effectively bring about the UV assisted reaction of ozone with organic contaminants, it is important that mass transfer of ozone and its reactants through the gas-liquid interface be maximized.  Also, to increase ozone solubility in water, temperatures should be maintained as low as possible and pressures as high as possible.  However, conditions such as high temperature, high pH, and high UV light flux favor ozone decomposition.  Under these conditions reactivity rates may increase, although costs may also increase due to less efficient use of ozone.  Decisions will have to be made on a case-by-case basis to establish the most effective operating conditions.

Several commercial designs are available for the conduct of gas/liquid reactions which bring reactants into contact as effectively as possible.  The types of reactor designs available range from mechanically agitated reactors to more complex spray, packed, and tray type towers.  Their advantages and limitations are discussed in detail in many standard texts and publications (for example, see References 5 through 7).

The process of UV/ozone treatment operates in the following manner.  The influent to the system is mixed with ozone and then enters a reaction chamber where it flows past numerous ultraviolet lamps as it travels through the chamber (see Figure 6.3.1).  Flow patterns and configurations are designed to maximize exposure of the total volume of ozone-bearing wastewater to the high energy UV radiation.  Although the nature of the effect appears to be influenced by the characteristics of the waste, the UV radiation enhances oxidation by direct dissociation of the contaminant molecule or through excitation of the various species within the waste stream.  In industrial systems, the system is generally equipped with recycle capacity.  Gases from the reactor are passed through a catalyst unit, destroying any volatiles, replenished with ozone, and then recycled back into the reactor.  The system has no gas emissions.

Regardless of the reaction mechanisms, there appears to be no doubt that the combination of ozonation with UV leads to increased oxidation rates. Typical design data for a 40,000 gal/day UV/ozone treatment process are shown in Table 6.3.3.  The plant is designed to reduce a 50 ppm PCB feed concentration to a 1 ppm effluent.

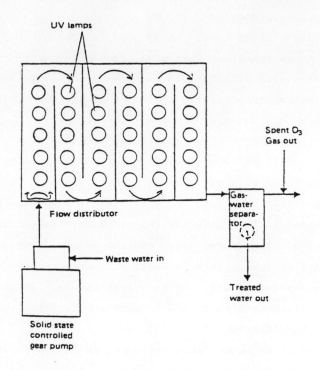

Figure 6.3.1.   Schematic of top view of ULTROX pilot plant by General Electric
(ozone sparging system omitted) (Edwards, B. H., 1983).

Source:   Reference 6.

TABLE 6.3.3.    DESIGN DATA FOR A 40,000 GPD
(151,400 L/DAY) ULTROX PLANT

Reactor

Dimensions:

| | |
|---|---|
| Meters (LxWxH) | 2.5 x 4.9 x 1.5 |
| Wet volume, liters | 14,951 |

UV lamps:

| | |
|---|---|
| Number of 65 watt lamps | 378 |
| Total power, KW | 25 |

Ozone Generator

Dimensions:

| | |
|---|---|
| Meters (LxWxH) | 1.7 x 1.8 x 1.2 |
| gms ozone/minute | 5.3 |
| kg ozone/day | 7.7 |
| Total power, kW | 7.0 |
| Total energy required (KW/day) | 768 |

Source:    Reference 16.

6.3.1.3  Post-Treatment Requirements--

Post-treatment of industrial wastewaters that have been contacted with ozone will involve elimination of residual ozone, usually by passing the effluent through a thermocatalytic unit as shown in Figure 6.3.1.  Some by-product residuals may be formed in the feed water and some contaminants, if present, will not undergo reaction.  Compounds considered unreactive probably will include many chlorinated aliphatic compounds.  If these compounds are present in the waste, technologies other than ozonation should be considered.

6.3.1.4  Treatment Combinations--

Apart from the employment of UV excitation with the ozonation process, ozonation can be considered as a finishing step for waste streams which have been treated by other technologies, principally biotreatment systems.  It has also been tested with some success as a means of enhancing biotreatability.  Although the use of ozonation in combination with other technologies such as biological treatment is a possible halogenated organic waste treatment alternative, it is not a demonstrated technology for industrial wastewaters, despite its extensive use and success in treating and disinfecting relatively clean drinking waters.

6.3.2  Demonstrated Performance

The limited information available deals primarily with pesticides, although some information regarding halogenated solvents and PCB/dioxins is presented in the solvent[3] and dioxin[4] TRDs, respectively.  Pesticides reportedly susceptible to ozonation alone include Aldrin, Benzene Hexachloride, DDT, and Dieldrin.[2]  Greater than 99.9 percent destruction of DDT using ozone at the 10 ppm level over a 30 minute time interval was reported in Reference 17.  However, the data were obtained at the bench scale level using concentrations in the ppb range.  Typical results of UV/Ozone as reported in Reference 18 are provided in Table 6.3.4.  It was noted that in 37 tests on wastewaters containing PCBs the maximum effluent concentration measured was 4.2 ppb.  Equivalent destruction efficiencies for dioxins were also reported in Reference 4.  Other halogenated compounds identified as economically treated include aldrin, chlorinated phenols, dieldrin, endrin, ethylene chloride, kepone, methylene chloride, and pentachlorophenol.

TABLE 6.3.4.   TYPICAL RESULTS OF UV/OZONE IRRADIATION

| Pesticide | Solution temperature (°C) | Ozone concentration (ppm) | UV light concentration (w/l) | Apparent percentage detoxification | Starting pesticide concentration (ppm) | Residual pesticide concentration (ppm) | Total time elapsed (min) |
|---|---|---|---|---|---|---|---|
| DDT | 30 | 10 | 1.32 | >99.1 | 0.057 | <0.0005 | 30 |
| PCP | 30 | 10 | 1.32 | >99.3 | 71.0 | <0.5 | 15 |
| Malathion | 25 | 50 | 1.32 | >99.8 | 55.0 | <0.1 | 30 |
| PCB | 22 | 10,000 | -- | 99.6 | 0.046 | 0.0002 | -- |

Source:   Reference 18

### 6.3.3   Cost of Treatment

Table 6.3.5 lists the costs for a 40,000 gpd UV/Ozone plant for which design data were shown in Table 6.3.3.   Cost estimates were based on waste-water containing 50 ppm PCB, designed to achieve an effluent PCB concentration of 1 ppm.   Costs were considered to be competitive with activated carbon.   The unit cost for treatment of the waste is greatly affected by whether or not the cost for a monitoring system is included.   The cost of PCB destroyed is in excess of $10/pound.   PCB data were used for costing purposes because of its availability.   However, the costs will increase substantially if ozonation is to be used as treatment for a waste containing 1 percent organic contaminants. This is 200 times the concentration used to develop the costs in Table 6.3.5. Assuming capital equipment costs follow a simple "sixth-tenths" factor scaling relationship, the costs of the reactor and generator would be about $3,000,000 (or 24 times the costs shown in Table 6.3.4) for treatment of this higher con-centration.   Scale factors would be variable for the operating and maintenance cost items listed in Table 6.3.4.   The net result of scale-up to handle the more concentrated waste would drastically increase the cost/1,000 gallons treated, but would also result in far lower costs when calculated on the basis of the amount of contaminant destroyed.   Costs of roughly $10/pound of contam-inant destroyed would be reduced to an estimated $1/pound, assuming comparable efficiencies.   Destruction efficiencies, however, may be adversely affected at higher concentrations due to mass-transfer and other considerations.   Thus, the cost benefits per pound of contaminant destroyed, as stated above, may not be fully achievable.   Ozone usage and the corresponding costs are dependent on the concentration of oxidizable species in the waste stream.   The amount of UV radiation used depends on quantum yield which can vary widely depending upon waste characteristics and process condition.   An optimal tradeoff must be made on the basis of pilot-scale or full scale test results.

### 6.3.4   Overall Status of Process

#### 6.3.4.1   Availability--

Ozonation equipment is available commercially from several manufacturers within the United States.   The Chemical Engineers' Equipment Guide published by McGraw Hill lists nine manufacturers of ozone generators and 10 manufac-

TABLE 6.3.5.   EQUIPMENT PLUS OPERATING AND MAINTENANCE
                COSTS; 40,000 GPD UV/OZONE PLANT

| | |
|---|---|
| Reactor | $ 94,500 |
| Generator | 30,000 |
| | $124,500 |

**O & M Costs/Day**

| | |
|---|---|
| Ozone generator power | $4.25 |
| UV lamp power | 15.00 |
| Maintenance | 27.00 |
| (Lamp Replacement) | |
| Equipment Amortization | |
| (10 years @ 10%) | 41.90 |
| Monitoring labor | 85.71 |
| TOTAL/DAY | $173.86 |
| Cost per 1,000 gals (3,785 liters) with monitoring labor | $4.35 |
| Cost per 1,000 gals without monitoring labor | $2.20 |

Source:   Reference 16.

turers of ozonators. The latter classification includes firms that usually provide the ozone generator, the reactor, and auxiliaries such as the catalytic unit for destruction of ozone from the treated stream. The status of UV/ozonation is far less advanced. Processes such as the Ultrox process[6] have been concerned with highly refractory compounds such as PCBs. Equipment specifically designed and available for UV/Ozonation of industrial wastewaters, is not available as a standard commercial item.

6.3.4.2  Application--

Ozonation, alone, generally cannot be used as a sole treatment technology for wastes which are resistant to oxidation such as chlorinated aliphatic hydrocarbon wastes, and for wastes containing contaminants which form stable intermediates that are resistant to total oxidation. Ozonation appears best suited for treatment of very dilute waste streams, similar to those streams treated by the ozone based water disinfection processes now used in Europe. It does not appear to be cost competitive or technically viable for most industrial waste streams where organic concentration levels are 1 percent or higher. However, it may be viable for certain specific wastes with high levels of a contaminant of special concern and high reactivity e.g. Dioxins. The combination of ultraviolet light and ozone does appear to greatly enhance destruction efficiency, but available data are not sufficient to identify specific waste streams that are treatable. Pesticide waste streams appear most likely to be treatable.

6.3.4.3  Environmental Impact--

Assuming adequate destruction of a contaminant by ozonation, the principal environmental impact would appear to be associated with ozone in the effluent vapor and liquid streams. However, thermal decomposition of ozone is effective and is used commercially to destroy ozone prior to discharge. Unreacted contaminants or partially oxidized residuals in the aqueous effluent may be a problem necessitating further treatment by other technologies. Presence of many such residuals will generally result in selection of a more suitable alternative technology.

6.3.4.4  Advantages and Limitations--

There are several factors which suggest that ozonation may be a viable technology for treating certain dilute aqueous waste streams:[1,6]

- Capital and operating costs are not excessive when compared to incineration provided oxidizable contaminant concentration levels are less than 1 percent.

- The system is readily adaptable to the onsite treatment of hazardous waste because the ozone can and must be generated onsite.

- It can be used as a preliminary treatment for certain wastes; e.g., preceeding biological treatment.

REFERENCES

1.   Prengle, H. W., Jr.  Evolution of the Ozone/UV Process for Wastewater
     Treatment.  Paper presented at Seminar on Wastewater Treatment and
     Disinfection with Ozone., Cincinnati, Ohio, 15 September 1977.
     International Ozone Association, Vienna, VA.

2.   Harris, J. C.  Ozonation.  In:  Unit Operations for Treatment of
     Hazardous Industrial Wastes.  Noyes Data Corporation, Park Ridge, N.J.
     1978.

3.   Breton, M. A. et al.  Technical Resource Document; Treatment Technologies
     for Solvent-Containing Wastes.  U.S. EPA HWERL.  Contract
     No. 68-03-3243.  August 1986.

4.   Arienti, M. et al.  Technical Resource Document:  Treatment Technologies
     for Dioxin-Containing Wastes.  Prepared for U.S. EPA, HWERL, Cincinnati,
     under Contract No. 68-03-3243, Work Assignment No. 2.  August 1986.

5.   Rice, R. G.  Ozone for the Treatment of Hazardous Materials.  In:
     Water-1980; AIChE Symposium Series 209, Vol. 77.  1981.

6.   Edwards, B. H., Paullin, J. N., and K. Coghlan-Jordan.  Emerging
     Technologies for the Destruction of Hazardous Waste - Ultraviolet/Ozone
     Destruction.  In:  Land Disposal: Hazardous Waste.  U.S. EPA
     600/9-81-025.  March 1981.

7.   Ebon Research Systems, Washington, D.C.  In:  Emerging Technologies for
     the Control of Hazardous Waste.  U.S. EPA 600/2-82-011.  1982.

8.   Rice, R. G., and M. E. Browning.  Ozone for Industrial Water and
     Wastewater Treatment, an Annotated bibliography.  EPA-600/2-80-142,
     U.S. EPA RSNERL, Ada, OK.  May 1980.

9.   Rice, R. G., and M. E. Browning.  Ozone for Industrial Water and
     Wastewater Treatment, A Literature Survey.  EPA-600/2-80-060.  U.S. EPA
     RSKERL, Ada, OK.  April 1986.

10.  International Ozone Institute, Inc., Vienna, VA.  First International
     Symposium on Ozone for Water and Wastewater Treatment.  1975.

11.  International Ozone Institute, Inc., Vienna, VA.  Second International
     Symposium on Ozone Technology.  1976.

12.  Hackman, E. Ellsworth.   Toxic Organic Chemicals-Destruction and Waste
     Treatment.   Park Ridge, NJ, Noyes Data Corp., 1978.

13.  Sundstrom, D. W., et. al.   Destruction of Halogenated Aliphatics by
     Ultraviolet Catalyzed Oxidation with Hydrogen Peroxide.   Department of
     Chemical Engineering, The University of Connecticut.   Hazardous Waste and
     Hazardous Materials, 3(1):   1986.

14.  Chillingworth, M. A., et al.   Industrial Waste Management Alternatives
     for the State of Illinois, Volume IV - Industrial Waste Management
     Alternatives and their Associated Technologies/Processes, prepared by GCA
     Technology Division, Inc.   February 1981.

15.  U.S., HWERL, Cincinnati Innovative and Alternative Technology Assessment
     Manual, Ozone Oxidation.   EPA 430/9-78-004.   February 1980.

16.  Arisman, R. K., and R. C. Musick.   Experience in Operation of a UV-Ozone
     Ultrox Pilot Plant for Destroying PCBs in Industrial Waste Effluent.
     Paper presented at the 35th Annual Purdue Industrial Waste Conference.
     May 1980.

17.  Dillon, A. P., Editor.   Pesticide Disposal and Detoxification Pollution
     Technology Review No. 81, Noyes Data Corporation, Park Ridge, NJ.   1981.

18.  Arienti, et. al.   Technical Assessment of Treatment Alternatives for
     Wastes Containing Halogenated Organics, Final Report to OSW, Washington
     D.C. under Contract No. 68-01-6871, Work Assignment No. 9, August 1984.

## 6.4   CHEMICAL DECHLORINATION

### 6.4.1   Process Description

Chemical dechlorination methods have been developed as possible alternatives to incineration or land disposal for halogenated organic compounds such as PCBs.[1]   Researchers have found that in order to decrease the degree of toxicity, as well as the chemical and biological stability of chlorinated compounds, it is not necessary to totally break down the molecular structure.[1]   Instead, the formation of a compound considered harmless and environmentally safe can be achieved through a reaction system that will result in the cleavage of C-Cl bonds or the rearrangement of the chlorinated molecule.   Although several different dechlorination methods exist, all of the processes are based primarily on two technologies; the "Goodyear process" developed by Goodyear Tire and Rubber, and the NaPEG system developed by the Franklin Research Institute.   Thus far, they have been studied as technologies for the destruction of highly toxic wastes such as PCBs and dioxins.   However, they should be applicable to the destruction of all halogenated organics.   The characteristics of these processes are summarized in Table 6.4.1.

The Goodyear Process was originally developed to reduce PCB laden heat transfer fluids from slightly above 500 ppm to less than 10 ppm.   The reaction chemistry is based on the use of a sodium-naphthalene reagent to form sodium chloride and an inert, combustible sludge.   The reagent is produced by disolving molten sodium and naphthalene in tetrahydrofuran.[2]   However, the reactivity of metallic sodium with water necessitates the use of an air free anhydrous reaction vessel to prevent rapid generation of hydrogen or loss of reagent through the formation of NaOH.

Since Goodyear has decided not to pursue the marketing of this process, several companies such as SunOhio, Acurex (now being marketed by Chemical Waste Management), and PPM Inc, have entered the field.   Generally, they have modified the process by substituting proprietary reagents for naphthalene, which is a priority pollutant.   These processes are also intended for treatment of PCB contaminated oils (50-500 ppm) and require pretreatment to remove water and inorganics such as soil.   Typically, these processes cannot handle PCB concentrations greater than 10 percent, and most are not suitable for sludges, soils, sediments, and dredgings.

TABLE 6.4.1. DECHLORINATION PROCESSES

| Process description | Compounds and forms of waste treated | Destruction Capabilities | Residuals | Comments |
|---|---|---|---|---|
| **PCB (SunOhio)** <br>• proprietary sodium reagent used to strip away chlorine <br>• oil is mixed with reagent and sent to reactor <br>• mixture is then centrifuged degassed and filtered | • liquid hydrocarbon streams, i.e. PCB contaminated oil from transformers <br>• cannot be used on aqueous or soil wastes | • 250 ppm PCB to 1 ppm <br>• 3000 ppm to below 2 with several passes | • metal chlorides <br>• polyphenyls <br>• treated oil | • mobile, continuous process <br>• moisture and contaminant removal required as pre-treatment <br>• moderate temperature and pressure <br>• pure PCBs destroyed at 150 ml/min |
| **Acurex (CMM)** <br>• proprietary sodium reagent used to strip chlorine <br>• contaminated oil is filtered, and mixed with reagent <br>• reaction takes place in processing tank | • can be used on PCB contaminated oils and soils <br>• also effective on transformer oil contaminated with 2,3,7,8-TCDD | • PCB feeds as high as 10% effectively treated <br>• 2,3,7,8-TCDD reduced from 200-400 ppt to 40 ppt | • treated oil <br>• sodium hydroxide effluent <br>• polyphenol sludge | • mobile, batch operation <br>• pretreatment needed to remove water, aldehydes and acids from transformer oils <br>• non-toxic solvent used to extract PCBs from soil |
| **APEG** <br>• sodium polyethylene glycol reagent (NaPEG) used for PCB <br>• Potassium Polyethylene Glycol used for TCDD <br>• reagent is added to contaminated material in the presence of air, and can be sprayed on | • PCB oils and soils <br>• TCDD contaminated soils <br>• also tested on hexachlorocyclohexane, hexachlorobenzene, PCP, DDT, KEPONE, Tri- and Tetra-chlorobenzenes | • PCB destruction 99.9% <br>• 2,3,7,8-TCDD reduced from 330 ppb to 101 ppb <br>• Other halogenated organics >99.9% | • sodium chloride <br>• oxgenated biphenyls <br>• decontaminated material <br>• hydrogen gas | • involves the application of reagent in the presence of air or oxygen <br>• water increases reaction times and decreases the degree of chlorination <br>• temps. above 100°C required for fast destruction |
| **PPM** <br>• proprietary sodium reagent used for chlorine stripping <br>• reagent is added to contaminated oil and left to react <br>• solid polymer formed is filtered out | • PCB contaminated oil <br>• TCDD detoxification will be investigated soon <br>• aqueous waste and soil not treated | • 200 ppm PCB reduced to below 1 ppm | • solid polymer <br>• decontaminated oil | • mobile, batch process, 700 gal/hr <br>• polymer is produced at a rate of 55 gal per 10,500 gal oil treated <br>• polymer is regulated and must be landfilled |

Source: Reference 3.

A more promising technology is the NaPEG or APEG process, originally developed in 1980 by Pytlewski, et al., at the Franklin Research Institute.[4] The intent was to devise a reaction system that would decompose PCBs and halogenated pesticides in an exothermic and self-sustaining manner. The dechlorination reagent was formed by reacting alkali metals such as sodium with a polyethylene glycol (M.W. 400) in the presence of heat and oxygen.[5] The reaction mechanism involves a nucleophilic substitution/elimination and the oxidative degradation of chlorine through the generation of numerous free radicals. The process reactivity can be "tuned" or directed at various aliphatic or aromatic systems by varying the molecular weight of the polyethylene glycol.[6] Two emerging technologies based on the APEG system are currently under development at the Galson Research and the Sea Marconi Corporations. The Galson Research process involves a series of processes for the degradation of chlorinated benzenes, biphenyls and dioxins from contaminated soils. The system, which was developed under EPA sponsorship, is based on the more reactive KPEG (potassium-based) reagent, in conjunction with a sulfoxide catalyst/cosolvent. A probable reaction scheme is presented in Figure 6.4.1.

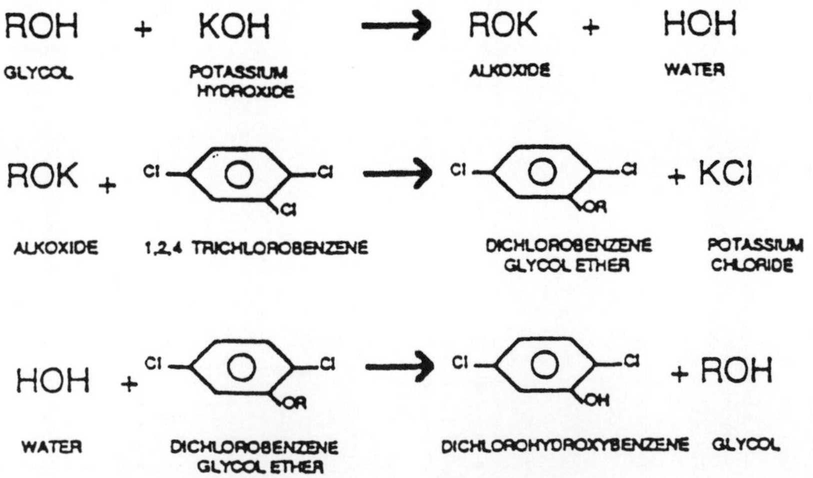

Figure 6.4.1.  Probable reaction mechanism.
Source:  Galson Research Corporation.

The Sea Marconi's chemical process, called CDP-Process, was first developed for the decontamination of PCB-laden mineral oils. However, the system has been more recently applied to materials and surfaces exposed to contaminants coming from fire or explosion of PCB equipment. The chemistry involves reaction with high-molecular weight polyethylene glycol in the presence of a weak base and a peroxide. No application to aqueous media can be expected from these processes due to their sensitivity to water.

The mode of operation of each of the above processes is basically the same with some slight variations. Each of the processes is a batch process except for the PCBs process which is continuous. All of the processes have been designed to be mobile. Waste throughput for these processes is generally in the range of 500 to 1,000-gallons per hour for a PCB contaminated oil.[2] For contaminated soils and other types of waste, the throughput would be different.

Another approach to dechlorination of halogenated organics is employed in the light activated reduction of chemicals (LARC). The LARC processes, which uses UV light and hydrogen gas to degrade extracted chlorinated hydrocarbons, was developed by Atlantic Research Corporation (U.S. Patent 4,144,152 awarded to Judith Kitchens of ARC). However, despite its initial promise as demonstrated by its ability to destroy halogenated organics such as Aroclor 1254, kepone, and tetrabromophthalic anhydride,[7] the process has not been actively pursued, primarily because of economic considerations.[8]

The processes have been designed primarily for the treatment of oils contaminated with 500 to 5,000 ppm of PCBs.[2] Some work has also been done on the treatment of wastes with low concentrations of dioxin[3]. All of the processes, except for the APEG process, require pretreatment to remove water and inorganics such as soil from the waste. The Acurex process can treat contaminants in a suitable solvent at virtually any concentration. The Acurex process is reported to be able to treat wastes with up to a 10 percent concentration of contaminants. If wastes are more concentrated than this, they are diluted before addition of the reagent.[9] In addition, compounds with phosphorus, sulfur, alcohols and acids are troublesome. The alcohols and acids interfere with the free radical reaction, upon which the dechlorination process is based. The PCBs and PPM processes are more limited in their applicability. Both cannot be used to treat soil contaminated wastes or wastes with any moisture content.

6.4.2   Demonstrated Performance

These processes should be applicable to most types of highly halogenated
compounds.  The sodium polyethylene glycol reagent used in the APEG process
was also used on the following compounds in place of PCBs:  hexachlorocyclo-
hexane, hexachlorobenzene, tri- and tetrachlorobenzenes, pentachlorophenol,
DDT, kepone, and chloroethylsulfide.  These compounds were dechlorinated
rapidly and completely as noted in the proceedings of the sixth annual
symposium on the treatment of hazardous waste.[4]

The destruction efficiency of PCB contaminated material is in the
99 percent range for each of the processes as can be seen in Table 6.4.1.
Equal or greater efficiencies should be achievable for most other halogenated
organic compounds in liquid streams.  Further detail regarding the performance
of the processes in degrading toxic compounds such as the PCBs and dioxins can
be found in Reference 10.

6.4.3   Cost of Treatment

At this time, costs are very well established for the decontamination of
PCB contaminated oils.  These costs are dependent of several variables:

- concentration of pollutant;

- quantity and characteristics of the material to be treated;

- reagent costs; and

- the resale value of the treated material.

The cost of treating bulk quantities of PCB-contaminated oil using the
SunOhio PCBs process will about $3.00 per gallon.  Costs will vary depending
upon contamination level, onsite or offsite treatment, transportation, and
ultimate disposition of the oil.  Costs for treating transformer oil will be
higher (5 to 9 dollars or more per gallon) with a minimum charge of $25,000
per transformer.  The average cost in early 1980 for the Acurex process was
$2.40 per gallon or $0.70 per kilogram of oil treated.[12]

Based upon the APEG laboratory field research that has been conducted over the past several years, a preliminary economic evaluation of this dechlorination process has been attempted[13]. Specifically, Galson Research Corporation, in conjunction with the U.S. EPA-HWERL, has roughly estimated the costs for APEG dechlorination using two hypothetical field scenarios. These costs, as shown in Table 6.4.2 below, indicate that there is approximately a $205/ton difference between the in situ process (operating on a 1 acre-3 feet deep contaminated area) and the slurry process (with excavation and 3 reactor systems operating). This difference comes from the fact that in the slurry APEG process, reagent recovery is possible which reduces the total cost of the process by approximately 65 percent.

TABLE 6.4.2.   PRELIMINARY ECONOMIC ANALYSIS OF IN SITU AND SLURRY PROCESSES
            (Peterson, R.L., et al., 1985)

| Cost item | Cost, $/ton soil | |
| --- | --- | --- |
| | In situ | Slurry |
| Capital recovery | 31 | 17 |
| Setup and operation | 65 | 54 |
| Reagent | 200 | 20 |
| Total costs | 296 | 91 |

## 6.4.4  Status of Development

### 6.4.4.1  Availability/Application--

The non-APEG processes are available both as fixed and mobile units, from the developers or operators. These firms are Chemical Waste Management, SunOhio, PPM Inc. and Goodyear. These processes are used exclusively for the treatment of PCB-contaminated oils, although the processes should be applicable to moisture free liquid organic wastes containing halogenated organic constituents.  Similarly the APEG process, now being tested by Galson Research Corporation and EPRI,[10] should be capable of destroying virtually all halogenated organics.

The limiting factor in all cases will be cost.

6.4.4.2  Environmental Impacts--

The residuals from these processes are listed in Table 6.4.1.  Each of these processes result in the production of a polyphenol type material which is nontoxic and can be readily handled.  Acurex has estimated that it takes about 2,000 gallons of treated oil to produce one drum of sludge and associated liquid.[2]  The chlorine atoms which have been removed from the PCB compound leave as sodium chloride in the sodium hydroxide effluent.[3]  Air emissions from this process are minimal because the destruction process occurs under an inert nitrogen atmosphere.  PPM and PCBs result in similar polyphenol/salt type residues which contain less than 2 ppm of PCB and can therefore be readily handled.

The residuals from the APEG process, however, are somewhat different because oxygen is involved in the reaction.  Oxygenated biphenyls are formed along with sodium chloride and hydrogen gas.  The products are reportedly nontoxic.  Incineration would appear to be the preferred disposal method for all solid wastes.

6.4.4.3  Advantages/Limitations--

The dechlorination processes discussed here represent proven technology for the dehalogenation of organic compounds such as PCBs and dioxins.  Equivalent performance can be expected for other halogenated compounds.  However, the costs are high and residuals are produced that will require further treatment, probably incineration for the solid residuals.

REFERENCES

1.   Weitzman, L. et al., "Disposing Safely of PCBs: What's Available, What's
     on the Way." Power, February 1981.

2.   Gin, W. et al., Technologies for Treatment and Destruction of Organic
     Wastes as Alternatives to Hand Disposal, State of California Air
     Resources Board, August 1982.

3.   Nunno, T. et al., Technical Assessment of Treatment Alternatives For
     Wastes Containing Halogenated Organics. GCA Report to U.S. EPA/OSW, under
     Contract 68-01-6871, WA No. 9.   1985.

4.   U.S. EPA, Treatment of Hazardous Waste, Proceedings of the Sixth Annual
     Research Symposium, pg. 72. EPA-600/9-8-011, March 1980.

5.   Ibid pg. 197.

6.   Telephone Conversation with Charles Rogers, U.S. EPA, Cincinnati, Ohio,
     May 15, 1986.

7.   Valentine, R. S.  LARC-Light Activated Reduction of Chemicals.  Pollution
     Engineering.  February 1981.

8.   Kitchens, J.  Atlantic Research Corporation.  Telephone conversation with
     M. Arienti of GCA.  August 1986.

9.   Wallbach, D.  Acurex Corporation.  Telephone conversation with M. Arienti
     of GCA.  October 2, 1984.

10.  Arienti, M. et. al.  Technical Resource Document - Treatment Technologies
     for Dioxin-Containing wastes.  Prepared for HWERL Cincinnati under
     Contract No. 68-03-3243, Work Assignment No. 2.  August 1986.

11.  Fisher, M. SunOhio.  Telephone conversation with N. Surprenant of GCA.
     August 1986.

12.  Weitzman, L.  Acurex Corporation, Cincinnati, Ohio.  Telephone
     conversation with M. Jasinski.  June 4, 1986.

13.  Peterson, R. L. et. al.  Chemical Destruction/Detoxification of
     Chlorinated Dioxins in Soils.  Paper presented at Eleventh Annual
     Research Symposium on Incineration and Treatment of Hazardous Waste.  EPA
     600/9-85-028.  September 1985.

# 7. Biological Treatment Methods

## 7.1 PROCESS DESCRIPTION

Biological treatment involves the degradation of organic compounds by
microorganisms. Conventional biological treatment processes include activated
sludge, aerated lagoons, aerobic and anaerobic digestion, trickling filters,
rotating biological contactors, and composting. For a detailed description of
these processes, one can refer to a number of texts on wastewater
treatment.[1,2] In each of these processes, there exists a population of
microorganisms which are either suspended in a liquid medium, as in the
activated sludge process, or attached to some solid surface, as in the
trickling filter process. The microorganisms metabolize the organic
constituents of the waste to carbon dioxide and water if the process is
aerobic, or carbon dioxide and methane if the process is anaerobic. When
treating toxic compounds, such as many halogenated organics, only partial
degradation may occur and the end products may be as toxic as the initial
compound. In addition, many of these compounds may be removed by some
mechanism other than biodegradation. Processes that require aeration may
cause volatile constituents in the waste stream to be air stripped prior to
degradation, while hydrophobic compounds with a high octanol/water partition
coefficient may be adsorbed by biological matter or other solids in the
wastestream and subsequently be removed as part of the settled sludge.
Therefore, compounds that may appear to have been biodegraded are sometimes
merely transferred from the liquid to the air or solid phase.

In treating wastes containing halogenated organic compounds, the
effectiveness of the system in removing these compounds is dependent primarily
on the microorganisms that are present. Most of these compounds are manmade
and, therefore, natural microorganisms did not originally have the ability to
degrade these compounds. Through exposure to the compounds, however, some

232

groups of microorganisms have developed enzymatic systems resistant to the toxic compounds and with a capability to degrade them at a slow rate.[3,4] Microorganisms that have demonstrated an ability to degrade halogenated organic compounds are listed in Table 7.1.   Treatment systems that are innoculated with these types of microorganisms may have the ability to remove these compounds.

## 7.2   DEMONSTRATED PERFORMANCE

### 7.2.1   Removal in Conventional Systems

In the past, the primary function of biological treatment systems has not been to remove toxic organic pollutants, but to remove the conventional, easily biodegradable organic compounds.   Recently, however, the biodegradation of toxic compounds has received increasing attention due to its potentially lower cost versus other treatment technologies, and a number of studies have been conducted.   In one study, a municipal wastewater stream treated by a conventional activated sludge process was spiked with a number of priority pollutants including toxaphene, lindane, pentachlorophenol, and heptachlor.[6]   The concentration of each of these compounds was 50 µg/L except for toxaphene which was 150 µg/L.   The treatment system included a primary clarifier, an aeration basin, and a secondary clarifier.   The primary sludge, return activated sludge, and final effluent were sampled over a 312 day period to determine whether the compound was biodegraded, air stripped, or adsorbed as it passed through the treatment system.   As shown in the table below, each compound responded differently.   The only similarity was that biodegradation was not the primary removal mechanism for any of the compounds.

| Compound | Percent adsorbed | Percent biodegraded or stripped | Percent in final effluent |
|---|---|---|---|
| Toxaphene | 58 | 40 | 2 |
| Heptachlor | 68 | 25 | 7 |
| Lindane | 20 | 25 | 55 |
| Pentachlorophenol | 28 | 0 | 72 |

TABLE 7.1. EXAMPLES OF MICROORGANISMS THAT CAN DEGRADE HALOGENATED ORGANIC COMPOUNDS

| Compound | Organisms | Condition | Remarks/Products |
|---|---|---|---|
| **Aliphatics** | | | |
| Trichloroethane | Marine bacteria | ae | |
| Trichloromethane | Sewage sludge | ae | |
| Trichloroethane, trichloromethane, methyl chloride, chloroethane, dichloroethane, vinylidiene chloride, trichloroethylene, tetrachloroethylene, methylene chloride, dibromochloromethane, bromochloromethane | Soil bacteria | an | Anoxic conditions |
| Trichloromethanes, trichloroethylene, tetrachloroethylene | Methanogenic (7) culture | an(8) | |
| Trichloromethane, trichloromethann, tetrachloromethane, dichloroethane, dibromochloromethane, 1,1,2,2-tetrachloroethane, bis-(2-chloroisopropyl) ether, bromoform, bromodichloromethane, trichlorofluoromethane, 1,1-dichloroethylene, 1,2-dichloroethylene, 1,3-dichloropropylene, 1,2-transdichloroethylene | Sewage sludge | ae(?) | |
| **Aromatic compounds** | | | |
| 1,2-; 2,3-; 1,4-dichlorobenzene; p-; m-; o-chlorobenzoate; 3,4-; 3,5-dichlorobenzoate, 3-methyl benzoate; 4-chlorophenol | Sewage sludge, Pseudomonas sp.(1), sewage | ae, ae | Plasmid transfer led to ability to attack a number of these compounds simultaneously; sole energy and carbon source |
| | Pseudomonas sp.B13(WR1) | ae | |
| Hexachlorobenzene, trichlorobenzene | Sewage sludge | ae | |
| 1,2,3- and 1,2,4-trichlorobenzene | Soil microbes | ae | 2,6-; 2,3-dichlorobenzene; 2,4- and 2,5-dichlorobenzene; $CO_2$; slow |
| Pentachlorophenol | Soil microbes, Flavobacterium | an, ae | tetra-, tri-, di-, and $m$-chlorophenol (8) Complete mineralization |
| Monochlorophenol, monochlorobenzoate | Nocardia, Mycobacterium (5) | ae | Cometabolism |
| Pentachloronitrobenzene (PCNB) | Aspergillus niger, Fusarium solani, Glomerella congulata, Helminthosporium victoriae, Myrothecium, Penicillium, Trichoderma viridae (4) | ae | Only during active growth |

(continued)

**TABLE 7.1 (continued)**

| Compound | Organisms | Condition | Remarks/Products |
|---|---|---|---|
| Methoxychlor | Nocardia sp., Streptomyces sp. (5) | ae | |
| | Aerobacter aerogenes (1) | ae/an | 1,1-dichloro-2,2-bis(p-methoxyphenol)ethylene; 1,1-dichloro-2,2-bis(p-methoxyphenol)-ethane |
| Acrolein | Site water (microbes) | ae | β-Hydroxypropionaldehyde |
| Aldrin | Site water (microbes) | ae | Dieldrin by epoxidation |
| | Sewage sludge | an | (8) |
| Endosulfan | Fungi, bacteria, soil antinomycetes | ae | Endosulfen (2), endodiol (1), endohydroether (5) |
| Endrin | Pseudomonas sp. Micrococcus sp., yeast (4) | ae | Soil organisms, aldehydes and ketones with 5 to 6 chlorine atoms (1) |
| | Sewage sludge | an | (8) |
| Chlorodimeform | Chlorella (2), Oscillatoria (3) | ae | 4-Chloro-o-formotoluidiene, 4-chloro-o-toluidiene, 5-chloroanthranilic acid, n-formyl-5-chloroanthranilic acid, suspected mutagens |
| Kepone | Treatment lagoon sludge | an | (8), Cometabolism |
| Polycyclic aromatic hydrocarbons (halogenated) | | | |
| PCBs (mono- and dichlorobiphenyls) | Pseudomonas, Vibrio, Spirillum, Flavobacterium | ae | Biodegradation appears to be inversely related to extent of chlorination |
| | Achromobacter | | High level dehalogenase, major end product $CO_2$ |
| | Chromobacter Bacillus (1), Nocardia (5) | | |
| 4-chlorobiphenyl; 4,4′-dichlorobiphenyl; 3,3′-dichlorobiphenyl | Fungi | ae | 4-Chloro-4′hydroxybiphenyl; 4,4′-dichloro-3-hydroxybiphenyl; chlorinated benzoic acid |
| Pesticides | | | |
| 2, 4-D | Pseudomonas Alcaligenes Eutrophus | ae | Full mineralization |
| 2, 4, 5-T | Pseudomonas cepacia (AC1100) | ae | |

(continued)

## TABLE 7.1 (continued)

| Compound | Organisms | Condition | Remarks/Products |
|---|---|---|---|
| Toxaphene | Corynebacterium pyrogenes (1) | an | |
| Heptachlorobornane | Micromonospora chalces (5) | ae | |
| | Bovine rumen fluid (7) | an | Hexachlorobornane (8) |
| Lindane | Chlorella vulgaris (2) | ae(?) | |
| | Chlamydamonas reinhardtii | ae | Pentachlorocyclohexane (non-toxic) (8) |
| | Chlosteridium sp., Pseudomonas (1) | an | |
| | Soil bacteria | an | -3,4,5,6-tetrachloro-1-cyclohexane, -BHC (7)(8) |
| | P. Chrysosporium | ae | Complete Mineralization |
| | Sewage sludge | an | |
| Dieldrin | Anacystis nidulans (3) | an | Photodieldrin |
| | Agmeneloum quardiplicatum (3) Pseudomonas (1) | an | Toxic epoxide moiety reduced to olefin |
| | Rumen fluid (7) | an | |
| | Actinomycetes | an/ae | Chlordene (8) chlordene epoxide (oxidation) |
| DDT (1,1'-bis(p-chlorophenyl)-2,2,2-trichloroethane) | Klebsiella pneumoniae (1) E. coli, Aerobacter aerogenes, Pseudomonas, Clostridium, Proteus vulgaris | an | More than 20 species of bacteria are reported to be able to reductively dechlorinate DDT. Aerobic conditions are sometimes reported but apparently do not promote much dechlorination. Anaerobically DDT goes mainly to DDD (TDE) while aerobically it appears to be transformed to DDE. |
| | Sewage | ae | TDE & DBP major products |
| | Soil bacteria | an | 7 possible metabolites, simplest reported was p-chlorobenzoate |
| | Rumen bacteria (7) | an | DDE, TDE, DDMU (8) |
| | Yeast | ae | TDE (8) |
| | Trichoderma viridae (4) | an | TDE, DDE (8) |

(continued)

## TABLE 7.1 (continued)

| Compound | Organisms | Condition | Remarks/Products |
|---|---|---|---|
| DDT | P. Chrysosporium | ae | Complete Mineralization |
| | Fusarium oxysporum (4) | ae | Complete mineralization; no DDT in 10-14 days |
| | Mucor alterans (4) | ae | Cometabolism |
| | Cylindrotheca, Closterium (2) | ae | DDE (slow) (9) |
| | Dunaliella (2) | ae | TDE, DDE, DDMS, DDOH (8)(9) |
| | Anaerobic sludge (7) | an | TDE rapid |
| | Nocardia, Streptomyces (5) | ae | DDE (9) |
| | Hydrogenomonas (1) | an/ae | 10 products, simplest was PCPA; 9 products, simplest was DBP |

ae—aerobic (may be ae/an)
an—anerobic (either anoxic or fastidious)

(1) Bacteria
(2) Algae
(3) Blue-green algae
(4) Fungi
(5) Actinomycetes
(6) Photosynthetic bacteria
(7) Consortium of anaerobices
(8) Reductive dechlorination
(9) Dehydrodochlorination

DDT--(only p,p'-DDT considered here)
DDE(TDE)--1,1'-bis(p-chlorophenyl)-2,2-dichloroethane
DDE--1,1'-bis(p-chlorophenyl)-2,2-dichloroethylene
DBP--4,4'-dichlorobenzophenone
DDMS--1,1'-bis(p-chlorophenyl)2-chloroethane
PCPA--p-chlorophenyl acetic acid
DDMA--1,1'-bis(p-chlorophenyl)-2-chloroethylene
DDOH--1,1'-bis(p-chlorophenyl)-2-hydroxyethane
BHC--1,2,3,4,5,6-hexachlorocyclohexane

Source: Reference 5.

One point that should be made, however, is that no attempt was made to optimize the system for the removal of the priority pollutants.  The primary purpose of an activated sludge treatment system such as this is to reduce the Chemical Oxygen Demand (COD) and the Total Suspended Solids (TSS) of the waste stream.  These removals averaged, respectively, 89 and 95 percent.  If the system had been inoculated with a microbial population that had been acclimated to the toxic constituents, and a longer solids retention time had been used, the removal by biodegradation may have been greater.  As it was, the majority of the toxaphene and heptachlor were removed by adsorption, and the majority of the lindane and pentachlorophenol were not removed at all.

In another study, the concentration of several pesticides in the influent and effluent of a municipal wastewater treatment system was measured.[7] These measurements indicated that biodegradation was the removal mechanism for 16 to 55 percent of the influent 2,4-D that ranged in concentration from 8 to 40 µg/L.  The measurements also indicated that DDT was transformed to DDE or DDD, and that aldrin degradation produced the compound dieldrin.  This illustrates the fact that the products of biodegradation are sometimes as toxic as the original compounds.

## 7.2.2   Research on Specific Microorganisms

In addition to research on the fate of compounds in conventional biological treatment systems, a number of researchers have studied the removal of halogenated and other toxic organic compounds by acclimated bacterial populations and other microorganisms.  The results of some of these projects are summarized in Table 7.2.  In most of these projects, natural bacterial populations from sewage, river water, or soil were slowly exposed to waste containing halogenated compounds (HOCs) after which a population was developed that could utilize the HOC as a food and energy source.

For example, in the first project listed in Table 7.2, microorganisms from several waste dumping sites (such as Love Canal) were collected and placed in a chemostat along with other microorganisms of known capabilities. 2,4,5-T was added to the chemostat in gradually increasing concentrations over a period of 8 to 10 months until the bacterial population could utilize 2,4,5-T as its sole food and energy source.  The bacteria isolated from this culture were identified as Pseudomonas cepacia, strain AC1100.   Table 7.3

TABLE 7.2.   RESEARCH ON MICROBIAL DEGRADATION OF HALOGENATED ORGANIC COMPOUNDS

| Compound | Microorganism | Degradation achieved | Comments | Reference |
|---|---|---|---|---|
| 2,4,5-T | Pseudomonas Cepacia, AC1100 | 98% reduction from starting concentration of 1,000 ppm in 1 week. | experiments conducted on contaminated soil | Chatterjee et al, 1982;[8] Kilbane et al, 1983[9] |
| 2,4-D | Pseudomonas (from sewage sludge) | 100 mg/L reduced to less than 1 mg/L in six days; 100 μg/L reduced to less than 20 μg/L in 3 days. | 2,4-D was the only carbon and energy source; 3-day lag period before rapid growth. | Kim and Maier, 1986[10] |
| 3,5-DCB | Pseudomonas (from sewage sludge) | 100 mg/L reduced to less than 10 mg/L in 6 days; 100 μg/L reduced to less than 20 μg/L in 3 days. | same as above | same as above |
| Pentachlorophenol | Flavobacterium (from River water) | 100 ppm in soil mineralized in 1 week; degradation did not occur at initial concentration of 500 ppm. | Bacteria adapted to PCP degradation after a 2-3 week exposure in saturated stream sediments; optimum conditions: 20% water, Temp = 24-35°C | Martinson et al, 1986[11] |
| 2,4-D | Pseudomonas, Alcaligones Eutrophus | Percent degradation not presented, but both types of bacteria showed rapid degradation of 2,4-D after an extended (80-120 hr) lag period | No degradation of 2,4-D occurred when glucose was added as an alternate substrate | Roy and Mitra, 1986[4] |
| Lindane, DDT 2,3,7,8-TCDD, 3,4,3',4'-Tetra-chlorobiphenyl | Phanerochaete Chrysosporium (White Rot Fungus) | 90% degradation of DDT after 50 days; 9.3% of DDT mineralized to $CO_2$ after 60 days. | | Bumpus et al, 1985[12] |
| Halogenated Benzoates | Methanogenic bacteria from lake sediment and sewage sludge | Mineralized halogenated aromatics to $CO_2$ and $CH_4$ | Dehalogenation did not depend on number of halogen atoms but on the type of halogen and position on aromatic ring | Suflita et al, 1982.[13] |
| Chlorinated Benzenes | Developed from primary sewage | 90-100% degradation of several chloro-benzenes from initial concentrations of 10 μg/L. | Microbial population was grown on either glass beads or granular activated carbon in a 25 cm upflow column | Bouwer and McCarthy, 1982[14] |

TABLE 7.3.   CONCENTRATIONS OF 2,4,5-T IN SOIL TREATED WITH
PSEUDOMONAS CEPACIA AC100[a] ($\mu$g/g of soil)

| Sample | Concentration of 2,4,5-T ($\mu$g/g of soil) | | | | |
|--------|---------|--------|---------|---------|---------|
|        | Initial | 1 Week | 2 Weeks | 3 Weeks | 6 Weeks |
| 1      | 1,000   | 30     | 20      | 20      | 60      |
| 2      | 2,500   | 35     | 20      | 20      | 20      |
| 3      | 5,000   | 280    | 70      | 25      | 20      |
| 4      | 10,000  | 9,000  | 3,500   | 2,050   | 488     |
| 5      | 20,000  | 17,000 | 11,900  | 4,300   | 1,410   |
| 6[b]   | 1,000   | 940    | 1,100   | 1,100   | 1,120   |
| 7[b]   | 10,000  | 11,200 | 10,800  | 10,800  | 9,095   |

[a]Samples treated with $5 \times 10^7$ micro-organisms per g of soil.

[b]Controls--no micro-organisms added.

Source:   Reference 9.

shows the effectiveness of this bacteria in degrading 2,4,5-T in soil at concentrations of up to 20,000 ppm.  The researchers plan to develop bacterial strains for degrading other compounds such as 2,3,7,8-TCDD, using the same method, which they call plasmid assisted molecular breeding.[8,9]

In other studies listed in Table 7.2, research was conducted to determine the effect of factors such as contaminant concentration and the presence of alternate substrates on the degradation of HOCs.  Kim and Maier of the University of Minnesota investigated the degradation of 2,4-D and 3,5-dichlorobenzene (DCB) by acclimated cultures derived from municipal sewage sludge.[10]  They found that the acclimated cultures could degrade the compounds over a wide concentration range (10 µg/L to 100 mg/L).  They also found that when a nutrient broth was added (to act as an alternate substrate), the rate of degradation of the HOCs was increased except at low concentrations.  In all cases, however, both substrates were utilized concurrently.

In a similar study,[4] when 2,000 mg/L of glucose was used as an alternate substrate along with 1,500 mg/L of 2,4-D. the glucose was degraded at a normal rate, but no degradation of 2,4-D occured.  In this study, as in the previous study, the bacterial population was derived by selective enrichment of municipal sewage sludge.  In this case, however, the degradation was undertaken by a pure culture of bacteria of the genus Pseudomonas, while in the previous case the degradation was by a mixed culture that was predominantly, but not solely Pseudomonas.  This illustrates another point, namely that degradation by mixed cultures is frequently more successful than by pure cultures.  This point was further demonstrated when pure cultures of Pseudomonas and Alcaligenes Eutrophus were mixed.  Degradation of 2,4-D by this mixed culture occurred without the initial 5 day lag period that existed when pure cultures of either of the bacteria were used.  The researchers attribute this to the fact that the enzyme systems in the two microorganisms are not identical, and instead they complement each other.[4]

The superior degradation of HOCs by diversified microbial populations is further demonstrated by a study that compared the degradative capabilities of a municipal mixed liquor with three commercial bacterial populations.  In this study, phenol, 2-chlorophenol, and 2,4-D were added to batch reactors which had been seeded with one of four different bacterial populations.  As shown in

Table 7.4, the rate of degradation of each of the compounds was higher in the reactor that had been seeded by the municipal mixed liquor than it was in any of the reactors that had been seeded with the commercial preparations.  The three commercial preparations, which had supposedly been preconditioned to the compounds, did not achieve high rates of degradation primarily because of their narrow population diversity.  Conversely, the mixed liquor contained a diverse population of microorganisms that could quickly adjust to varying substrates.  The highest degradation rates were achieved when either of the three commercial preparations were mixed with the municipal mixed liquor.[15]

One point that was emphasized in another study[11] is that in order to enhance biodegradation, particularly in soils, physical parameters such as water content, pH, oxygen availability, and bacterial concentration must be optimized.  This particular study involved degradation of pentachlorophenol in soil by Flavobacterium.  One of the most important factors affecting the rate of PCP removal was the water content of the soil.  The greatest amount and rate of PCP removal occured at a water content in the range of 15 to 20 percent.  The research also indicated that the optimum temperature range was 24 to 35°C.  No significant mineralization was observed at 12 or 40°C.  In addition, PCP degradation did not occur when PCP concentration exceeded 500 ppm.

Not all of the microorganisms identified as having the ability to degrade HOCs are bacteria.  Recent investigations of the white rot fungus, Phanerochaete chrysosporium, have demonstrated the ability of this lignin degrading organism to mineralize several toxic compounds including lindane, DDT, 2,3,7,8-TCDD, and 3,4,3'4'-Tetrachlorobiphenyl.[12]  As shown in Table 7.5, the rate of degradation was highest for lindane.  Over 20 percent was fully mineralized in 60 days.  In model degradation studies P. Chrysosporium degraded 50 percent of the DDT initially present within the first 30 days.  Four percent of the initial DDT had been fully mineralized to $CO_2$ while the remainder had either been incorporated in the organism, or was present as an intermediate in the pathway between $CO_2$ and DDT.  These intermediates were identified as being DDD, dicofol, and 4,4'-dichlorobenzo-phenone.  At 30 days, additional glucose was added to the 10 mL culture, and after 18 more days, more than 90 percent of the initial DDT had been degraded.  The researchers feel that this microorganism would be extremely

TABLE 7.4.  ZERO-ORDER RATE CONSTANTS FOR COMMERCIAL AND MIXED LIQUOR BACTERIAL POPULATIONS (ppm/hr)* (values in parentheses are the rate constants divided by MLSS-in ppm/hr/1,000 ppm MLSS)

| | Livingston (mixed liquor) | Hydrobac | BI-CHEM | LLMO | Livingston +Hydrobac | Livingston +BI-CHEM | Livingston +LLMO |
|---|---|---|---|---|---|---|---|
| phenol | 82(34) | 25(230) | 52(37) | 9(8) | 130(57) | 131(57) | 101(44) |
| 2-chlorophenol | 11(5) | 2(2)* | 4(3)** | 0.4(0.4) | 21(9) | .28(12) | 47(20) |
| 2,4-D | 0.6(0.5) | -- | -- | -- | 0.7(0.5) | 0.2(0.2) | 0.4(0.3) |

*Average values
**For runs III&IV

TABLE 7.5.  DEGRADATION OF ORGANOPOLLUTANTS BY P. CHRYSOSPORIUM

| | Initial rate of degradation to $^{14}CO_2$ (pmoles/day) | Radiolabeled substrate evolved as $^{14}CO_2$ (pmoles) | | Percent of Radiolabeled substrates evolved as $^{14}CO_2$ in 60 days |
|---|---|---|---|---|
| | | 30 days | 60 days | |
| Lindane | 11.3 | 190.8 | 267.6 | 21.4 |
| Benzo(a)pyrene | 7.5 | 117.2 | 171.9 | 13.8 |
| DDT | 2.7 | 48.0 | 116.4 | 9.3 |
| TCDD | 1.2 | 27.9 | 49.5 | 4.0 |
| 3,4,3',4'-TCB | 0.7 | 13.8 | 25.1 | 2.0 |
| 2,4,5,2',4',5'-HCB* | 2.4 | 44.2 | 86.0 | 1.7 |

*Substrate concentration was 1.25 nmoles/10 ml for all $^{14}$c-radiolabeled compounds except s,4,5,2',5'-HCB.  Because of its low specific radioactivity a concentration of 5.0 nmoles/10 mL was used for 2,4,5,2',4',5'-HCB.

useful in degrading organopollutants that are physically adsorbed to soils and sediments.   In this state, the pollutants may not be available for uptake and metabolism by microorganisms.   However, P. chrysosporium secretes extracellular fungal enzymes that degrade large, insoluble particles to smaller substituents which may then be internalized and biodegraded.   In addition, P. chrysosporium is highly nonspecific in its degradative capabilities and degradation will occur even under low substrate conditions. Each of these characteristics is desirable for treating contaminated soils and sediments.

## 7.3   COST

Biological treatment has conventionally been used to treat aqueous wastes containing biodegradable organic compounds.   A typical activated sludge plant will consist of an equilization basin, an activated sludge basin with surface aerators, a clarifier with activated sludge recycle, and some type of sludge dewatering unit such as a vacuum filter or centrifuge.   The cost of a system such as this will be a function of the organic content of the wastewater (BOD) and the flow rate.   Capital and operating costs for a 1 mgd and a 5 mgd system are presented in Table 7.6.

The cost of treating wastes containing toxic constituents such as halogenated organic compounds is difficult to estimate due to the lack of experience in such projects.   If specific bacterial populations are required to attain full degradation of the waste, a major portion of the cost may go into research and development as opposed to equipment and construction.   In addition, hazardous wastes are frequently not liquid waste streams, but instead they are contaminated soils or sludges.   In this case, conventional biological processes such as activated sludge or trickling filters cannot be employed unless the contaminant is first leached from the soil.   Otherwise, biological treatment will involve the adaption of indigenous bacteria to the contaminants, or innoculation of the soil with microorganisms that have been acclimated to the contaminant.   In either case, it will be necessary to maintain optimal temperature, water content, pH, oxygen availability and nutrient concentration in the soil to attain maximum biodegradation.

TABLE 7.6.   COSTS FOR BIOLOGICAL TREATMENT[a]

| Item | Facility size | |
| --- | --- | --- |
| | 1 mgd | 5 mgd |
| Direct Capital ($1,000) | $2,753 | $ 7,010 |
| Indirect Capital ($1,000)[b] | 2,552 | 6,433 |
| Total Capital | $5,305 | $13,443 |
| Direct O&M ($1,000/yr) | $   145 | $   529 |
| Indirect O&M ($1,000/yr)[c] | 340 | 878 |
| Total O&M | $   485 | $ 1,407 |
| Closure ($1,000) | $   482 | $ 2,303 |
| Annual Revenue Requirement ($1,000/yr) | $   871 | $ 2,376 |
| Unit Cost ($/lb BOD) | $ 0.37/lb | $  0.21/lb |

[a]Updated from June 1983 to October 1986 using ENR construction cost index.

[b]92% of direct capital costs.

[c]5% of total capital costs plus 10% of total annual cost.

Source:   Reference 16.

## 7.4   OVERALL STATUS

Biological treatment of wastes containing halogenated organic compounds is not a common practice because many of the compounds are not readily degraded by naturally occurring microorganisms.  Instead, this type of waste is more commonly incinerated or treated by carbon adsorption.  However, recent research has led to the development of microbial populations that do have the ability to degrade some of these compounds.  It is possible that, in the near future, these microbes will be used in biological treatment systems to remove halogenated organic compounds.  At the present time, however, biological processes as the sole method of treatment of these wastes is only in a developmental stage.  Biological treatment is still useful, in reducing the overall organic content of a wastestream.  For example, in one case, leachate from a landfill containing toxic industrial wastes was being treated by activated carbon adsorption.  The carbon removed the toxic contaminants, but the carbon usage rate was exceedingly high due to the presence of many nontoxic organic compounds.  Therefore, a biological treatment system was installed to remove the biodegradable organic compounds and reduce the organic load to the carbon adsorbers.  This significantly reduced the carbon exhaustion rate and the overall cost of treatment.[17]  This example demonstrates that although biological treatment may not always be used for the removal of specific halogenated compounds, it is still be useful in removing conventional pollutants.

## REFERENCES

1.  Metcalf and Eddy, Inc. <u>Wastewater Engineering:  Treatment/Disposal/ Reuse</u>. McGraw-Hill, Second Edition. 1979.

2.  Nemerow, Nelson, L. <u>Industrial Water Pollution:  Origins, Characteristics and Treatment</u>. Addison-Wesley. 1978.

3.  Ghosal, D. et al. <u>Microbial Degradation of Halogenated Compounds</u>. Science, 228, 4696, 135-142. 1985.

4.  Roy, D., and S. Mitra. Biodegradation of Chlorophenoxy Herbicides. In:  Incineration and Treatment of Hazardous Wastes:  Proceedings of the Eleventh Annual Research Symposium. EPA/600-9-85-028.

5.  Kobayashi, H., and B.E. Rittman. Microbial Removal of Hazardous Organic Compounds. Environmental Science and Technology, 16(3) 170A-181A. 1982.

6.  Patrasek, A.C., et al. Fate of Toxic Organic Compounds in Wastewater Treatment Plants. J. Water Pollut. Control Fed. 55, 1286-1295. 1983.

7.  Saleh, F.Y., et al. Selected Organic Pesticides, Occurence, Transformation, and Removal from Domestic Wastewater. J. Water Pollut. Control Fed. 52, 19-29. 1980.

8.  Chatterjee, D.K., and A.M. Chakrabarty. Genetic Rearrangements of Plasmids Specifying Total Degradation of Chlorinated Benzoic Acids. Mol. Gen. genetics, 188, 279-285.

9.  Kilbane, et al. Detoxification of 2,4,5-Trichlophenoxyaxetic Acid from Contaminated Soil by <u>Pseudomonas</u> <u>Cepacia</u>. Applied and Environmental Microbiology, 45. May 1983.

10. Kim, C.J., and W.J. Maier. Acclimation and Biodegradation of Chlorinated Organic Compounds in the Presence of Alternate Substrates. J. Water Pollut. Cont. Fed. 58, 157-163. 1986.

11. Martinson, M.M., et al. Microbial Decontamination of Pentachlorphenol in Soils, Surface Waters, and Ground waters. Presented before the Division of Environmental Chemistry, American Chemical Society, New York. April 1986.

12.   Bumpus, J.A., et al.   Biodegradation of Environmental Pollutants by the
      White Rat Fungus Phanerochaete Chrysosporium.   In:   Incineration and
      Treatment of Hazardous Waste:   Proceedings of the Eleventh Annual
      Research Symposium.   EPA 600/9-85-028.

13.   Suflita, J.M., et al.   Dehalogenation:   A Novel Pathway for the Anaerobic
      Biodegradation of Haloaromatic Compounds.   Environ. Sci. Technol. 218,
      1115-1116.   1982.

14.   Bouwer, E.J., and P.C. McCarty.   Removal of Trace Chlorinated Organic
      Compounds by Activated Carbon and Fixed-Film Bacteria.   Environ, Sci.
      Technol.   16, 836-843.   1982.

15.   Lewandowski, G, et al.   Biodegradation of Toxic Chemicals Using
      Commercial Preparations.   Environmental Progress 5, 212-217.   1986.

16.   ICF, Incorporated.   The RCRA Risk--Cost Analysis Model Phase III Report.
      Submitted to the U.S. EPA, Office of SOlid Waste, Economic Analysis
      Branch, March 1, 1984.

17.   Ying, Wei-Chi, et al.   Biological Treatment of a Landfill Leachate in
      Sequencing Batch Reactors.   Environmental Progress, 5, 41-50.   1986.

# 8. Incineration Processes

Incineration is a principal alternative to land disposal for wastes containing halogenated organics. Despite the low heat of combustion and high thermal oxidation stability (TOS) exhibited by many of the halogenated organics, incineration represents an established, if costly, means of achieving effective thermal destruction. A number of incineration technologies, including liquid injection incinerators, rotary kilns, fixed and multiple hearth incinerators, and fluidized-bed incinerators have demonstrated the ability to meet RCRA incineration standards for the destruction of halogenated organics. As noted in the solvent TRD,[1] many industrial boilers and process furnaces are also capable of achieving RCRA destruction and removal efficiency (DRE) standards for halogenated solvents, including chlorinated aromatics. However, pollution control equipment may be required to meet RCRA particulate and HCl emission standards if significant quantities of wastes are introduced as feed.

Incineration facilities permitted to operate by EPA under RCRA are required to achieve at least a three-tiered environmental standard.[2]

1. They must achieve a destruction and removal efficiency (DRE) of 99.99 percent for each principal organic hazardous constituent (POHC);

2. They must achieve a 99 percent HCl scrubbing efficiency or emit less than 4 lb/hr of hydrogen chloride; and

3. They must not emit particulate matter in excess of 0.08 grains/dscf, (0.18 grams/dscm) corrected to 7 percent oxygen.

Other standards which may affect the decision to incinerate halogenated organic wastes include limitations on the generation of CO, $SO_x$, $NO_x$, and toxic air pollutant; e.g., toxic metals. To become permitted, an incineration facility must submit to a full scale evaluation of design and performance. This evaluation includes a trial burn monitored by EPA, demonstrating the ability to perform to expected levels for various wastes. Most large incinerators are equipped with control systems to limit both particulate matter and acid gas emissions. The latter control system is needed to remove halogen acids such as HCl, a product of chlorinated organic compound incineration, in cases where the chlorine content of the feed is sufficient to exceed the 4 lb/hr HCl emission limitation. The need for particulate control systems will depend in large measure on the ash content of the feed.[3] The emission standard of 0.18 g/dscm (63 ng/joule for a 19,000 Btu/lb fuel) is well below the EPA emission factors of 6 and 23 ng/joule for distillate and residual fuel oils,[4] respectively. Thus, particulate standards appear obtainable for many wastes, particularly when blended with conventional fuels. Most incinerators operating in 1981 were not equipped with air pollution control devices, probably because these facilities handled only low ash, nonhalogenated liquid[5] wastes for which control measures were not necessary.

Costs of incineration are higher than most hazardous waste management alternatives because of the large energy input requirements and cost of environmental controls. Costs vary widely depending upon waste characteristics, incinerator design, and various operational considerations. Costs of commercial incineration were found to vary from approximately $0.10/lb to $3.00/lb.

## 8.1 OVERVIEW

### 8.1.1 Incineration Theory

The term incineration normally refers to the destruction of organic wastes by combustion or thermal oxidation. The reaction sequence which takes place in the destruction of a hazardous waste involves a complex series of pyrolysis, free radical, and oxidation reactions. Several intermediate stages occur before a halogenated organic waste feed is oxidized into its final product, depending upon the chemical composition of the waste and the design and operation of the incinerator. Extensive discussion of incinerator theory

may be found in several recent publications, including the proceedings of
several conferences and symposia (References 5 through 13).

Efficient oxidation depends upon three parameters:  temperature, time,
and availability of oxygen.  Combustion temperature affects the rate of
reaction, and thus the time needed to achieve the desired destruction
efficiency.  Residence time or dwell time is determined by the incinerator
size, combustion gas flow rate, turbulence and the rate at which waste is
processed.  Availability of oxygen, or adequate contact of waste and oxygen is
achieved by turbulence and the use of excess air (addition of more air than
the stoichiometric amount needed for combustion).

Oxidation occurs when organic materials are heated in the presence of
oxygen.  A second type of reaction, pyrolysis, is chemical change resulting
from heat alone (in the absence of oxygen).  It normally involves the breaking
of chemical bonds and the formation of new, smaller organic molecules.
Because oxidation causes a more complete destruction of waste, most
incinerators promote this reaction, with pyrolysis occurring incidentally.
The exception occurs with pyrolytic incinerators, in which pyrolyzed products
are generated intentionally.  Usually, the products of pyrolysis are
subsequently oxidized to reclaim their heat content.[14]

The typical chemical reaction scheme for incineration of hazardous wastes
containing chlorinated organics is shown below:

$$C_x H_y Cl_z + M \qquad\qquad\qquad NO_2$$

waste mixture          +          stoichiometric volume
including noncombustible                of oxygen
solid (M)

$$\longrightarrow \quad CO_2 + H_2O + HCl + M$$

most common reaction products

$$+ \; [Cl_2 + C_a H_b Cl_c + \text{others}]$$

plus other species including incompletely combusted species and
noncombustible species

In practical operations, the incinerator is operated to minimize the
formation of the second group of products listed above.[6]  The formation of
organics as byproducts is considered a consequence both of inefficient

operation and the contribution of organics from fuels and reformation.  The formation of $Cl_2$ gas, which is highly toxic, is very undesirable because it is relatively difficult to remove from stack gases by conventional air pollution control systems.  In practice, almost all chlorine is emitted as HCl as a result of auxillary fuel addition.  Auxiliary fuel is often utilized as much for its contribution of hydrogen to suppress $Cl_2$ formation as for its contribution to the overall heat value of the combustion mixture.[6]

Although the kinetics of incineration are highly complex, the overall high rate of reaction allows the general reaction scheme to be described in terms of first order kinetics:[7]

$$\frac{-d(C_A)}{dt} = k\ (C_A)$$

where    $C_A$ = concentration of constituent A in the waste

    $k$ = reaction rate coefficient

    $t$ = time


The reaction rate coefficient is a function of waste and operating characteristics, as indicated below:

$$k = Ae^{-E/RT}$$

where    $A$ = Arrhenius coefficient (characteristic parameter)

    $E$ = activation energy (characteristic parameter)

    $R$ = universal gas constant

    $T$ = absolute temperature


Thus, the most significant factors impacting the destruction of wastes in an incinerator include the temperature, time, turbulence, and concentration of principal constituents.  This observation has been supported by practical experience, although there is no absolute level of these factors that has been correlated with DRE or formation of products of incomplete combustion (PICs).[5]

8.1.2  Applicability of Incineration to Wastes Containing Halogenated Organics

The determination of the applicability of incineration  and specific
incineration technologies to the management of hazardous wastes is based upon
waste physical and chemical characteristics.  The overall ability to
incinerate a specific waste is a function of the relative ease with which the
materials may be fed to the combustion system, the ignitability and
combustibility of the materials during the oxidation/pyrolysis process, the
relative hazardousness of potential combustion byproducts (dictating
post-combustion handling and control), and the general impact on the system
from their incineration.  Several chemical and physical characteristics of the
wastes must be considered in determining whether incineration is technically
and/or economically feasible, what incinerator design will handle a waste most
effectively, and what form of pretreatment should be performed to enhance
performance.  As a result of their halogen content, halogenated organic
hazardous wastes, in general, are considered difficult to incinerate.  Apart
from system/emission problems posed by HCl generation, halogenated organics
have lower heats of combustion and normally higher thermal oxidation stability
(TOS) values than their nonhalogenated counterparts.  The following tentative
guidelines can be drawn for the TOS of halogenated organics:[15]

1.    Chlorinated aromatics such as chlorobenzene, PCB and
      chloronaphthalene appear to have the highest TOS values of any of
      the Appendix VIII constituents.

2.    Halogen substitution seems to increase TOS in the following order:
      F > Cl > Br.

3.    Halogenation does not increase the TOS of all organic compounds.
      Halogenation of straight-chain olefins, for example, may result in a
      TOS decrease.

4.    For aromatic compounds, TOS appears to increase with increasing
      halogen substitution.

Hazardous wastes to be burned in an incinerator, including halogenated
organic hazardous wastes, may be classified into two basic categories relative
to their incinerability, as follows:[9]

1.  Combustible wastes which sustain combustion without the use of
    auxiliary fuels (i.e., heat content above 8500 Btu/lb); and

2.  Noncombustible wastes which will not sustain combustion without the
    use of auxiliary fuels.

All combustible wastes are obviously applicable to incineration, but this may
not be the best disposal option for such substances. Instead, combustible
wastes may be better handled in fuel burning devices such as industrial
boilers specially designed to burn hazardous wastes, which would make more
effective use of the recoverable heat energy from these substances.   The
primary focus of this discussion will be on noncombustible wastes.
Non-combustible wastes exhibit characteristics which limit their
combustibility.  Whether or not these limitations will present a technological
or economic barrier to incineration must be determined.

The primary waste characteristics which determine relative abilities of
wastes to be incinerated include the following:

● Physical form;

● Heat content/heat of combustion;

● Autoignition temperature/thermal stability;

● Moisture content.

These are discussed below in terms of their affect on the incineration process.

Physical Form--
The physical form of a waste is the primary factor in the selection of an
appropriate incineration technology.  Although some technologies, such as
rotary kilns, can handle all physical forms, others such as liquid injection
incineration and fluidized-bed incineration cannot.  For certain wastes,
pretreatment by filtration, size reduction, heating, or blending may be
sufficient to ensure applicability of the last two technologies.

Heat of Combustion--
The heat of combustion of a halogenated organic is the amount of heat
energy produced when the substance is totally oxidized.  Wastes with a higher
heat of combustion usually produce a higher flame temperature when burned

which will, in turn, produce a higher destruction efficiency.  Although the concept of using the heat of combustion to rank relative incinerability has been questioned,[9e,16-18] apparently with some validity, heats of combustion above a minimum value are needed to stabilize flames in conventional liquid burners.[19]

Wastes with a heat content above 8,500 Btu/lb are considered fuels, and can be burned in facilities regulated under RCRA Subpart D.  These wastes can sustain combustion in most furnaces.  Between approximately 2,500 and 8,500 Btu/lb wastes may require auxiliary fuels to sustain combustion.  Below 2,500 Btu/lb, wastes require auxiliary fuels and, in many cases, other forms of pretreatment before incineration.[20]  A good example of low Btu content wastes are those with high moisture contents, which sometimes require dewatering before incineration can be conducted.  High moisture and chloride contents both limit incinerability.  Heat of combustion for several halogenated organics are listed in Table 8.1 in the order of their physical state and chloride content.  The correlation between heat of combustion and chlorine content is pronounced.

Autoignition Temperature/Other Incineration Indicators--

Autoignition Temperature (AIT) as well as several other temperature-based experimental parameters, have been used as indicators of relative ease of incineration.  AIT is defined as the temperature at which a waste will first sustain combustion.  In theory, the lower the AIT of a material, the lower the required combustion temperature and, thus, the easier it will be to incinerate.

Results of field tests at various facilities were compared with the results predicted by heat of combustion, AIT, and several other indices of thermal destruction.  The most useful predictive procedure proved to be a gas phase thermal stability method using an oxygen deficient pyrolytic environment.  Field stability rankings could be predicted from laboratory data in 70 percent of the cases evaluated.[9]  However, additional study will be required to fully assess means of predicting TOS.

TABLE 8.1. HEAT OF COMBUSTION BASED ON PHYSICAL STATE AND CHLORINE CONTENT

| RCRA waste code | Compound name | Molecular formula | Molecular weight | Halogen content (% by weight) | Heat of combustion (Btu/lb) |
|---|---|---|---|---|---|
| Liquid compounds (@25°C) | | | | | |
| U062 | Diallate | $C_{10}H_{17}Cl_2NOS$ | 270.2 | 13 Cl | 10,120 |
| U048 | 2-Chlorophenol | $C_6H_5ClO$ | 128.6 | 28 Cl | 12,400 |
| P028 | Benzyl chloride | $C_7H_7Cl$ | 126.6 | 28 Cl | 11,120 |
| U030 | 1-Bromo-4-phenoxy benzene | $C_{12}H_9B_2O$ | 249 | 32 Br | 10,510 |
| P036 | Dichlorophenyl arsive | $C_6H_5AsCl_2$ | 222.9 | 32 Cl | 4,160 |
| U097 | Dimethyl carbamoyl chloride | $C_3H_6ClNO$ | 107.6 | 33 Cl | 9,140 |
| U042 | 2-Chloroethylvinyl ether | $C_4H_7ClO$ | 106.6 | 33 Cl | 9,340 |
| U041 | Epichlorohydrin | $C_3H_5ClO$ | 92.5 | 38 Cl | 9,340 |
| P027 | 3-Chloropropionitrile | $C_3H_4ClN$ | 89.5 | 40 Cl | 8,100 |
| U024 | Bis(2-chloroethoxy) methane | $C_5H_{10}Cl_2O_2$ | 173.1 | 41 Cl | 8,300 |
| U027 | Bis(2-chloroisopropyl) ether | $C_6H_{12}Cl_2O$ | 171.1 | 42 Cl | 8,870 |
| U046 | Chloromethoxymethane | $C_2H_5ClO$ | 80.5 | 44 Cl | 6,260 |
| P023 | Chloroacetaldehyde | $C_2H_3ClO$ | 78.5 | 45 Cl | 5,260 |
| U006 | Methyl chloride | $C_2H_3Cl)$ | 98.9 | 45 C | 4,990 |
| U025 | Bis(2-chloroethyl) ether | $C_4H_8Cl_2O$ | 143 | 50 Cl | 6,080 |
| U023 | Benzotrichloride | $C_7H_5Cl_3$ | 195.5 | 54 Cl | 7,020 |
| P017 | Bromoacetone | $C_3H_5Br$ | 137 | 58 Br | 4,790 |
| U034 | Trichloroacetaldehyde | $C_2HCl_3O$ | 147.4 | 72 Cl | 1,440 |
| U130 | Hexachlorocyclopendadiene | $C_5Cl_6$ | 272.8 | 78 Cl | 3,780 |
| U128 | Hexachlorobutadiene | $C_4Cl_6$ | 260.8 | 82 Cl | 3,820 |

(continued)

TABLE 8.1 (continued)

| RCRA waste code | Compound name | Molecular formula | Molecular weight | Halogen content (% by weight) | Heat of combustion (Btu/lb) |
|---|---|---|---|---|---|
| U066 | 1,2-Dibromo-3-chloropropane | $C_3H_5Br_2Cl$ | 236.4 | 68 Br 83 total<br>15 Cl | 2,660 |
| U184 | Pentachloroethane | $C_2HCl_5$ | 202.3 | 88 Cl | 954 |
| Solid Compounds (@25°C) | | | | | |
| P026 | o-(1-chlorophenyl) thiourea | $C_7H_7ClN_2$ | 187 | 19 Cl | 9,540 |
| U047 | 2-Chloronaphthalene | $C_{10}H_7Cl$ | 162.6 | 22 Cl | -- |
| U035 | Chlorambucil | $C_{14}H_{19}Cl_2NO_2$ | 304.2 | 23 Cl | 10,670 |
| U039 | p-chloro-m-cresol | $C_7H_7ClO$ | 142.6 | 25 Cl | 9,140 |
| P057 | Fluoroacetamide | $C_2H_4FNO$ | 77 | 25 F | 5,830 |
| U158 | 4,4'-Methylene-bis-2-chloroaniline | $C_{13}H_{12}Cl_2N$ | 267.2 | 27 Cl | 8,710 |
| P024 | p-chloroaniline | $C_6H_6ClN$ | 127.6 | 28 Cl | 11,052 |
| U192 | Pronamide | $C_{12}H_{11}Cl_2NO$ | 256.1 | 28 Cl | 10,300 |
| U237 | Uracil mustard | $C_8H_{11}Cl_2N_3O_2$ | 252.1 | 28 Cl | 7,200 |
| U073 | 3,3'-dichlorobenzidine | $C_{12}H_{10}Cl_2N_2$ | 253.1 | 28 Cl | 10,300 |
| D014, U247 | Methoxychlor | $C_{16}H_{15}Cl_3O_2$ | 345.7 | 31 Cl | 10,060 |
| D016, P035 | 2,4-D | $C_8H_6Cl_2O_3$ | 221 | 32 Cl | 6,520 |
| D017, U233 | 2,4,5-TP | $C_9H_7Cl_3O_3$ | 269.5 | 40 Cl | 10,040 |
| U232 | 2,4,5-T | $C_8H_6Cl_3O_3$ | 255.5 | 42 Cl | 5,160 |
| U060 | DDD | $C_{14}H_{10}Cl_4$ | 320.1 | 44 Cl | 9,250 |
| U082 | 2,6-Dichlorophenol | $C_6H_4Cl_2O$ | 163 | 44 Cl | 6,860 |

(continued)

TABLE 8.1 (continued)

| RCRA waste code | Compound name | Molecular formula | Molecular weight | Halogen content (% by weight) | Heat of combustion (Btu/lb) |
|---|---|---|---|---|---|
| U081 | 2,4-Dichlorophenol | $C_6H_4Cl_2O$ | 163 | 44 Cl | 6,860 |
| U061 | DDT | $C_{14}H_9Cl_5$ | 354.5 | 50 Cl | 8,120 |
| U132 | Hexachlorophene | $C_{13}H_6Cl_6O_2$ | 406.9 | 52 Cl | 6,880 |
| P050 | Endosulfan | $C_9H_6Cl_6O_3S$ | 406.9 | 52 Cl | 4,190 |
| U231 | 2,4,6-Trichlorophenol | $C_6H_3Cl_3O$ | 197.5 | 54 Cl | 5,180 |
| U230 | 2,4,5-Trichlorophenol | $C_6H_3Cl_3O$ | 197.5 | 54 Cl | 5,180 |
| P037 | Dieldrin | $C_{12}H_8Cl_6O$ | 380.9 | 56 Cl | 6,230 |
| D012, PU51 | Endrin | $C_{12}H_8Cl_6O$ | 380.9 | 56 Cl | 6,230 |
| P004 | Aldrin | $C_{12}H_8Cl_6$ | 365 | 58 Cl | 6,750 |
| U185 | Pentachloronitrobenzene | $C_6Cl_5NO_2$ | 295.4 | 60 Cl | 2,920 |
| U212 | 2,3,4,5-Tetrachlorophenol | $C_6H_2Cl_4O$ | 231.9 | 61 Cl | 4,010 |
| U207 | 1,2,4,5-Tetrachlorobenzene | $C_6H_2Cl_4$ | 215.9 | 66 Cl | 4,700 |
| P059 | Heptachlor | $C_{10}H_5Cl_7$ | 373.4 | 67 Cl | 5,330 |
| P090, U242 | Pentachlorophenol | $C_6HCl_5O$ | 266.4 | 67 Cl | 3,760 |
| U036 | Chlordane | $C_{10}H_6Cl_8$ | 409.8 | 69 Cl | 4,880 |
| D015, P123, U224 | Toxaphene | $C_{10}H_{10}Cl_8$ | 413.8 | 69 Cl | 4,500 |
| U183 | Pentachlorobenzene | $C_6HCl_5$ | 250.3 | 71 Cl | 3,690 |
| U142 | Kepone | $C_{10}Cl_{10}O$ | 490.7 | 72 Cl | 3,870 |
| U127 | Hexachlorobenzene | $C_6Cl_6$ | 284.8 | 75 Cl | 3,220 |

Moisture Content--

Moisture reduces the incinerability of a waste.  In the combustion
process, water will absorb heat energy and vaporize, but will not oxidize or
pyrolyze.  This will tend to reduce the heat energy available to assist
combustion.  Water may also absorb combustion intermediates and waste
components and thus limit their availability for combustion.

The requirement to drive off moisture increases the overall stress on
incineration systems and their operating costs.  Certain incinerator designs,
including fixed hearth furnaces and rotary kilns, are not equipped to handle
high moisture content wastes.  Moisture content may be reduced by dewatering
pretreatments, but these tend to be expensive.  The most common way of dealing
with high moisture content wastes is to blend them with solid wastes or other
high heat content materials.[1]


Ash Content (Solids/Metals/Thermally Inert Materials Content)--

Ash content is a major factor in determining the type of incinerator, air
pollution control equipment, and ash recovery system required, and is often
directly used in incineration pricing structures.  Rotary kiln and hearth type
incinerators are, in general, more applicable to wastes with higher ash
content, while liquid injection and fluidized-bed incinerators are less
applicable.  Fluidized-bed incinerators have a particular limitation to wastes
containing sodium salts which tend to fuse within the bed, leading to process
failure.[21]  The costs of incinerating wastes with higher ash content are
higher primarily due to increased air pollution control and ash recovery
costs.  Many incineration facilities appear to use ash content as a factor in
determining the price of incinerating wastes.  One facility contacted, for
example, indicated that they charged an extra 1 cent per pound per each
percent of ash content.[22]

It is conceivable that blends of halogenated organic containing wastes
with fuel oils may meet RCRA particulate emission standards without the need
for particulate control devices.  As noted previously, the RCRA particulate
emission standard for incinerators, 0.18 g/dscm, represents an emission level
that is above the EPA AP-42 emission factor levels for distillate and residual
fuel oils.  If a hazardous waste fuel blend with a heating value not
noticeably different from fuel oils, is burned, the RCRA particulate emission

factor will be of the order of 50 ng/J (versus 6 and 37 ng/J for the
distillate and residual fuel oils, respectively).  The ash content
corresponding to the 0.18 g/dscm RCRA standard assuming a similar blend and
stoichiometric emission of ash as particulate is 0.28 percent at 7 percent
oxygen.

Chloride Content--

The chloride content of a waste is directly related to the formation and
emission of HCl.  The RCRA emission standard for HCl of 4 lb/hr will be
exceeded by most incinerators burning wastes containing more than nominal
levels of chlorine.  For example, the emission standard will be exceeded by a
$3.8 \times 10^6$ Btu/lb input unit burning a 19,000 Btu/lb waste with a chlorine
content of 2 percent.  Thus, most incinerators burning chlorinated organic
wastes will require a high level of blending or installation of acid gas
scrubbers to meet RCRA standards.

The chloride concentration is also related to the overall corrosivity of
combustion byproducts.  As a result, most incinerators establish a limiting
chloride concentration for their systems.  It is common for this limit to fall
under 3 percent by weight.  Most incinerators also appear to have established
surcharges for chloride content.  One facility stated that an additional
charge of 0.2 cents per pound per each 1 percent of chloride was common
practice in the industry.[22]

Viscosity--

Liquid injection incinerators require wastes feedstocks which are
pumpable and atomizable.  Common limits for pumping and low pressure atomizing
of fuel oils are 10,000 ssu or less and 100 ssu or less, respectively.  These
same limits can be applied to waste incineration and are commonly met by
heating of the waste, blending with fuel oil, or both.

Viscosity limits also affect the compatibility of wastes with other
incinerator types.  Fluidized-bed incinerators also require that liquid wastes
be pumpable in order to achieve effective dispersion in the bed.  Conversely,
low viscosity wastes may pass through the heating zone of a multiple hearth
incinerator too rapidly for effective destruction.

8.1.3   Strategy for Assessment of Incinerability

    Incineration is a potential option for the disposal of halogenated organic-containing wastes.  An approach to assessing incineration as an option and identifying the best incineration technology for the specific waste of concern is provided in Reference 20.  It involves the following steps:

1.    Determine whether the waste can be physically introduced to the combustion zone as is, or if pretreatment is required. . This determination is based upon physical form and viscosity.  For example, if the waste is a liquid with low viscosity, it can be atomized and, thus, may best be incinerated in a liquid injection system.

2.    Determine the overall physical effect of the waste on the incineration system.  This consideration is primarily based upon the physical form, solids content, and corrosivity of a waste.  These factors may be such that the incineration of the waste will rapidly lead to process failures due to debilitation of equipment. Refractory linings, for example, are highly susceptible to chipping and cracking by large solid particles.

3.    Determine if auxiliary fuel should be used.  This determination is commonly made solely on the basis of the heat of combustion.  For example, wastes with a heat of combustion below 2,500 Btu/lb are almost always mixed with a fuel or blended with a high Btu waste.

4.    Determine the temperature and residence time requirement for effective combustion.  This determination is largely based on characteristics such as moisture content.  Many incinerator designs operate with a specified residence time or temperature range.

5.    Determine the disposal or handling method required for combustion byproducts other than gaseous products.  This consideration is largely based upon the solid/metal/thermally inert material concentration of the waste.  Wastes with a high ash concentration, for example, may require a continuous ash removal system.

6.    Determine if air pollution controls are required.  This consideration is largely based upon the chloride and ash content of a waste.  Most wastes containing more than a very small amount of chloride will require a scrubber to remove acid gases.  Need can be calculated assuming emissions are directly related to input.

7.    Determine if relevant environmental standards can be met.  This determination, again, is based upon chloride and/or ash concentration.  Most incinerators operate with a chloride

concentration limit.  If the chloride content is too high, the air pollution control system will not be adequate to limit emissions to the applicable standard.

8.    Determine the relative costs of the various incinerator options.  It is important to note that the technology with the lowest base cost may not be the most cost-effective alternative, should one of the factors listed above come significantly into play.

Although incineration is a potential option for disposal of halogenated organics, the wide spectrum of properties found in these compounds of concern require that each waste stream be carefully characterized and appraised. Generally, many of the aromatic, highly halogenated compounds are not ideal candidates for incineration for the following reasons:

● Many are solid compounds which would require special handling or blending if they are to be used in liquid injection incinerators;

● The high chloride content will limit commercial facility availability and normally require air pollution control systems;

● Heating values will be low and thermal oxidation stabilities will be high, generally increasing the difficulty of combustion and the formation of products of incomplete (PIC) combustion; and

● High levels of blending will almost invariably be required to ensure destruction and protect equipment, thus increasing the size of the facility and its cost.

Factors which favor incineration as an option include the resistance of these halogenated compounds to destruction by other options, including biological treatment.

## 8.2  PROCESS DESCRIPTION

There are numerous incineration system designs available to handle the wide variation of chemical and physical characteristics found in hazardous wastes.  Hazardous waste incineration technologies range from those with widespread commercial application and many years of proven effective performance, to those currently in development.  As many as 67 companies may be involved in the design and development of hazardous waste incinerators,[23] with more expected as limitations on land disposal of hazardous wastes increase.  Several incineration technologies have been demonstrated to be effective for a wide range of hazardous wastes.  They comprise about 80 percent (by number) of the U.S. market.[11a,23]  They include:

- Liquid injection incinerators

- Rotary kilns

- Fluidized-bed incinerators

- Fixed hearth incinerators, particularly the starved air or pyrolysis type units, and

- Multiple hearth incinerators.

The first two (and the fixed hearth units) are the most widely used for the disposal of hazardous wastes.  A description of the first three types of units listed above will be provided here, following a brief discussion of basic components common to all incinerators.  The hearth type incinerators, particularly the fixed hearth unit, are also used extensively, but data on their ability to handle hazardous wastes have not been widely published in the literature.  Discussions of the design and operation of these systems can be found in the literature.

## 8.2.1  Basic Components of Incineration Systems

All incineration systems are designed in consideration of the four basic elements of combustion:  temperature, time, turbulence, and concentration. Temperature is the most important element of an incineration system.[24]  The heat requirements govern the method by which heat energy is supplied and sustained within the combustion chamber, and governs many of the pretreatment operations conducted.  Residence time requirements impact the size of the combustion chamber, as the volume of the combustion zone must be sufficient to allow for completion of thermal destruction.  Turbulence is strictly a function of incinerator design.  Elements such as baffles, rotation, or changes in direction within the combustion chamber increase turbulence (and, therefore, enhance mixing of wastes and oxygen to allow for more effective performance). Concentration governs the oxygen input requirement, as sufficient air must be supplied to insure complete combustion of hazardous constituents.

There are essentially five component parts common to any incineration facility, as shown in Figure 8.1 and discussed below.

1.  Material Storage and Preparation--Waste materials are received, analyzed, stored and prepared for input into the incinerator. In this initial step of the incineration process, the waste characteristics which may affect the performance of the incinerator are identified. If necessary, pretreatment operations are conducted to mitigate these characteristics. In some cases, wastes are rejected for incineration when pretreatment will not render them "incinerable".

    Common methods of pretreatment include preheating, chemical neutralization, filtration/sedimentation of suspended solids and water, and distillation.

2.  Waste Feed Mechanism--The waste feed mechanism is the means by which waste materials are input into the combustion chamber of an incinerator. Feed mechanisms also control the volume of waste present in the chamber at any moment, and thus control waste residence time. Feed mechanisms also play a key role in creating surface area, to increase combustion rate, and in developing turbulence within the combustion system. Dispersion of wastes is particularly critical in liquid injection and fluidized-bed incinerators.

3.  Combustion System--Combustion systems perform three functions: 1) heating of waste materials to vaporize and pyrolyze them; 2) mixing of wastes with combustion air; and 3) oxidation and subsequent formation and separation of combustion products.

4.  Heat Recovery--Heat recovery systems are often employed with incineration of hazardous wastes in order to achieve greater cost effectiveness. Generally, heat recovery is accomplished by either standard heat exchange equipment or waste heat boilers which burn the waste byproducts. There are generally two limitations in heat recovery. First, the cost benefits of heat recovery must justify the expense of the heat recovery system, including design, installation, and maintenance. Second, heat recovery systems should not be used if they lead to a more difficult waste management problem; i.e., form new pollutants of concern, or require difficult maintenance such as cleanup of waste byproducts plugging the heat recovery system.

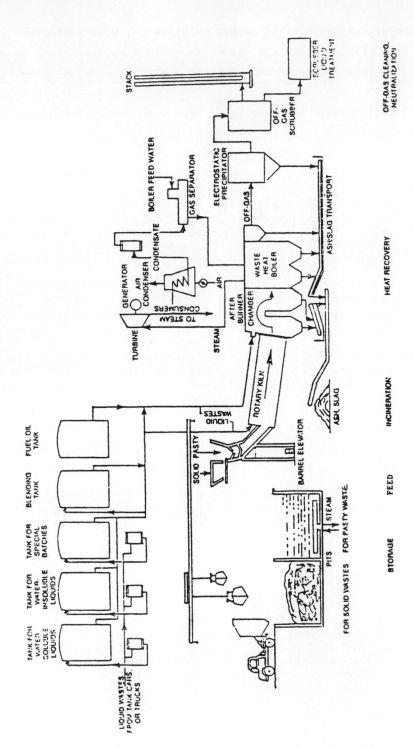

**Figure 8.1.  Flow sheet of an incineration plant for hazardous wastes.**

Source:    Babcock Krauss-Maffei Industrieanlagen GMBH
(Revised by P. Adie).25

5.  Solid and Liquid Waste Control--Air pollution control devices are
    required if the combustion process produces air pollutants at levels
    exceeding applicable emissions standards.  Most commonly, the
    primary pollutants of concern generated by incineration of hazardous
    wastes are particulate matter and hydrochloric acid (HCl) vapor.
    Air pollution control is often, but not always, used at hazardous
    waste incinerators.  Incineration processes produce solid and liquid
    waste streams which must be managed.  These streams are usually not
    hazardous themselves.  Ash produced in combustion is collected
    either continuously (e.g., a screw conveyor built into the bottom of
    the combustion system), or periodically by manually cleaning the
    combustion chamber.  Sludges can be produced by air pollution
    control or heat recovery systems, and are removed periodically from
    the process systems.  Liquid wastes are produced by air pollution
    control or heat recovery systems, and are removed periodically from
    the process systems.  Liquid wastes produced by air pollution
    scrubbers or quench towers are continuously treated.  In most cases,
    ash may be disposed of in a landfill, as may dried sludges.  Liquid
    wastes may be subject to wastewater treatments before discharge.

## 8.2.2  Liquid Injection Incinerators

Liquid Injection (LI) incinerators are the most widely used hazardous
waste incineration systems in the United States, accounting for 64 percent of
the total number of waste incinerators currently in use.[23]  LI systems may
be used to incinerate virtually any liquid hazardous waste, due to their very
basic design and high temperature and residence time capabilities.  Liquid
injection incinerators generally represent the most effective system available
for hazardous wastes that can be processed to produce a pumpable and
atomizable feedstock, from a both a technical (i.e., destruction efficiency)
and economic perspective.

Liquid injection incinerator systems typically employ a basic, fixed
hearth combustion chamber.  Pretreatment systems to blend wastes and fuels, to
remove solids and free water, and to lower viscosity through heating, are
often used in conjunction with liquid injection incinerators.  Ash recovery
systems may not be required, at least on a continuous basis, because many
liquid hazardous wastes fired in an LI system contain low volumes of ash or
suspended solids.[8]

The liquid waste feed system is the key element of the LI process. Liquid injection incinerators operate as "suspension burners", whose combustion efficiency (and hence destruction efficiency for constituents of hazardous wastes) is dependent upon the extent to which the feed mechanism can disperse the liquid waste within the combustion chamber and provide sufficient area for contacting waste with combustion air.  There are two atomizer designs commonly employed in LI systems, denoted as fluid systems and mechanical systems.  Typical characteristics of several atomizer designs are described in detail in Reference 26.

Once liquid wastes enter into the liquid injection incinerator and are ignited at the burner, efficient combustion is achieved by proper mixing of combustion air and waste to create a turbulent flow of waste throughout the combustion chamber.  Combustion temperature capabilities of the systems can be very high, reaching over 3000°F in many cases.  Table 8.2 summarizes operating parameters for typical hazardous waste liquid injection systems.

Applicability of hazardous wastes to liquid injection incinerators is generally limited by the extent to which they may be atomized, and the physical effect they may have on the incinerator equipment (most notably, on the atomizer).  The primary restrictive waste characteristics of interest are the liquid viscosity, solids content, and corrosivity.  Wastes with low heat value may also be restricted from burning in a liquid injection incinerator.

In some cases, the applicability of an LI incinerator may be extended by the use of multiple injection systems.  In this way, an injector may be fitted to more specific ranges in waste characteristics allowing a broader range of overall usage without requiring pretreatment.  As discussed earlier, certain atomization device designs are better suited to more viscous or high suspended solids containing wastes than others.  In addition, the use of multiple injection points may allow for coincineration of incompatible wastes.

8.2.3  Rotary Kiln Incinerators

Rotary kiln (RK) incinerators have found widespread application in the U.S. for management of hazardous wastes, both at chemical manufacturing and at hazardous waste facilities.  MITRE estimated that rotary kilns comprised 12.3 percent of the total number of hazardous waste incinerators in

TABLE 8.2.    OPERATING PARAMETERS OF HAZARDOUS WASTE LIQUID INJECTION
INCINERATORS

| | |
|---|---|
| Form of Waste Feed: | Liquid wastes only |
| | Limiting liquid viscosity for atomization is typically 16,000 centistokes |
| | Limiting solids content may be as high as 10% by weight undissolved solids |
| | Limiting solid particle size may be as high as 1/8 inch diameter |
| Heat Input Capacity Range: | $5 - 150 \times 10^6$ Btu/hr |
| Heat Release: | 25,000 Btu/hr.ft$^3$ (typical) |
| | 1,000,000 Btu/hr.ft$^3$ (maximum) |
| Operating Temperature Range: | 1200 - 3000°F |
| Residence Time Range: | 0.5 - 2.0 secs |
| Excess Air: | 20% (typical) |
| | For nitrogen-containing wastes, excess air requirements may be 65-95% |
| Pressure: | 0.5 - 4 in $H_2O$ (typical) |

Source:  MITRE, 1982 (Reference 23).

operation.[23]   Rotary kiln systems are considered the most versatile of the
established incinerator technologies.   Liquid, solid, and slurried hazardous
wastes may all be burned in rotary kilns, without extensive adaptation of the
design for specific waste types.

Rotary kiln systems employ a fairly basic design concept.   As depicted in
Figure 8.2, the typical rotary kiln system involves two-stage combustion of
waste materials with primary combustion occurring in the rotary kiln followed
by secondary combustion of gaseous byproducts.   Heat recovery, ash recovery,
and air pollution control devices are usually included in the overall system.
Combustion byproducts are most often scrubbed for both particulate matter and
acidic byproducts; e.g., HCl.   Heat recovery is employed in the majority
(~70 percent, according to recent estimates) of cases.[23]

Pretreatment of hazardous wastes is not often required for incineration
in a rotary kiln, because of the great versatility of the system.   The most
common preparatory operations conducted at rotary kiln incinerators include
size reduction, mixing of liquid wastes with solid wastes, and chemical
neutralization.   Wastes with an average heating value of 4,500 Btu/lb are
reported adequate to sustain combustion at kiln temperatures between 1600 and
1800°F.   In those cases where auxiliary fuel is required, No. 2 fuel oil is
used most often.   Size reduction of solid wastes, via crushing and grinding
operations, is a common preparatory operation.   This is often done both to
preserve the life span of the kiln refractory lining and to increase the
combustion efficiency of the system.   Mixing of liquid wastes with solid
wastes helps to increase the liquid waste residence time and thus enhance
destruction efficiency.   Highly corrosive wastes are often neutralized by
chemical treatment before being fed to the rotary kiln.   This helps preserve
the working life of the kiln refractory.[23]

Waste materials, following pretreatment, are fed to the elevated end of
the rotary kiln.   Waste feed mechanisms employed are typically simple hoppers
which feed a regulated amount of material to the kiln.   Waste materials flow
through the rotary kiln as a consequence of the rotation and the angle of
inclination.   The kiln is often designed with baffles, which serve to regulate
the flow rate through the unit, generally resulting in increased residence
times.   The rotation of the kiln serves to enhance the mixing of waste with
combustion air and provides continuously renewed contact between waste

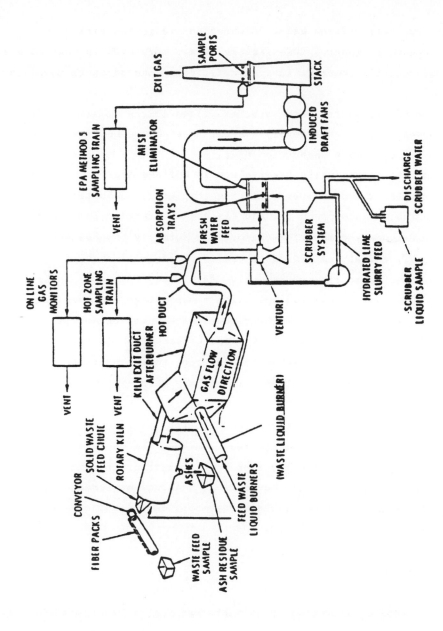

Figure 8.2.  Rotary kiln incinerator with liquid injection capability.

material and the hot walls of the kiln. Combustion air is fed either concurrently or countercurrently. One feature of a rotary kiln is that it may be operated under substoichiometric (oxygen deficient) conditions to pyrolyze certain wastes.

As combustion of the waste progresses, ash flows to the bottom of the unit and is conveyed to the ash recovery system. Gaseous combustion products are exhausted to the secondary combustion unit.

Secondary combustion generally takes place in a fixed hearth type unit, where gaseous products of combustion, including completely combusted waste components, combustible waste products, and fly ash are fired. The gaseous products from the secondary combustion chamber are normally then passed through heat recovery and air pollution control systems, while ash is collected and transported to the ash recovery facility.

Most rotary kiln systems are equipped with a multistage scrubber system to control particulate matter and acid byproducts of combustion. Heat recovery systems are often used not only for the conservation of energy, but also to reduce temperatures to allowable levels prior to introduction to the scrubbers. Typical operating parameters for a rotary kiln system are shown in Table 8.3.

Rotary kilns are generally large systems, and thus require a large capital expenditure. Due to their energy requirements, the operating costs associated with rotary kiln systems may also be higher than other incinerator systems. Their versatility may lead, however, to benefits measurable in overall reduced costs for hazardous waste management; cost considerations are further discussed later in this section.

### 8.2.4  Fluidized-Bed Incinerators

Fluidized-bed (FB) incineration systems represent a new incineration technology which has not yet made a significant commercial impact in the established incinerator market. Although fluidized-bed processing units were developed nearly 50 years ago and have found extensive application both in chemical processing and, more recently, in sewage sludge incineration, the development of FB systems capable of destroying wastes containing hazardous components is still in its early stages. As indicated in MITRE's 1982

TABLE 8.3.   OPERATING PARAMETERS FOR ROTARY KILNS

| | |
|---|---|
| Form of Wastes Fed | Liquid, solid, slurry.  Virtually any waste may be fired to a Rotary Kiln. |
| Thermal Capacity | $1 - 150 \times 10^6$ Btu/hr (Rotary Kiln) 20,000 Btu/hr (secondary combuster) |
| **Typical Overall System Flowrate** | |
| gas flow | 47,000 acfm @ 2200°F |
| pressure drop | 10 - 25 in $H_2O$ |
| solid feed rate | 10,000 lbs/hr |
| liquid feed rate | 3,000 lbs/hr |
| **Combustion Temperature** | |
| 1st chamber (Rotary Kiln) | 500 - 2300°F |
| secondary chamber | 1600 - 2800°F |
| **Residence Time** | |
| gases | 0.5 - 3.0 secs |
| solids | Highly variable, depending on viscosity, angle of inclination, rotation of kiln |
| Rotational Speed | 12 revolutions/hr (typical) |
| Length-to-Diameter | 2:10 (typical) |
| Excess Air | 60 - 150% |
| Refractory Life | 24 - 30 months |

Source:  MITRE, 1981 (Reference 23).

survey,[23] only nine fluidized-bed units, representing 2.6 percent of the total number of hazardous waste systems in operation, had been put into actual service at hazardous waste processing facilities.  The basic fluidized-bed system is depicted in Figure 8.3.  Fluidized-beds are always oriented vertically.  Feed and air flow are balanced to achieve fluidization in the bed.  The fluidized-bed promotes turbulence and serves as an excellent heat transfer medium, thus assisting combustion.  As will be discussed later, the fluidized-bed material can be chosen to react directly with combustion production such as HCl, thus minimizing subsequent air pollution control requirements.

Operating parameters for fluidized-bed incineration are shown in Table 8.4.  Operating temperatures are lower than those found in other types of incinerators.  However, the long residence times and the excellent distribution of thermal energy within the bed are sufficient to provide excellent destruction efficiency of organic solvents.

The usage of a fluidized-bed incinerator may be limited by certain chemical characteristics of a hazardous waste.  In general these restrictive waste characteristics are those properties which may affect the fluidity of the bed itself.  The key to the effectiveness of an FB incinerator is the ability of the bed to display certain liquid-like physical properties.  Those wastes with characteristics which lead to either an increase or decrease in bed particle mobility are not suitable for FB incineration.  The primary waste characteristics identified as potentially restrictive include sodium content, corrosivity, moisture content, and fusible ash content.

Sodium content has been identified as the most significant property of concern, in determining the applicability of fluidized-bed incinerators to the treatment of a facility's hazardous wastes.  Certain sodium salts, most notably sulfates and nitrates, may form eutectic complexes with other inorganic salts present in the bed which serve to bind bed particles together and thus destroy the fluidity of the bed.[27]

TABLE 8.4.    OPERATING PARAMETERS OF FLUIDIZED BED INCINERATORS

| | |
|---|---|
| Feed Materials | Granular solids, sludges, slurries are best; can handle liquids, bulk solids as well |
| Capacity | $2 - 200 \times 10^6$ Btu/hr heat input |
| Operating Temperature | $1600 - 1850°F$ in combustion zone |
| Residence Time gaseous | 5-10 secs. |
| solids | no limit |
| Pressure Drop | 90% of height of fluidized bed (in $H_2O$) |
| Excess $O_2$ | 30 - 50% |
| Air Flow Rate | 2.5 - 8.0 ft/sec |
| Typical Bed Thickness | 6 - 8 ft |
| Preheat Requirements | 4000 Btu/lb $H_2O$ for cold windbox |
| Air Pollution Control | Acid scrubber Particulate scrubber Quench tower |
| Startup and Shutdown | Rapid startups and shutdowns possible; continuous feed not necessary |
| Bed Particle Size | 20 - 80 mesh inert |

Source:    Reference 23.

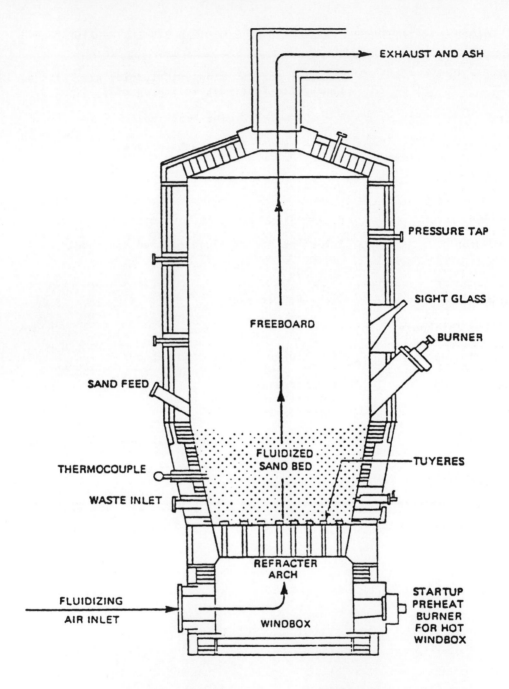

Figure 8.3. Cross-section of a fluidized-bed furnace.

Source: Reference 25.

Highly corrosive wastes pose a different threat to the integrity of the fluidized-bed. The fluidity of the fluidized-bed is dependent upon maintenance of a certain bed particle density and size distribution. Thus, reactions which alter these properties are detrimental to the effective operation of the bed. Corrosion of the bed may therefore lead to a loss of fluidization and result in significantly lower destruction efficiencies than are typically achieved by this type of incineration system.

Wastes with very high moisture content may reduce the overall effectiveness of the fluidized-bed system. Wastes containing more than 75 percent moisture, by weight, may require temperatures or residence times which are not practical for an FB system.[23] Pretreatment of wastes to reduce high moisture content is highly recommended for fluidized-bed incineration. Numerous standard dewatering techniques may be employed, including fractionation, filtration, and settling.

The consequences of a high concentration of fusible solid byproducts of waste combustion are very much the same for fluidized-bed incineration as those associated with the formation of inorganic salt eutectic mixtures described earlier. These materials may impair the fluidity of the bed by binding the granular solids into large, nonfluid solids.

## 8.3  PERFORMANCE OF HAZARDOUS WASTE INCINERATORS IN THE DESTRUCTION OF HALOGENATED ORGANIC WASTES

Most of the available incinerator performance data detail the destruction of low molecular weight halogenated organic wastes, which generally may be categorized as solvent wastes. These studies are discussed in detail in Reference 1, as well as in many sources available in the open literature. More recently, however, there has been some study of the applicability and performance of incineration technologies to the full range of halogenated organic wastes.[28-31]

The relative scarcity of data for nonsolvent halogenated waste is not surprising since this wastes category accounts for less than 10 percent of the total (solvent and nonsolvent) halogenated organic waste generated in the United States. Available nonsolvent halogenated organic waste incineration data may be classified in two areas: pesticide waste; and highly chlorinated halogens of high thermal stability; e.g., halogenated aromatic compounds such as hexachlorobenzene. Much of the data were generated during studies of the destruction of PCBs in incinerators. Although PCB wastes are not considered in this document as part of the nonsolvent halogenated organic waste category, their destruction characteristics in incinerators may be quite comparable to those wastes that do fall into this category. In addition, studies of PCB destruction have often tested compounds such as hexachlorobenzene as "PCB surrogates". The incinerability of certain halogenated organic wastes has also been subject to study because of their appearance in certain common industrial wastes streams. As a class they are regarded as thermally stable, and they are suspected of being high temperature precursors of PICs.[28]

Waste destruction efficiencies for a variety of halogenated pesticides, PCBs, and nonsolvent halogenated organics, determined in several full scale and pilot scale programs, were summarized in Reference 14. The data, shown in Table 8.5, demonstrate that 99.99 percent DREs can be achieved for all wastes in several types of incinerators.

Pesticide destruction levels, shown in Table 8.5, are relatively high, even for highly chlorinated compounds such as toxaphene and kepone. The data are similar to an extensive tabulation of destruction efficiencies measured for several chlorinated pesticides incinerated in a variety of pilot and commercial scale incinerators.[40] Halogenated organic pesticides identified in Reference 40 include aldrin; chlordane; dieldrin; 2,4-D; 2,4,5-T; herbicide orange, kepone, and toxaphene. These pesticides exhibited DREs in excess of 99.99 percent in all but 3 of 80 test data points reported as a result of several EPA, Canadian Government and industry-sponsored programs.

EPA's Combustion Research Facility (CRF) at Pine Bluff, Arkansas has been involved in studies of the destruction of hexachlorobenzene (HCB) and 1,2,4-trichlorobenzene (1,2,4-TCB) in a $1.8 \times 10^2$ Btu/hr rotary kiln using propane as a primary fuel. These two compounds were tested because they were recognized as thermally stable compounds suspected as being precursors of PICs. The CRF facility consistently produced DRE values above 99.99 percent

TABLE 8.5.    WASTE DESTRUCTION EFFICIENCIES ACHIEVABLE BY INCINERATION.

| Waste type | Incinerator type | Major operating parameters | | Waste destruction efficiency (%) | Reference and comments |
|---|---|---|---|---|---|
| | | Temperature (°C) | Residence time (sec) | | |
| Phenoxy herbicide | Catalytic | 480 | Not stated | "complete destruction" | Paulson (Reference 32). Detection limit used to measure "complete destruction" was not specified. |
| Kepone | Liquid injection (pilot scale) | 1090 | 2 | 99.99991 | Carnes (Reference 33). This test was conducted with the afterburner portion only of the rotary kiln system below. |
| Kepone, sewage sludge | Rotary kiln (pilot scale) | Kiln: 500 Afterburner: 1090 | 2 | 99.9999 | Carnes (Reference 33). |
| PCBs (1.7% in fuel oil) | Not specified | Not specified | 3 | 99.992 to 99.995 | Elliot (Reference 34). |
| DDT (20% emulsion in fuel oil) | Not specified | 870 to 980 | 3 | >99.9999 | Elliot (Reference 34). |
| PCBs (375 g/l of Aroclor 1242 in oil) | Rotary cement kiln | Not specified | Not specified | 99.99998 | Ackerman (Reference 35). Citing work done by the Swedish Water Air Pollution Research Institute. |
| PCBs (50 ppm in sewage sludge) | Multiple hearth | Maximum: 790 Gas exit temp.: 615 | 0.1 | 91.7 to 97.1 | Ackerman (Reference 35). |
| Vinyl chloride | Laboratory test equipment: vapor phase | 760 | 0.5 | 99.9 | Lee (Reference 36). Idealized incinerator (plug flow, isothermal). |
| Polyvinyl chloride • Polyvinyl chloride, vinyl chloride monomer • Sludge (72% water) | Rotary kiln | Kiln: 870 Afterburner: 980 to 1090 | 2 to 3 | >99.996 | TRW (Reference 37), pp. 5, 22, 23. |
| Nitrochlorobenzene waste • Nitrochlorobenzene • Liquid; crystalline, particulate | Liquid injection | 1310 to 1330 | 2.3 | 99.999 | TRW (Reference 37), pp. 5, 22, 23. |
| PCB capacitors (hammermilled) • PCBs, paper and plastics, ash • Solid waste | Rotary kiln | Kiln: 1250 Afterburner: 1330 | 3.2 | 99.999 | TRW (Reference 37), pp. 5, 22, 23. |

(continued)

## TABLE 8.5 (continued)

| Waste type | Incinerator type | Major operating parameters | | Waste destruction efficiency (%) | Reference and comments |
|---|---|---|---|---|---|
| | | Temperature (°C) | Residence time (sec) | | |
| 25% DDT (Emulsifiable concentrate) | Pilot scale, type not clearly specified | 1000 | 2 | 99.992[a] (vapor only) | MRI (Reference 38), pp. 3, 4 (source for next 5 entries). Range: 99.97 to 99.998% (includes residue).[b] |
| 10% DDT dust | Pilot scale, type not clearly specified | 1000 | 2 | 99.995[a] (vapor only) | Range: 99.99 to 99.997% (includes residue).[b] |
| 41% Aldrin (emulsifiable concentrate) | Pilot scale, type not clearly specified | 1000 | 2 | 99.992[a] (vapor only) | Range: 99.995 to 99.9995% (includes residue).[b] |
| 19% Aldrin (granular) | Pilot scale, type not clearly specified | 1000 | 2 | 99.998[a] (vapor only) | Range: 99.996 to 99.99998% (includes residue).[b] |
| 50% Toxaphene (emulsifiable concentrate) | Pilot scale, type not clearly specified | 1000 | 2 | 99.995[a] (vapor only) | Range: 99.995 to 99.992% (includes residue).[b] |
| 20% Toxaphene dust | Pilot scale, type not clearly specified | 1000 | 2 | 99.995[a] (vapor only) | Range: 99.995 to 99.997% (includes residue).[b] |
| Hexachlorocyclopentadiene waste <br> • Mixture of chlorinated toluenes <br> • Pentanes and benzenes liquid; suspended particulate | Liquid injection | 1350 to 1380 | 0.17 to 0.18 | >99.999 | TRW (Reference 39), pp. 5, 22, 23. |

[a]Based on the respective pesticide plus all related species in the incinerator off gas.

[b]The respective pesticide plus all related chemical species detailed in the incinerator off gas and solid residues. Results cover a range of operating parameters (temperature and residence time); this range was not specified.

for refractory POHCs (HCB and 1,2,4-TCB).   DRE values were higher for
1,2,4-TCB than for HCB under comparable residence time/temperature
conditions.   A large number of PICs, including the two POHCs, were identified
in the flue gas; under some conditions POHC output from the afterburner was
greater than the POHC input.[28]

As noted in Reference 1, all halogenated solvents of concern, including
the halogenated organics chlorobenzene and the dichlorobenzene have been
effectively destroyed by incineration processes.   A summary of recent
EPA-sponsored field tests of commercial and industrial incinerators is
provided in Table 8.6.   Although most of the halogenated organics identified
in the table could be considered as solvents, the DRE for all halogenated
organics present at levels above 1000 ppm in the waste feed exceeded
99.99 percent.   However, candidate POHCs present at levels less than 500 ppm
did not always meet the 99.99 percent DRE standard.   This concentration effect
is under study by the EPA.[7]

DRE values for specific halogenated organics were also obtained during
these field tests, as shown in Table 8.7.   Achievement of 99.99 percent DRE
was not always realized at the conditions of test.   An extensive discussion of
the results of the test data are provided in Reference 29 and summarized in
Reference 1.   As noted, the data show that:

"1.   Extensive analysis of organics emissions data provided the following
      insight into the factors affecting DRE:

      ● DREs for the incinerators tested were generally above
        99.99 percent.   The average DRE for volatile organic
        constituents was found to be 99.992 percent.

      ● DRE appears to be strongly correlated to concentration of the
        POHC in the waste feed.   POHCs at higher waste feed
        concentrations were observed to be destroyed or removed to a
        higher degree.   The phenomenon that caused this relationship
        was not identified.

      ● Analyses of data collected on this program showed no clear
        correlation between DRE and heat of combustion for POHCs.

      ● Data compiled from the eight tests were not sufficient to
        define parametric relationships between residence time,
        temperature, heat input, or $O_2$ concentrations and DRE.

TABLE 8.6.   INCINERATION FACILITIES TESTED

| Facility | Control device | Waste | DRE[a] (number of nines)[b] | HCl control (average) | Average particulate emissions (g/dscf) |
|---|---|---|---|---|---|
| Commercial rotary kiln-liquid incinerator (87 million Btu/hr) | Packed-tower adsorber, ionizing wet scrubber | Drummed, aqueous, liquid organic waste with carbon tetrachloride, TCE,[c] perchloroethylene, toluene, phenol | 5.3 | 99.4% | 0.67 |
| Commercial fixed-hearth, two-stage incinerator (25 million Btu/hr) | Electrified gravel bed filter; packed-tower adsorber | Liquid organic and aqueous aqueous waste with chloroform, carbon tetrachloride, TCE, toluene, perchloroethylene | 4.4 | 98.3% | 0.178 |
| Onsite two-stage liquid incinerator (6 million Btu/hr) | Packed-tower adsorber | Liquid organic waste with carbon tetrachloride, dichlorobenzene, TCE, chlorobenzene, chloromethane, aniline, phosgene | 4.4 | 99.7% | 0.027 |
| Commercial fixed-hearth two-stage incinerator (2 million Btu/hr) | None | Liquid organic waste with TCE, carbon tetrachloride, toluene, chlorobenzene | 4.7 | 4 lb/hr[d] | 0.089 |
| Onsite liquid injection incinerator (4.8 million Btu/hr) | None | Liquid organic waste with analine, diphenylamine, mono- and dinitrobenzene | 6.7 | 4 lb/hr[d] | 0.092 |
| Commercial fixed-hearth two-stage incinerator (10 million Btu/hr) | None | Aqueous and organic liquid waste with carbon tetrachloride, TCE, benzene, phenol, perchloroethylene, toluene, methylethyl ketone | 4.8 | 4 lb/hr[d] | 0.40 |
| Onsite rotary kiln with liquid injection (35 million Btu/hr) | Venturi scrubber with cyclone separators and packed-tower adsorbers | Liquid organic, paint waste and filter cakes with methylene chloride, chloroform, benzyl chloride, hexachloroethane, toluene, TCE, carbon tetrachloride | 5.3 | 99.9% | 0.01 |
| Commercial fixed-hearth two-stage incinerator (75 million Btu/hr) | Venturi scrubber | Aqueous and organic liquids and sold waste with methylene chloride, chloroform, carbon tetrachloride, hexachlorocyclopentadiene, toluene, benzene, TCE | 4.6 | 98.3% | 0.075 |

[a]Destruction and removal efficiency (mass weighted average for all POHCs).

[b]For example, 99.995% DRE = 4.5 nines.

[c]TCE = trichloroethylene.

[d]No HCL control device; waste is low in total organic chlorine content.

Source:   Reference 5.

TABLE 8.7. SUMMARY OF RESULTS OF INCINERATOR TEST PROGRAMS

| Facility | No. of runs | Average waste feed rate (lbs/hr) | Waste constituents | Waste characteristics | | | Average incinerator value | | | | Performance | | |
|---|---|---|---|---|---|---|---|---|---|---|---|---|---|
| | | | | Ash % | Chloride % | Moisture % | Temperature °C | Residence time sec | Heat input 10⁶ lbs/hr | O₂, stack % | DRE % | Particulate Emissions | HCL emission removal |
| Plant C Upjohn | 3 | 243 | Chlorobenzene<br>m-Dichlorobenzene<br>o-Dichlorobenzene<br>p-Dichlorobenzene<br>1,2,4-Trichlorobenzene<br>Phosgene | 0.19 | 21 | NA | 1116 | 5.2 | 6.2 | 8.2 | 99.934<br>99.920<br>99.997<br>99.9977<br>99.67<br>99.997 | 130 mg/dscm | 99.75% |
| Plant G | 3 | DCB Coke 126 | Benzyl Chloride<br>Hexachloroethane | -- | -- | -- | -- | -- | -- | -- | 99.9995<br>99.99 | -- | -- |
| Plant H | 4 | Solid Wastes 542.1 | Benzene<br>Tetrachloroethylene<br>Toluene<br>Chlorobenzene<br>Hexachlorocyclopentadiene<br>Chlordane<br>Hexachlorobutadiene | -- | 2.5 | 3.0 | -- | -- | -- | -- | 99.9848<br>99.9017<br>99.9947<br>99.897<br>99.99<br>99.999<br>99.98 | -- | -- |
| Plant I* | 4 | Liquid waste and No. 2 fuel | Hexachlorocyclopentadiene | -- | 25. | -- | 1430 - 1870 | 0.17 - 0.18 | 7.4 | 6.4 | 99.99 | NA | NA |

NA = Not available.

Source: Reference 29.

*From Reference 30.

- The data from the eight tests suggest that POHC levels in scrubber water and ash were generally very low or nondetectable. These data suggest that the majority of POHCs are destroyed rather than merely transferred to another media in the incineration process.

- Some Appendix VIII compounds detected in the stack (primarily trihalomethanes) appear to be stripped from the scrubber water by the hot stack gas. Trihalomethanes detected in the scrubber inlet water were not detected in the effluent water. The effect can be lower measured/calculated DREs even though the destruction mechanisms may not be affected.

2. Evaluation of organic emissions data for compounds classified as Products of Incomplete Combustion (PICs are Appendix VIII compounds detected in the stack which were not found in the waste feed in concentrations above 100 µg/g) led to the following observations:

- Stack gas concentrations of PICs were typically as high as or higher than those for POHC compounds in the stack.

- PIC output rate infrequently exceeded 0.01 percent of POHC input rate. (The 0.01 percent criterion was proposed in FR Vol. 45, No. 197, October 8, 1980.)

- The three likely mechanisms that explain the presence of most PICs are:

  a. Poor DREs for Appendix VIII compounds present at low concentration (<100 µg/g) in the waste feed;

  b. Input of Appendix VIII compounds to the system from sources other than waste feed (e.g., scrubber water); and

  c. Actual intermediate products of combustion reactions or products of complex side reactions including recombination.

- Data from the tests suggest that benzene, toluene, chloroform, tetrachloroethylene, and naphthalene have a high potential for appearing as byproducts of the combustion of organic wastes.

- A summary of the PICs detected in this study are given in Table 8.8.

3. Compliance with the particulate standard of 180 mg/Nm was not achieved at half of the sites tested. Particulate control devices were operated at five of the eight facilities, and two of these five failed to achieve the standard. Two of the three facilities without control devices also failed the particulate standard. Data from this study suggest that any facility firing wastes with ash content greater than 0.5 percent will need a particulate control device to meet the standard.

TABLE 8.8.   PICs FOUND IN STACK EFFLUENTS

| PIC | Number of sites | Concentrations (ng/L) |
|---|---|---|
| Benzene | 6 | 12-670 |
| Chloroform | 5 | 1-1,330 |
| Bromodichloromethane | 4 | 3-32 |
| Dibromochloromethane | 4 | 1-12 |
| Bromoform | 3 | 0.2-24 |
| Naphthalene | 3 | 5-100 |
| Chlorobenzene | 3 | 1-10 |
| Tetrachloroethylene | 3 | 0.1-2.5 |
| 1,1,1-Trichloroethane | 3 | 0.1-1.5 |
| Hexachlorobenzene | 2 | 0.5-7 |
| Methylene chloride | 2 | 2-27 |
| o-Nitrophenol | 2 | 25-50 |
| Phenol | 2 | 4-22 |
| Toluene | 2 | 2-75 |
| Bromochloromethane | 1 | 14 |
| Carbon disulfide | 1 | 32 |
| Methylene bromide | 1 | 18 |
| 2,4,6-Trichlorophenol | 1 | 110 |
| Bromomethane | 1 | 1 |
| Chloromethane | 1 | 3 |
| Pyrene | 1 | 1 |
| Fluoranthene | 1 | 1 |
| Dichlorobenzene | 1 | 2-4 |
| Trichlorobenzene | 1 | 7 |
| Methyl ethyl ketone | 1 | 3 |
| Diethyl phthalate | 1 | 7 |
| o-Chlorophenol | 1 | 2-22 |
| Pentachlorophenol | 1 | 6 |
| 2,4-Dimethyl phenol | 1 | 1-21 |

Source:   Reference 29.

4.    HCl emissions were generally easily controlled to meet one of the
two criteria specified in the regulations--less than 1.8 kg HCl/hr
or greater than 99 percent removal efficiency."

In addition to the effluent data discussed above, the study included
analyses of two other residuals, ash and the scrubber liquor from the air
pollution control device, at four sites.    The results of the analyses indicate
that both ash and scrubber liquor contain concentration levels that are at
acceptable levels for the compounds analyzed.

Summaries of additional tests of full scale incinerators are reported in
the solvent TRD.[1]    Data were provided for two test programs conducted at
rotary kiln incinerators using wastes containing nonsolvent chlorinated
organics.    Incinerator standards of 99.99 percent DRE were met for all test
compounds as follows:

| Compound | Range of DREs Reported |
|----------|------------------------|
| Hexachloroethane | 99.99 |
| Trichlorobenzene | 99.992-99.995 |
| 2,4-Dichlorophenol | 99.999 |
| 2,4,6-Trichlorophenol | 99.999 |

Correlations between residence time and temperature with DRE were observed, as
expected, in both studies.

Very little data were found, other than shown in Table 8.6, for the
destruction of nonsolvent halogenated organics in fixed hearth and multiple
hearth incinerators and fluidized-bed incinerators although data for
halogenated solvents are available from several test programs.    Available data
do indicate that 99.99 percent DREs are achievable for halogenated solvents.
Similar DREs can be anticipated for nonhalogenated organics.

Several recent EPA sponsored studies have examined the destruction
efficiency of halogenated organics in industrial boilers and process kilns
used in the lime, cement, and aggregate industries.[5]    Extensive data have
been obtained for halogenated solvents indicating that, in most cases,
incinerator DRE standards can usually be met.    Most units for which data are
available also achieved the particulate emission standard of 0.08 grams/scf or

less although ash content data were not available.[5,31]   Particulate
emissions appear to increase with an increase in chlorine content.   These
increases were attributed to a lowering of ESP efficiency due to changes in
both the electrical resistivity of the particle and the particle size
distribution of the particulate emissions.[31]

## 8.4   COSTS OF HAZARDOUS WASTE INCINERATION

The overall costs associated with the incineration of hazardous wastes
are high relative to other hazardous waste treatment or disposal methods.
Incineration facilities require large capital costs due to the size and
complexity of the systems involved, and the requirements associated with the
handling of hazardous wastes and their combustion products.  Operating costs
are high, primarily due to the large energy input required, and also as a
consequence of large raw material costs and stringent environmental control
requirements.   Incineration costs are difficult to specify, in general,
because in each situation the number of factors impacting costs is large.
These factors may be classified fundamentally as follows:

- Waste characteristics;
- Facility design characteristics; and
- Operational characteristics.

The general significance of many of the factors affecting incineration costs
will be discussed in detail below.

## Waste Characteristics

The chemical and physical properties of a waste considered for
incineration govern the type of incinerator selected, the processing capacity,
environmental controls employed, pretreatment employed, required maintenance
and equipment lifespan, and operational parameter levels.  Several waste
characteristics which significantly affect the costs of incineration are
described below:

- Physical State—Physical state dictates the type of incinerator and the type of waste feed mechanism selected. Liquid injection incinerators, for example, are applicable only to liquid wastes. Limited data available on prices charged by commercial incinerators suggest that solid and sludge wastes are more expensive to incinerate than liquid wastes.

- Heat Value—Heat value is used as a measure of auxiliary fuel requirement. The higher the heat value of a waste, the less fuel is required to sustain combustion.

- Rheological Characteristics—The way in which liquid viscosity of a waste changes with temperature is important in determining the need for preheating, waste feed mechanism, and incinerator type. Some of the wastes are easily handled at higher temperatures, while others maintain viscosities which render them nonpumpable and/or nonatomizable over practical limits of temperature.

- Water Content—Water content of a waste strongly affects temperature and destruction efficiency of the combustion system. In some cases, dewatering of wastes is conducted as a pretreatment operation.

- Chloride Content—The chloride content of a waste has strong bearing on the air pollution control methods employed at an incinerator. High levels of chlorine necessitate acid gas scrubbing and also require combustion methods which prevent the formation of toxic chlorine gas.

- Ash Content/Heavy Metals Content—The amount of ash which will be formed in combustion, and the nature of the ash is related to the inorganic salt and heavy metal content of a waste to be incinerated, and greatly affects the particulate matter air pollution control requirement and the ash collection and disposal system design.

- Volatile Content—The presence of volatile low flash point components should be considered. If such materials are present in significant amount special pretreatment or precautions must be taken.

The impact of various waste characteristics on incineration costs may, in some cases, be measured directly. A survey was conducted of a cross-section of hazardous waste incineration facilities operating commercially in the United States, and it revealed that pricing structures are often established based on certain waste characteristics.[11,23,41-45] As shown in Table 8.9 chloride content and ash content commonly are used to establish surcharges based on additional air pollution control requirements. The physical form of the waste may also be seen as leading to price differentials. In general, solid and sludge wastes cost more to incinerate than liquid wastes.

TABLE 8.9.    SURVEY OF HAZARDOUS WASTE INCINERATORS - COSTS OF INCINERATION AND COST IMPACTING FACTORS

| Facility | Incineration system | Costs to incinerate hazardous wastes | | Additional costs | |
|---|---|---|---|---|---|
| | | Type of waste | Cost ($) | Basis | Cost ($) |
| A | Liquid Injection/ Rotary Kiln | Blendable[a] aqueous | 0.18/lb | Phase separation | 0.3475/lb |
| | | Blendable organic | 0.2675/lb | | |
| | | Directly-burned aqueous | 0.2050/lb | | |
| | | Directly-burned organic | 0.2850/lb | | |
| | | Directly-burned sludges or solids | 0.5/lb | | |
| B | Liquid Injection/ Rotary Kiln | Liquids | 39/55 gal. drum | Chloride content or ash content | N/A |
| | | Solids and sludges | 125/55 gal. drum | | |
| C | Liquid Injection/ Rotary Kiln | Any | 0.25/lb | Chloride content or ash content | 0.002/lb per each 1% of chloride or ash content |
| D | Liquid Injection | Liquids | 0.86/gallon | Suspended solids | 0.01/gal per 1% content |
| | | | | Metals (e.g. chromium) | 0.0005/100 ppm/gallon |
| | | | | Chlorine | 0.02/gal per 1% chlorine |
| E | Liquid Injection/ Rotary Kiln | Bulk liquids | 1.93/gallon | Handling fee | 25/drum |
| | | Drummed liquids | 230/drum | "Approval" charge | 150/job |
| | | Drummed solids | 300/drum | | |

[a]Blendable defined as a waste with a viscosity of below 10,000 SSU.

Source:  References 11, 23, 24, 41-45.

## Facility Characteristics

Facility characteristics, i.e., the design and size of incineration unit equipment, are measured in terms of capital costs. Capital costs for incineration facilities are high relative to many other hazardous waste management technologies, which are generally less complex and less sensitive to thermal and mechanical tolerances. For each incineration technology, there is a large variation in the designs available commercially, and great differences in the pricing policies of manufacturers. As a result, it is difficult to specify a range of costs for any particular type of system.

To determine the cost of a hazardous waste incineration facility, several key factors must be assessed. Several of the key factors influencing capital costs are:

- Size requirements--flow rates, heat input capacities, exhaust rates, etc.;

- Equipment lifespan;

- Pretreatment requirements;

- Heat recovery;

- Environmental control requirements;

- Feed mechanisms; and

- Equipment availability.

The size requirements of the system have the most bearing on capital costs, while the environmental control equipment costs may be the largest element of the overall capital costs. The capital costs of a particular hazardous waste incineration system are strongly affected by the overall availability of that technology. Certain systems, such as liquid injection incinerators, are manufactured by many companies. Other technologies, most notably the newer type systems, are manufactured by a few or, in many cases, only one company.

Capital cost data available for hazardous waste incineration systems were limited. Several manufacturers of incineration systems were contacted, but most were reluctant to specify costs for their systems because the cost for a

specific application is dependent upon so many different factors.  A study
conducted by McCormack, et al.,[3] provided cost estimation curves for several
of the established hazardous waste incineration technologies:  liquid
injection, rotary kilns, and hearth incinerators.  These cost curves,
including estimation curves for heat recovery systems (waste heat boiler) and
acid gas scrubbing systems, are presented in Figures 8.4 through 8.9.  This
information was generated in 1982, and has been updated to reflect the changes
in the Chemical Engineering Plant Index between May 1986 and the date for
which costs were estimated in Reference 3.  A study conducted by MITRE
Corporation in 1981, in which several visits were made to incinerator
manufacturers, generated additional cost data summarized in Table 8.10.

   In general, it may be noted that certain hazardous waste technologies are
considered to be more expensive in terms of capital costs than are others.
Rotary kilns are most expensive.  Relative capital costs for the five
established incineration technologies are as follows in order of decreasing
cost:

- Rotary Kiln

- Fluidized-Bed

- Multiple Hearth

- Liquid Injection

- Fixed Hearth

## Operating Characteristics

   Numerous factors impact the operating costs of a hazardous waste
incineration facility.  The most significant factor governing operating costs
is energy usage.  Energy is used in incineration to heat wastes during
combustion, and to operate materials transport and control systems.  In many
cases, the energy usage of an incineration system is large enough to justify
the costs of installing and operating heat recovery systems.  A summary of
other important operating characteristics affecting the costs associated with
incineration is presented below:

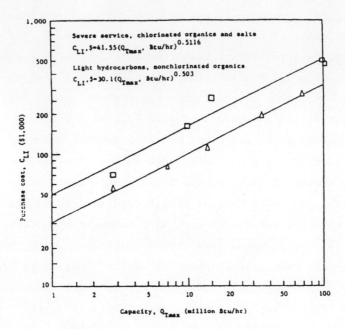

Figure 8.4. Purchase cost of liquid injection system (May 1982).

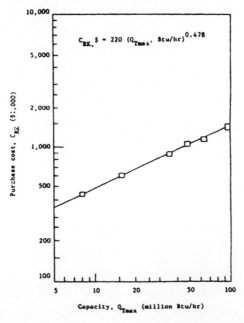

Figure 8.5. Purchase cost of rotary kiln system (May 1982).

Source: Reference 3.

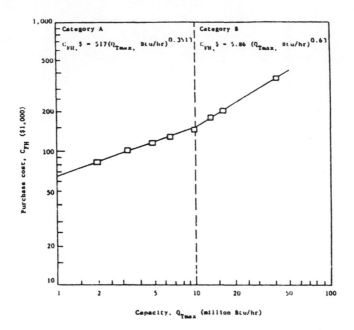

Figure 8.6.    Purchase cost of hearth incinerators (mid 1982).

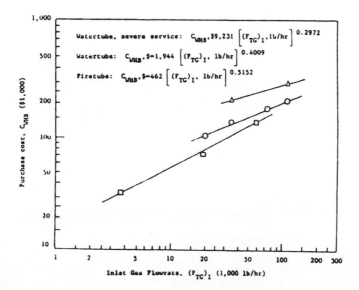

Figure 8.7.    Purchase cost of waste heat boilers (July 1982).

Source:    Reference 3.

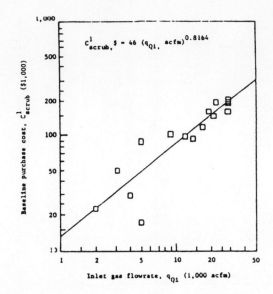

$$C_{scrub}^{1}, \$ = 46 \, (q_{Q1}, \, acfm)^{0.8164}$$

Figure 8.8.    Purchase cost of scrubbing systems receiving 500 to 550°F gas (July 1982).

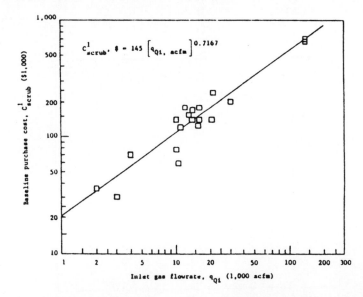

$$C_{scrub}^{1}, \$ = 145 \left[ q_{Q1}, \, acfm \right]^{0.7167}$$

Figure 8.9.    Purchase costs for typical hazardous waste incinerator scrubbing systems receiving 1800 to 2200°F gas (July 1982).

Source:    Reference 3.

TABLE 8.10.    SUMMARY OF COST DATA COMPILED BY MITRE CORPORATION, 1981

| Facility | Incineration technology | Capacity (mmBtu/hr) | Capital cost ($) | Description of cost factors |
|---|---|---|---|---|
| 1 | Fluidized Bed | 10 | 700,000 | Without energy recovery. |
| 2 | "Packaged" Rotary Kiln | --- | 40-50,000/(100 lbs/hr) | Scale-up factor for cost estimation is 0.6 exponent. Installed cost, including heat recovery and air pollution control. |
| 3 | Rotary Kiln | 37.5 | 800,000 | Not installed. Includes one item of air pollution control. Estimated installation cost was 20 percent. |
| 4 | Rotary Kiln | 80-150 | $10\text{-}15 \times 10^6$ | Total installed cost. |
|  |  | 0.5 | 600,000 | Total installed. |
|  |  | 1.02 | $1.9 - 2.2 \times 10^6$ | All not installed. |
|  |  | 1.24 | $2.34 - 2.66 \times 10^6$ |  |
|  |  | 14.1 | $2.66 - 3.04 \times 10^6$ |  |
|  |  | 17.0 | $3.25 - 3.65 \times 10^6$ |  |
| 5 | Rotary Kiln | 90 | $8.5 \times 10^6$ | Total installed. |

(continued)

TABLE 8.10 (continued)

| Facility | Incineration technology | Capacity (mmBtu/hr) | Capital cost ($) | Description of cost factors |
|---|---|---|---|---|
| 6 | Liquid Injection | 5 | 150,000 | Base cost, not installed, no APC, heat recovery. |
|  |  |  | 300,000 | Total installed with APC. |
| 7 | Fixed Hearth | 5 | 150,000 | Base cost, no APC or heat recovery, not installed. |
|  |  |  | 300,000 | Installed with APC. |
| 8 | Liquid Injection | 18 | 500,000 | Not installed. |
|  |  | 70 | $1.5 \times 10^6$ | Total installed cost. |
|  |  |  |  | Scale-up factor is exponent – 0.65. |
| 9 | Combined Liquid Injection and Rotary Kiln | 150 | $2.2 \times 10^6$ | Not installed. No APC, heat recovery. |
|  |  |  |  | Estimated cost of APC given is $1.2 \times 10^6$. |
| 10 | Liquid Injection | 30 | 400–500,000 | With boiler and scrubber, not installed. |
| 11 | Pyrolysis | 3,000 lbs/hr | $1 \times 10^6$ | Including heat recovery, no APC installed. |
|  |  | 6,000 lbs/hr | $4 \times 10^6$ |  |

Source: Reference 23.

- Residence Time--Residence time affects the volume of the combustion chamber, secondary combustion requirement, and the exhaust rate. Residence time may be increased by employing devices such as baffles or recirculation blowers.

- Temperature--Temperature affects the volume and type of the incinerator refractory lining, volume of insulation for other systems, and the need for heating and cooling systems.

- Raw Materials Usage--A variety of raw materials are used in incineration systems, including chemical agents, fluidized-bed granular material, scrubber and cooling tower water, sorbents, and oxygen. The consumption of these materials leads to additional costs.

- Maintenance--Maintenance requirements for incineration systems are considered high due to the number of systems involved and the thermal and mechanical stresses under which they operate. The maintenance of refractory linings, for example, is considered a particularly significant cost consideration.

- Disposal—Disposal of solid and liquid combustion byproducts can be very expensive depending on the characteristics of the materials produced. In some cases, systems are limited in applicability based on disposal costs of, for example, heavy metals containing wastes.

Because of the uncertainties in many of the above items, it is difficult to assign meaningful values to elements of operational costs. These factors, however, have been considered by operators in assigning differential costs based on waste characteristics (see Table 8.9).

## 8.5  STATUS OF DEVELOPMENT

### 8.5.1  Hazardous Waste Incinerator Manufacturing Industry

Several surveys have been conducted to determine the number of companies currently involved in the development, manufacture, and installation of hazardous waste incineration systems.[23,25,46,47] Investigation of the current hazardous waste incinerator market in 1982, indicates that there are approximately 67 companies known to be actually involved. This number may not necessarily include the number of companies who are developing newer, more innovative thermal technologies. The conventional technologies offered by

these commercial companies are summarized in Table 8.11.   In general, the
following conclusions, drawn by the MITRE Corporation in 1982, are supported
by these data:

- "About 342 incinerators have been put into hazardous waste service
  since January 1969.  These units were manufactured by 29 companies,
  all of which were based in the United States at the time the units
  were delivered.  Within the past year one of these companies (BSP
  Envirotech) was purchased by a West German firm, the Lurgi
  Corporation.  The count of 342 units is believed to be reasonably
  accurate, but cannot be exact for the following reasons:

  - A number of small vendor companies have disappeared since
    1969.  These companies have probably manufactured a few
    incinerators which are still in use, but their existence could
    not be determined.

  - Incinerators originally sold for hazardous waste disposal or
    for nonhazardous wastes could be operating, at least part time,
    on the other waste.

  - Some incinerators have been manufactured strictly in accordance
    with a customer's specifications and the manufacturing company
    has no knowledge of, or declines to speculate on, the nature of
    the purchaser's wastes.

  - A few incinerators which have been manufactured since January
    1969 are probably no longer in use.  A vendor will not
    generally know this.

  - A few incinerators manufactured since 1969 cannot fulfill the
    design function and are not operating.  Vendors will not
    voluntarily acknowledge these.

- The most common type of hazardous waste incinerator is liquid
  injection, representing 64.0 percent of all hazardous waste
  incinerators in service.  This type of incinerator is not designed
  to operate on liquids containing any significant amount of salts or
  other suspended or dissolved solids.

- The next most common types of hazardous waste incinerators are the
  fixed hearth (FH) and the rotary kiln (RK), with 17.3 and
  12.3 percent, respectively, of the total manufactured.  Both of
  these types of units will dispose of solid wastes, liquid wastes,
  and/or fumes.

- Although there are nine companies offering fluidized-bed (FB)
  incinerators, only nine such units are in hazardous waste service.
  Apparently most of these nine companies believe that the market is
  potentially good for this technology.

TABLE 8.11.    NUMBER OF HAZARDOUS WASTE INCINERATORS IN SERVICE IN THE U.S.A.

| Type | No. of companies offering | No. in H.W. service | Range of capacities | Numerical share of market (%) |
|---|---|---|---|---|
| Liquid Injection | 23 | 219-231 | 3-300 mmBtu/hr | 64 |
| Rotary Kiln | 17 | 42 | 1-150 mmBtu/hr | 12.3 |
| Fixed Hearth[a] | 15 | 64 | 200-2500 lbs/hr | 18.5 |
| Multiple Hearth | 2 | 7 | 1000-1500 lbs/hr | 2.0 |
| Fluidized Bed | 9 | 9 | N/A | 2.6 |

[a]Includes other hearth-type systems including Pulse Hearth (2), Rotary Hearth (2), Reciprocating Grate (1).

N/A - Information not available.

Source:  MITRE, 1982 (Reference 23).

- Two companies are actively marketing fused salt bath technology, but there are no units in service or under construction yet.

- Of about 219 liquid injection units in service, about 129 (59 percent) were produced by two companies, John Zink and Trane Thermal. The data furnished by Zink are not well verified. Of 23 companies marketing LI incinerators, 8 have sold no units to date. However, several of the 8 indicate that sales are imminent.

- Of the 17 companies offering rotary kiln incinerators, 8 have sold none to date.

- Of the 9 companies offering fluidized-bed incinerators, 5 have sold none to date.

- Of a total of 57 companies offering 14 types of incinerators, 28 have sold no units in the United States. (Several companies represent European technologies and all have sold at least one unit each in Europe.

- The fact that 28 (of the total of 57 companies) have not sold any units to date is indicative of the extent of: 1) new technology being made available in the United States by both U.S. and foreign companies; 2) the formation of new corporate ventures in the field of technology; and 3) efforts by European companies to invade the U.S. market. It is therefore believed that the market, or technology, is not static at this point in time.

- Two companies are allegedly developing new technology, which they would not describe at this time. It is known that other companies are researching other techniques for hazardous waste incineration, but these techniques are not described in this report. The new processes included plasma, microwave plasma, and several unusual fluidized-bed techniques."[23]

A comparative assessment of the available incinerator technologies and a discussion of their advantages and limitations has been provided in Reference 1 and in many of the other references cited which provide an overview of hazardous waste incineration. Manufacturers are identified in the survey studies such as Reference 23; listings of equipment manufacturers can also be found in McGraw Hill's Chemical Engineering Equipment Buyers' Guide.

## 8.5.2   Environmental Impacts of Incineration

Incineration processes potentially affect the environment through generation of air emissions, and liquid, sludge, and solid wastes. As a result, EPA has established environmental standards of performance for incinerators in the RCRA permit process.  Most incinerators must be equipped with appropriate air pollution control systems, leading to higher capital and operating costs.  Environmental impacts associated with incineration are, therefore, a significant factor in the determination of the appropriateness of incineration as a management option for hazardous wastes.

## 8.5.2.1  Air Emissions--

Air emissions of pollutants produced in incineration are a primary area of environmental concern.  Emissions may be generated from the incinerator stack and from fugitive emission sources.  Emissions from incinerators primarily consist of the following "criteria" pollutants:  oxides of nitrogen and sulfur, and particulate matter.  Other air pollutants of concern include undestroyed organics such as benzene, toxic heavy metals (in particles), hydrochloric acid, and other acid gases.

As part of the RCRA permit process, incinerators must demonstrate their ability to achieve various performance levels established by EPA.  Among the performance criteria are emission standards for hydrochloric acid gas and particulate matter.  These standards are:

1.  Hydrochloric acid emissions are limited to a rate of 4 lb/hr or, if acid gas scrubbing is employed, a scrubbing efficiency of 99 percent or greater; and

2.  Particulate matter emissions are limited to 0.08 grains per dry standard cubic feet of flue gas at 7 percent oxygen (180 milligrams per dry standard cubic meter).

Emissions from incinerators are also regulated under Federal NESHAPS and state air toxics program standards.  These may affect, in particular, the emissions of heavy metals such as lead or mercury vapors.

Available technologies for the control of emissions from hazardous waste incinerators include devices to control emissions of particulate matter, acid gases, oxides of sulfur, and possibly oxides of nitrogen.  Gaseous pollutant control devices include various wet and dry scrubbers.  Both wet and dry scrubbing systems are effective in removing acid gases, although the dry scrubbing systems are newer and, as a result, not as well established as the wet systems. Oxides of nitrogen emissions can sometimes be minimized by the use of combustion modifications which reduce the peak flame temperature in an incinerator.

For control of particulate matter, the primary candidates are the wet and dry electrostatic precipitators (ESPs), ionizing wet scrubbers, and baghouses. Conventional scrubbers are not very effective in the removal of fine particulate matter. Particulate matter control devices must be compatible with the acid removal device. A wet acid scrubber is more compatible with a wet ESP or the ionizing wet scrubber than with a baghouse. Since baghouses are compatible only with a dry gas system, the use of baghouses on hazardous waste incinerators is not as prevalent as ESP usage. Properly designed baghouses and ESPs are both effective particulate matter control devices. Discussion of the various types of emission control devices used on hazardous waste incinerators and their control efficiency capabilities can be found in numerous texts and publications dealing with air pollution control and incineration.

8.5.2.2  Liquid and Solid Wastes Generation--

Wastes formed both in the combustion unit and in pretreatment and air pollution control systems constitute a potential environmental hazard which must be properly managed. Presence of hazardous materials in the incinerator ash, scrubbing liquor, and scrubber sludges is primarily dependent upon two factors: composition of wastes fed; and destruction effectiveness of the incinerator. The primary constituents of concern in these residues are thermally inert materials such as toxic heavy metals. Toxic organic compounds are generally not a significant contaminant of these streams, owing to good destruction efficiencies.

Ash from incineration--Incinerator ash formed during the combustion reaction consists almost entirely of thermally inert materials (metals and other inorganics) introduced in the waste feed. Ash, not emitted with the combustion flue gas, generally collects at the bottom of the incinerator units. Many incinerator designs include a conveyor system which continuously removes ash from the bottom of the unit for subsequent disposal. Contaminated ash is now commonly disposed of in a Class I landfill. As noted in

Reference 29, ash residuals from incineration have been found to be suitable tor landfill disposal. Alternatives to direct landfilling, if required, could include encapsulation/solidification treatments.

Scrubber liquor/scrubber sludges—Scrubber systems, which directly contact the gaseous byproducts of combustion with liquid (or solid) media, most commonly water, may produce contaminated liquid or solid waste streams. The primary contaminants of such streams are toxic solid particles carried as fly ash, acids such as hydrochloric acid formed in combustion, and various organic products of incomplete combustion. The quantity, quality, and types of liquid wastes formed from the control equipment is dependent on the constituents of the waste feed, destruction efficiency, and collection efficiency.

Liquid effluents from scrubber and quench systems usually will undergo neutralization and removal of solids before they are discharged to local sewage systems. A very common practice is to discharge these streams to settling ponds (volatilization of organics from these ponds is not considered a significant problem). Sludges are commonly treated in a sewage sludge incinerator, or are dewatered and directly landfilled. Residual analysis of scrubber liquor and sludges[29] have indicated that they are essentially free of organic materials.

## 8.5.3  Summary

The advantages and disadvantages of the various incineration technologies available for the destruction of hazardous wastes are presented in Table 8.12. In general, most of the common incineration technologies might be of limited applicability to halogenated organic wastes, depending upon the individual characteristics of the waste.

## TABLE 8.12.   SUMMARY OF INCINERATION TECHNOLOGIES

| Incineration method | Limitations | Advantages | Disadvantages | Approximate costs | |
|---|---|---|---|---|---|
| | | | | Capital | Operating |
| Liquid Injection | Feedstock must be atomizable; relatively free of particulates | Can process all types of hazardous liquids | Requires pretreatment to remove impurities, heat, and blend | $4-500,000 for 30 mmBtu/hr (installed, with heat recovery and APC 1982) | $1-250/1000 gal |
| Rotary Kiln | Requires large batch throughput to be practical or economical | Can process virtually any type of waste; can coincinerate various types of wastes | Requires air pollution controls | $40,000-50,000/(100 #/hr) $10-15 x 10$^6$ for 80-150 mmBtu (total installed, 1982) | $2500-100/ton/day |
| Fluidized Bed | Requires large batch throughput; limited to liquids or non bulky solids; no sodium salt wastes | Can process many wastes types; good temperature response in processing | Requires periodic bed replacement; requires air pollution controls | $700,000 for 10 mmBtu (total installed, no heat recovery, 1982) | N/A |
| Fixed Hearth | Requires afterburner; can't burn liquids if use continuous ash recovery | Can achieve very high combustion temperatures; low maintenance required | Not energy-efficient; requires higher tempera-tures and residence times | $3-400,000 for 10 mmBtu (installed, 1982) | $0.5/lb |
| Multiple Hearth | Requires afterburner; can't burn bulky solids, corrosives | Best for sludge incin-eration; low capital cost | Possible high maintenance costs; not energy efficient | N/A | N/A |

Source:  Engineering-Science (Reference 48).

## REFERENCES

1.  Breton, M., et al.  Technical Resource Document--Treatment Technologies for Solvent Containing Wastes.  Prepared for U.S. EPA, HWERL, Cincinnati, under Contract No. 68-03-3243.  Work Assignment No. 2.  August 1986.

2.  Federal Register 1982, 47, 27516-35.

3.  McCormack, R. J., et al.  Costs for Hazardous Waste Incineration. Capital, Operation and Maintenance, Retrofit.  Acurex Corporation, Mountain View, CA.  Noyes Publications, Park Ridge, NJ.  1985.

4.  U.S. EPA.  Compilation of Air Pollution Emission Factors.  Third Edition including Supplements 1 through 15.  Publication No. AP-42.  August 1982.

5.  Oppelt, E. T.  Hazardous Waste Destruction, Environmental Science and Technology, Vol. 20, No. 4.  1986.

6.  Manson, L., et al.  Hazardous Waste Incinerator Design Criteria. EPA-600/2-79-198.  TRW, Inc., Redondo Beach, CA.  Prepared for U.S. Environmental Protection Agency, Industrial Environmental Research Laboratory, Cincinnati, OH.  October 1979.

7.  Lee, K. C., H. J. Jahnes, D. C. Macauley.  Thermal Oxidation Kinetics of Selected Organic Compounds.  In:  71st Annual Meeting of the Air Pollution Control Association, Houston, TX.  June 23-30, 1978.

8.  Advanced Environmental Control Technology Research Center.  Research Plaing Task Group Study - Thermal Destruction.  EPA-600/2-84-025. Prepared for U.S. Environmental Protection Agency, Industrial Engineering Research Laboratory, Cincinnati, OH.  January 1984.

9.  U.S. EPA.  Incineration and Treatment of Hazardous Waste:  Proceedings of the 11th Annual Research Symposium.  EPA-600/9-85-028.  Articles cited include:

    (a)  Olexsey, R., G. Hoffman, and G. Evans.  "Emission and Control of By-Products from Hazardous Waste Combustion Processes";

    (b)  Gorman, P., and D. Oberacker.  "Practical Guide to Trial Burns at Hazardous Waste Incinerators";

    (c)  Westbrook, W., and E. Tatsch.  "Field Testing of Pilot Scale APCDs at a Hazardous Waste Incinerator";

(d)  Clark, W. D., et al.  "Emergency Analysis of Hazardous Waste
     Incineration:  Failure Mode Analysis for Two Pilot Scale
     Incinerators";

(e)  Dellinger, B., J. Graham, D. Hall, and W. Rubey.  "Examination of
     Fundamental Incinerability Indices for Hazardous Waste Destruction."

(f)  Kramlich, J., E. Poncelet, W. R. Seeker, and G. Samuelsen.  "A
     Laboratory Study on the Effect of Atomization on Destruction and
     Removal Efficiency for Liquid Hazardous Wastes";

(g)  Chang, D., and N. Sorbo.  "Evaluation of a Pilot-Scale Circulating
     Bed Combustor with a Surrogate Hazardous Waste Mixture";

(h)  Evans, G.  "Uncertainties and Incineration Costs:  Estimating the
     Margin of Error"; and

(i)  Graham, J., D. Hall, B. Dellinger.  "The Thermal Decomposition
     Characteristics of a Simple Organic Mixture".

10.  Lee, K. C., N. Morgan, J. L. Hansen, G. M. Whipple.  Revised Model for
     the Prediction of the Time-Temperature Requirements for Thermal
     Destruction of Dilute Organic Vapors, and Its Usage for Predicting
     Compound Destructability.  In:  75th Annual Meeting of the Air Pollution
     Control Association, New Orleans, LA.  June 20-25, 1982.

11.  Incineration and Treatment of Hazardous Waste:  Proceedings of the 8th
     Annual Research Symposium.  EPA-600/9-83-003.  Articles cited include:

     (a)  Frankel, J., N. Sanders, and G. Vogel.  "Profile of the Hazardous
          Waste Incinerator Manufacturing Industry":

     (b)  Vogel, G. A., Frankel, I., and N. Sanders.  "Hazardous Waste
          Incineration Costs":

     (c)  Staley, L. J., G. A. Volten, F. R. O'Donnell, and C. A. Little.  "An
          Assessment of Emissions from a Hazardous Waste Incineration
          Facility"; and

     (d)  Johnson, S. G., S. J. Yosium, L. G. Keeley, and S. Sudar.
          "Elimination of Hazardous Wastes by the Molten Salt Destruction
          Process."

12.  PEDCo, Inc.  Evaluation of the Feasibility of Incinerating Hazardous
     Waste in High-Temperature Industrial Processes.  EPA-600/2-84-049.
     Prepared for U.S. Environmental Protection Agency, Industrial
     Environmental Research Laboratory, Cincinnati, OH.  February 1984.

13.  Air Pollution Control Association.  Technical Conference on the Burning
     Issue of Disposing of Hazardous Wastes by Thermal Incineration.
     April 29-30, 1982.  Hilton Gateway, Newark, NJ.  Articles cited include:

(a)    Deneau, K. S.  "Pyrolytic Destruction of Hazardous Waste";

(b)    Austin, D. S., R. E. Bastian, and R. W. Wood.  "Factors Affecting
       Performance in a 90 million Btu/hr Chemical Waste Incinerator:
       Preliminary Findings";

(c)    Bierman, T. J. and J. C. Reed.  "Determination of Principal Organic
       Hazardous Constituents (POHCs) in Hazardous Waste Incineration"; and

(d)    Sesaverns, G. A., D. R. J. Roy, and W. B. Rossnagel.  "Air Pollution
       Control Technology:  For Hazardous Waste Incineration".

14.    Arienti, M., et al.  Technical Assessment of Treatment Alternatives for
       Wastes Containing Halogenated Organics.  Prepared for U.S. EPA, OSW Under
       Contract No. 68-01-6871.  Work Assignment No. 9.  October 1984.

15.    Clark, J. N. and J. J. Cudahy.  Impact of the Resource Conservation and
       Recovery Act on the Design of Hazardous Waste Incinerators.  In:
       Detoxification of Hazardous Waste.  Ann Arbor Science Publishers.  1982.

16.    Cudahy, J. J., W. L. Troxler and L. Sroka.  Incineration Characteristics
       of RCRA Listed Wastes.  U.S. EPA Contract No. 68-03-2568.  Work Directive
       T-7021.  Industrial Environmental Research Laboratory, Cincinnati, OH.

17.    Tsang, W., and W. Shaub.  Chemical Processes in the Incineration of
       Hazardous Materials.  Paper presented at the American Chemical Symposium
       on Detoxification of Hazardous Wastes.  New York, NY.  August 1981.

18.    Guidance Manual for Hazardous Waste Incinerator Permits.  Report SW-966.
       EPA, Washington, D.C., 1983.

19.    Senkan, S. M.  Combustion Characteristics of Chlorinated Hydrocarbons.
       In:  Detoxification of Hazardous Waste.  Ann Arbor Science Publishers.
       1982.

20.    Martin, E. J., and L. W. Weinberger, et al.  Practical Limitation of
       Waste Characteristics for Effective Incineration.  Presented at the
       Twelfth Annual Research Symposium on Land Disposal, Remedial Action,
       Incineration, and Treatment of Hazardous Waste.  Sponsored by U.S.
       Environmental Protection Agency, Hazardous Waste Engineering Research
       Laboratory, Cincinnati, OH.  April 21-23, 1986.

21.    Edwards, B. H., J. N. Paullin, K. Coughlan-Jordan.  Emerging Technologies
       for the Control of Hazardous Wastes.  Ebon Research Systems, Washington,
       D.C., Noyes Data Corporation, Park Ridge, NJ.  1983.

22.    Marti, Bruce.  Telephone Conversation with M. Kravett, GCA Technology
       Division, Inc.  Chemical Waste Management, Inc., Chicago, IL.  February
       1986.

23.    MITRE Corporation.  Survey of Hazardous Waste Incinerator Manufacturers,
       1981.  MITRE Corporation, METREK Division, McLean, VA.  1982.

24.  Cross, F. C.  Hazardous Waste Incinerators--Operational Needs and
     Concerns.  Cross/Tessitore and Associates, P.A., Orlando, FL.  In:
     Hazardous Waste and Environmental Emergencies--Management, Prevention,
     Cleanup, and Control.  March 12-14, 1984.

25.  State of California Air Resources Board.  Air Pollution Impacts of
     Hazardous Waste Incineration:  A California Perspective.  Technical
     Support Document.  A Report to the California State Legislature.
     Prepared by the California Air Resources Board.  December 1983.

26.  MITRE Corporation, Working Paper.  Liquid Injection Incinerator Burner
     Performance.  WP-83W00393.  MITRE Corporation, METREK DIvision, McLean,
     VA.  October 1983.

27.  Kiang, Y. H., and A. A. Metry.  Hazardous Waste Processing Technology.
     Ann Arbor Science Publishers, Inc.  Ann Arbor, MI.  1981.

28.  Lee, C. E. and G. L. Hoffman.  An Overview of Pilot-Scale Research in
     Hazardous Waste Thermal Destruction.  In:  International Conference on
     New Frontiers for Hazardous Waste Management.  U.S. EPA-600/9-85-025.
     September 1985.

29.  Trenholm, A., P. Gorham, and G. Sungclaus.  Performance Evaluation of
     Full Scale Hazardous Waste Incinerators.  EPA-600/2-84-181.  Midwest
     Research Institute, Kansas City, MO.  Prepared for U.S. Environmental
     Protection Agency, Office of Research and Development, Cincinnati, OH.
     November 1984.

30.  U.S. EPA.  Destroying Chemical Wastes in Commercial Scale Incinerators
     Facility Report No. 1.  SW-122c.1.  U.S. Environmental Protection Agency,
     Office of Solid Waste, Washington, D.C.  1977.

31.  Mournighan, R. E., et al.  Hazardous Waste Incineration in Industrial
     Processes:  Cement and Lime Kilns.  In:  International Conference on New
     Frontiers for Hazardous Waste Management.  U.S. EPA-600/9-85-025.
     September 1985.

32.  Paulson, E. G.  How to Get Rid of Toxic Organics.  Chemical Engineering.
     October 17, 1977.  pp. 21-27.

33.  Carnes, R. A.  Combustion Characteristics of Hazardous Waste Streams.
     Presented at the 71st Annual Meeting of the Air Pollution Control
     Association.  No. 77-19-1.  June 20-24, 1977.

34.  Elliot, W. H. and W. B. McCormack.  Incineration of Hazardous Substances.
     Presented at the 70th ANnual Meeting of the Air Pollution Control
     Association.  No. 77-19.1.  June 20-24, 1977.

35.  Ackerman, D. G., et al.  Guidelines for the Disposal of PCBs and PCB
     Items by Thermal Destruction (DRAFT) by TRW, for U.S. EPA, Industrial
     Environmental Research Laboratory, Contract No. 68-02-3174.  May 1980.

36.  Lee, Kun-Chieh, et al.  Thermal Oxidation Kinetics of Selected Organic
     Compounds, presented at the 71st Annual Meeting of the Air Pollution
     Control Association, 78-58.6, Houston, Texas.  June 25-30, 1978.

37.  Manson, L. and S. Unger.  Hazardous Material Incinerator Design Criteria,
     by TRW, for U.S. EPA, Office of Research and Development.
     EPA-600/2-79-198.  October 1979.

38.  Ferguson, T. L., et al.  Determination of Incinerator Operating
     Conditions Necessary for Safe-Disposal of Pesticides, by Midwest Research
     Institute, for U.S. EPA, Office of Research and Development.
     EPA-600/2-75-041.

39.  Ackerman, D. G., et al.  Destroying Chemical Wastes in Commercial Site
     Incinerators, Final Report--Phase II, by TRW and Arthur D. Little, Inc.
     for U.S. EPA, Office of Solid Waste, SW-155c.  November 1977.

40.  Dillon, A. P., Editor.  Pesticide Disposal and Detoxification.  Noyes
     Data Corporation, Park Ridge, N. J.   1981.

41.  ICF Incorporated.  RCRA Risk/Cost Policy Model - Phase III Report.
     Prepared for U.S. Environmental Protection Agency, Office of Solid Waste,
     Washington, D.C.  1984.

42.  Anderson, R.  Telephone Conversation with M. Kravett, GCA Technology
     Division, Inc.  IT Corporation, Martinez, CA.  April 1986.

43.  Warren, P.  Telephone Conversation with M. Kravett, GCA Technology
     Division, Inc.  Stablex Corporation, Rock Hill, SC.  April 1986.

44.  Garcia, G.  Telephone Conversation with M. Kravett, GCA Technology
     Division, Inc.  TWI, Inc., Sauget, IL.  April 1986.

45.  Bell, R.  Telephone Conversation with M. Kravett, GCA Technology
     Division, Inc.  Systech Corporation, Paulding, OH.  April 1986.

46.  A Profile of Existing Hazardous Waste Incineration Facilities and
     Manufacturers in the United States.  PB-84-157072.  EPA, Washington,
     D.C.  1984.

47.  National Survey of Hazardous Waste Generators and Treatment, Storage and
     Disposal Facilities Regulated Under RCRA in 1981.  U.S. Government
     Printing Office Order No. 055000-00239-8.  EPA, Washington, D.C.  1984.

48.  Engineering-Science.  Final Report.  Technical Assessment of Treatment
     Alternatives for Waste Solvents.  Prepared for U.S. Environmental
     Protection Agency, Technology Branch.  November 1983.

# 9. Emerging Thermal Treatment Technologies

With the passage of the 1984 amendments to RCRA banning the land disposal of hazardous wastes, thermal treatment of hazardous wastes has become an increasingly attractive option. Accordingly, there has been a great deal of interest shown in the development of new technological approaches to thermal treatment. HWERL has identified 21 "innovative thermal processes for treating or destroying hazardous organic wastes", many of which are applicable to halogenated organic wastes.[1]

Emerging thermal treatment technologies include modifications of conventional incineration technologies (e.g., the circulating bed incinerator) as well as more unconventional approaches to thermal destruction, e.g., the plasma arc pyrolysis system. The technologies included here, and discussed below, are:

1. Circulating Bed Combustion

2. Catalytic Fluidized Bed Incineration

3. Molten Glass Incineration

4. Molten Salt Destruction

5. Pyrolysis Processes

6. In Situ Vitrification

Discussions of these and other technologies can also be found in Reference 1 and the TRDs for solvents and dioxins.[2,3] In general, the thermal technologies discussed here can be applied to the destruction of any halogenated organic.

9.1   CIRCULATING BED COMBUSTION

Circulating bed combustion (CBC) systems constitute an innovation in
fluidized bed incineration technology.  These systems utilize high air
velocities and recirculating granular bed materials to maintain and achieve
combustion of waste under fluidized bed conditions.  The circulating bed
material also can be chosen for its chemical characteristics to bring about
reaction and neutralization of certain products of combustion such as sulfur
oxides and hydrochloric acid.  CBC systems are applicable to solids, liquids,
slurries, and sludges,[1] over a wide range of heat values and ash contents.
Numerous performance tests have been conducted which indicate that circulating
bed combustion can achieve very high destruction and removal efficiencies,
while limiting other pollutant emissions to acceptable levels.  CBC systems
can offer both technological and economic advantages over established
fluidized-bed incineration systems primarily due to the increased turbulence
of the system.  CBC systems operate at higher air velocities, and are not
limited, as are fixed bed units, to the narrow range of design velocities
needed to maintain fluidization.  At the same time, they limit entrainment and
carry over of bed material.

9.1.1   Process Description

The circulating bed combustion process, depicted in Figure 9.1.1,
represents a design innovation to standard fluidized bed (FB) incineration
systems.  The CBC system is designed to handle all forms of waste, including
solids, liquids, and sludges.

The primary operating unit, the circulating bed combustor, incorporates a
two-chamber design consisting of a combustion chamber and a hot cyclone
chamber, as shown in Figure 9.1.1.  The bed material used, limestone
($CaCO_3$), is fed to the system concurrently with the waste material.
Limestone is used because it readily reacts with sulfur and chlorine
combustion products to form relatively innocuous salts such as $CaCl_2$ and
$CaSO_4$.  The general reaction scheme for the CBC process is as shown in
Figure 9.1.2.  Waste material is fed to the system either before the
combustion loop for solids and sludges, or just at the start of the loop for

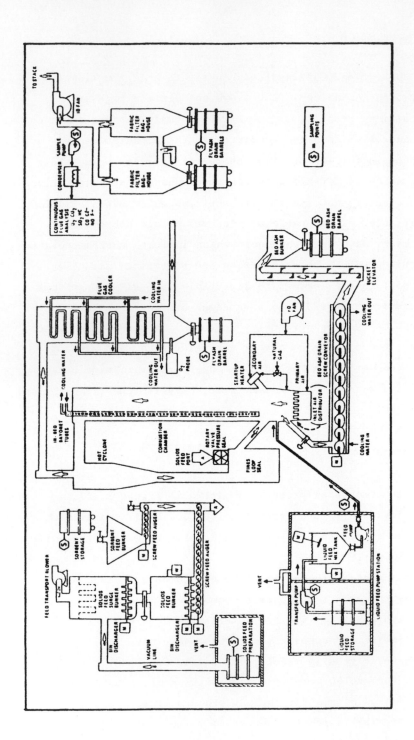

Figure 9.1.1.  CBC incineration pilot plant located at GA Technologies.

Source:  Reference 4.

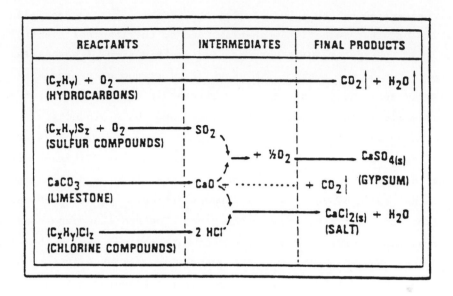

Figure 9.1.2.   Chemical reactions that occur in CBC combustion chamber.

Source:   Reference 4.

liquids.  As stated by the manufacturer, the CBC requires no specialized waste atomization or other injection mechanism, due to the inherently high level of turbulence in the system which ensures good distribution of waste feed.[5] During operation of the system, a high velocity stream of heated air (15 to 20 feet/sec.) entrains the material and carries it up the combustion column. As the waste flows upward, combustion occurs, and the byproducts are dispersed.  The gaseous products, primarily $CO_2$ and water vapor, flow out the top of the combustor.  The acidic byproducts such as HCl react with the limestone to form inorganic salts (generally these form as particulates). They and other solid byproducts flow downward through the hot cyclone, in which solids and gases are further separated.  The hot flue gases pass first to a heat exchange system, then to a particulate control device, before being vented through the exhaust stack.  Ash eventually settles within the combustion column and falls to a screw conveyor (as shown in Figure 9.1.1) where it is transported to ash recovery.

The operating conditions are as shown below.[1]

- Waste Feed:                Applicable to any physical form - granular
                             solids, liquids, sludges, slurries

- Temperature Range:         1400-1600°F (760-870°C)

- Residence Time:
    Gas Phase:               2-3 seconds
    Solids:                  10 seconds to 10 hours.

- Capacity (lbs/hr):         See Table 9.1.1

- Energy Type and Requirements:

    Thermal:                 Sensible and latent heat; self-sufficient
                             for wastes up to 85 percent water content

    Electrical:              Blower and feeder operation—approximately
                             30 HP for 2 MMBtu/hr incinerator

Capacities, shown in Table 9.1.1, are dependent upon the type of waste fed to the unit.  As noted, the data were furnished by the developer; commercial units covering the range of capacities have not yet been constructed.

TABLE 9.1.1.   CIRCULATING BED COMBUSTION UNITS

| Customer | Startup | Fuel | Output (MMBtu/hr) | Application |
|---|---|---|---|---|
| **USA – GA** | | | | |
| GA Technologies Inc. San Diego, CA | 1982 operating | Varied | 2 MW (t) | Pilot plant |
| **USA – Pyropower** | | | | |
| Gulf Oil Exploration Bakersfield, CA | 1983 operating | Coal, coke, and limestone | 50 | Enhanced oil recovery |
| California Portland Cement Co. Colton, CA | 1984 under construction | Coal and limestone | 209 | Cogeneration |
| B. F. Goodrich Henry, IL | 1985 | Coal and limestone | 123 | Process steam |
| Central Soya Chattanooga, TN | 1985 | Coal and limestone | 105 | Process steam |
| General Motors Corp. Pontiac, MI | 1986 | Coal, limestone, and plant wastes | 370 | Cogeneration |
| Colorado Ute Electric Utility Nucla, CO | 1987 | Coal and limestone | 1000 | Electrical generation |
| **Foreign – Ahlstrom** | | | | |
| Hans Ahlstrom Laboratory Karhula, Finland | 1976 operating | Varied | 2 MW (t) | Pilot plant |
| Pihlava Board Mill Finland | 1979 operating | Peat, wood, and coal | 50 | Cogeneration |
| Suonenjoki, Finland | 1979 operating | Peat, wood, and coal | 22 | District heating |
| Kemira Oy, Finland | 1980 operating | Zinciferous sludge | — | Sludge incineration |
| Kauttua, Finland | 1981 operating | Peat, wood, and coal | 220 | Cogeneration |

(continued)

TABLE 9.1.1 (continued)

| Customer | Startup | Fuel | Output (MMBtu/hr) | Application |
|---|---|---|---|---|
| Foreign (cont'd) | | | | |
| Hyvinkaa, Finland | 1981 operating | Coal, peat, oil, and municipal wastes | 85 | District heating |
| Skelleftea, Sweden | 1981 operating | Peat, wood, and coal | 22 | District heating |
| Ruzomberok, Czechoslovakia | 1982 operating | Sewage sludge | — | Sludge incineration |
| Hylte Bruk, Sweden | 1982 operating | Peat and coal | 157 | Cogeneration |
| Koskenkorva Distillery Finland | 1982 operating | Peat and oil | 63 | Process steam |
| Kemira Chemical Finland | 1982 operating | Peat and oil | 173 | Cogeneration |
| Zellstoff und Papierfabrik Frantschach AG Carinthia, Austria | 1983 operating | Bark, brown coal, and sludge | 188 | Cogeneration |
| Ahlstrom Varkaus, Finland | 1983 operating | Wood waste | 68 | Cogeneration—retrofit |
| Neste Lampo Oy Mantsala, Finland | 1983 operating | Coal-water mixture and coal | 10 | Heating—firetube design |
| Bord Na Mona Ballyforan, Ireland | 1984 | Peat and oil | 61 | Cogeneration |
| Oriental Chemical Co. Inchon, Korea | 1984 | Petroleum, coke, and coal | 330 | Cogeneration |
| Ostersunds Fjarrvarme AB Ostersund, Sweden | 1985 | Peat, wood chips, and coal | 85 | Heating |
| Papyrus AB Kopporfors, Sweden | 1985 | Bark, peat, and coal | 190 | Cogeneration |
| Metsaliiton Teollisuus Oy Aanekoski, Finland | 1985 | Wood waste, peat, coal, and oil | 258 | Retrofit |
| Kereva Power Company Kereva, Finland | 1985 | Coal and limestone | 102 | Utility-heating |

Source:   Reference 2.

Waste streams of primary environmental concern in the CBC process are: (a) the acidic byproducts and organic products of incomplete combustion (PICS); and (b) hazardous heavy metals or other solid byproducts remaining in the ash. To date, performance testing has indicated that acid or PICs in the flue gas stream are not usually significant. The ash will be handled as a solid waste. If hazardous materials exist, they will be disposed of in an appropriate manner.[5]

### 9.1.2  Demonstrated Performance

A pilot-scale CBC was tested by the California Air Resources Board in cooperation with the manufacturer, GA Technologies, in 1983.[6] The testing involved a surrogate waste mixture which had a heating value of 8000 Btu/lb and included organic compounds such as xylene, ethylbenzene, toluene, hexachlorobenzene, Freon, and carbon tetrachloride. The CBC unit operated at a capacity of 0.5 MMBtu/hr, and a temperature of below 1600°F (870°C).

A DRE of 99.99 percent or greater was achieved for all principal organic constituents. Emissions of hydrochloric acid met EPA standards by use of a limestone sorbent in the bed.

Some of the conclusions drawn by Chang and Sorbo in Reference 6 are presented below:

1. The DRE of volatile and semi-volatile POHCs under less than optimum combustion conditions met RCRA requirements (99.99 percent DRE).

2. Total volatile PIC formation was found to correlate well with CO and THC, normalized to fuel flowrate ($CO_2$). Penetration of volatile chlorinated PICs (based on total chlorine content of the fuel) exceeded $1 \times 10^{-4}$. PIC benzene appeared in substantial concentrations in several samples and was not correlated with any conventional combustion parameters.

3. The DRE dropped sharply when the bed temperature fell below 700°C. Temperature appeared to be a major factor in the destruction of the fluorinated compounds and a moderate correlation between sulfur hexafluoride, DRE and temperature was observed.

4. The CBC seemed to behave as a plug-flow reactor, susceptible to pockets of non-stoichiometric air/fuel mixtures passing through the bed causing increased PIC formation. This observation suggests the importance of the fuel feed system on CBC performance and should be evaluated carefully by permitting authorities.

GA Technologies, the developer, has reported more than 7,500 hours of performance testing conducted under the auspices of DOE, EPRI (Electric Power Research Institute), TVA, and a number of commercial sponsors. These tests were conducted at the company's $2 \times 10^6$ Btu/hr test unit in San Diego, CA. The system has been tested with a variety of fuels and wastes to establish the combustion efficiency and the pollutant removal efficiency of the system relative to specific waste types. For halogenated organic waste types, the system has generally shown a DRE above 99.99 percent, and an HCl capture of 99 percent and above. Particulate emissions downstream of the unit's fabric filter baghouse have been measured at 0.002 grains per standard cubic foot, well below the RCRA incinerator standard. In summary, DREs found for POHCs existing in the organic wastes are as shown below:

| Compound | DRE | Temperature (°F) |
|---|---|---|
| Ethylbenzene | 99.99 | 1600  (871°C) |
| Trichloroethane | 99.999 | 1600 |
| Vinyl chloride | 99.9999 | 1600 |
| 1,2-trans-dichloroethylene | 99.99 | 1600 |
| 1,2-dichloroethane | 99.99 | 1600 |
| PCBs | 99.9999 | 1800 (982°C) |
| Pentachlorophenol | 99.992 | ---- |

## 9.1.3  Cost of Treatment

The costs of circulating bed incinerators according to GA Technologies Inc., are equivalent to the costs of conventional fluidized bed systems and less than those for rotary bed incinerators. Additional cost savings will also result from control of pollutants, such as those resulting from chlorine and sulfur in the waste, through addition of dry limestone to the bed. Chlorine capture efficiencies are reported to exceed 99 percent, a condition that meets EPA incinerator requirements. EPA requirements for particulate emissions will require the use of pollution control equipment, e.g., ESPs or baghouses. Overall cost of incineration was estimated to range from $31 to $235 per metric ton for PCBs, pesticides, and halogenated solvents. Costs for nonsolvent halogenated organics should be similar.

9.1.4   <u>Status of Technology</u>

Circulating bed combustion systems are in operation worldwide, for many process applications.  However, there are no CBC incinerators operating specifically as hazardous waste incinerators.  Although, as the manufacturer points out, many of the wastes disposed of by currently operating CBCs contain hazardous constituents.  A listing of the operating units, submitted by the company, is shown in Table 9.1.1.  GA Technologies is the only manufacturer of CBC technology.[5]  In terms of market potential, the company provides the general comparison between existing technologies and CBC shown in Table 9.1.2.

While the CBC concept and available performance data are promising, additional data are needed to validate DREs and establish air emission levels for particulates, PICs, chlorine based pollutants, and other possible toxics.  As noted in Reference 6, plug-flow reactor behavior, if it occurs, could lead to incomplete combustion and high emission levels of contaminants in the feed.

TABLE 9.1.2.   CIRCULATING BED INCINERATOR VS. CONVENTIONAL INCINERATORS

| Item | Circulating Bed | Bubbling Bed | Rotary Kiln |
|---|---|---|---|
| **Cost** | | | |
| Capital | $ | $ + scrubber<br>+ extra feeders<br>+ foundations | $$ (double)<br>+ scrubber<br>+ afterburner |
| Operating | $ | $ + more feeder<br>maintenance<br>+ more limestone<br>+ scrubber | $$ + more auxiliary<br>fuel<br>+ kiln maintenance<br>+ scrubber |
| **Pollution control** | | | |
| POHCs | In minimum-temperature<br>combustor | In high-temperature<br>combustor or afterburner | In afterburner |
| Cl, S, P | Dry limestone in combustor | Downstream scrubber | Downstream scrubber |
| $NO_x$, CO | Low due to turbulence,<br>staged combustion | High: bubbles bypass and<br>poor fuel distribution | High $NO_x$: hot afterburner |
| Upset Response | Slump bed; no pollution | Bypass scrubber pollution<br>released | Bypass scrubber pollution<br>released |
| Effluent | Dry ash | Wet ash sludge | Wet ash sludge |
| **Feeding** | | | |
| No. of Inlets | 1-solid<br>1-liquid | 5-solids<br>5-liquids | 1-solids<br>2-liquids |
| Sludge Feeding | Direct | Filter/atomizer (5 each) | Filter/atomizer (2 each) |
| Solids Feedsize | Less than 1 in. | Less than ½ to ¼ in. | Larger, but shredded |
| Fly-Ash Recycle | Inherent (50 to 100 ×<br>feedrate) | Difficult mechanical/<br>pressure (10 × feedrate<br>max) | Not done |
| **Unit size** | | | |
| Land area | Smaller | Larger (over 2×) | Larger (over 4×) |
| **Efficiency** | | | |
| Thermal, % | >78 | <75 | <70 |
| Carbon, % | >98 | <90 | — |
| Feeder, hp | Minimum | High | High |

SI Conversion: mm = in. × 25.4

Source:   GA Technologies, Inc.

## 9.2   CATALYTIC FLUIDIZED BED INCINERATION

The combustion temperatures required for high level destruction of wastes and, hence, the operating costs associated with an incinerator may be significantly reduced through the use of catalysts.  Catalysts serve to increase the rate of oxidation by adsorbing reactants onto active sites, thus increasing their localized concentration.  By increasing the reaction rate, less energy input is required to achieve component destruction efficiencies which are substantially similar to conventional incinerator systems.  Thus, a catalytic incinerator operating at temperatures on the order of 500 to 600°F lower may destroy certain species as effectively as a conventional system (Reference 7).

Catalytic incinerators have been proven effective as a fume abatement (air pollution control) device for many years.  Such systems typically employ a static catalyst bed into which material flows and oxidation ensues (see Figure 9.2.1).  The catalyst bed may be either a single "honeycomb"-type unit, or a bed of pellets or uniquely-designed pieces (not unlike the packing materials used in packed-tower scrubber systems).  Such systems have not been proven effective, however, in the incineration of streams containing high particulate loadings or high concentrations of halogens.  As stated in Reference 9, the performance life of a catalyst may be seriously deteriorated by the effect of plugging or corrosion, especially by HCl.  Excessively high localized temperatures will also tend to deactivate the catalyst in a shorter time period.

Beginning in the 1970's, studies began to indicate that, while problems could not always be entirely eliminated, the continuous replacement of a portion of the catalyst bed during normal operation could allow continuing operation at high efficiency, even in the presence of poisoning agents.  This criterion could be satisfied by a fluidized bed reactor into which catalyst could be added, and from which withdrawals could be made during operation. The use of a fluidized bed could also provide some protection against catalyst poisoning because of the self-abrading action of the catalyst particles in the bed, which continuously cleans the surfaces available for activation.  Such

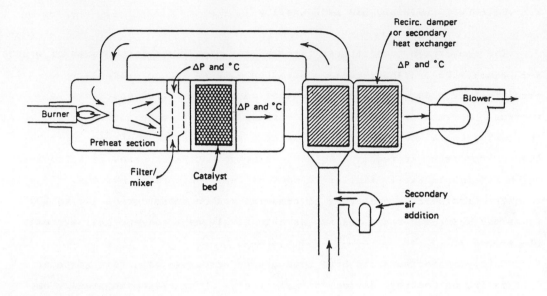

Figure 9.2.1.  Typical components of a catalytic incinerator.
ΔP and °C indicate pressure drop and heated
gas, respectively.  (Reference 8)

systems have been researched by several different manufacturers, both as an independent system or as a system to be used in combination with another waste treatment technology. Although the overall status of catalyzed fluid bed incineration is still in an early development stage some promising results have reportedly been observed relative to incineration of halogenated organic wastes.

## 9.2.1  Process Description

The catalytic fluidized bed incineration process would appear to be essentially identical to a standard fluidized bed incinerator process, with the exception of the bed materials. An example of this process is the "CATOXID" process designed by BF Goodrich, as shown in Figure 9.2.2.

It is likely that pretreatment of the waste stream would be even more important in catalytic fluidized bed incineration processes, given the negative effect that high particulate loadings and high halogen concentrations would have on the catalyst bed. It is also necessary to remove waste constituents such as sodium salts which may contribute to the clogging (defluidization) of the bed. Information detailing the specifics of the processes currently in development is limited in availability, particularly with respect to the catalyst materials being used. For more detailed information on the fluidized bed incineration process, the reader is advised to refer to Section 8.

## 9.2.2  Demonstrated Performance

Performance data relative to catalytic fluidized bed incineration in general, and their performance in the destruction of halogenated organic compounds in particular, are limited. The three primary studies which appear to have been conducted to date were summarized by Manning (Reference 7).

Hardison and Dowd (Reference 10) described the performance of a pilot-scale catalytic unit of 0.91m in diameter with a 15cm bed of propriety catalyst. In the disposal of a refinery waste stream, the unit oxidized

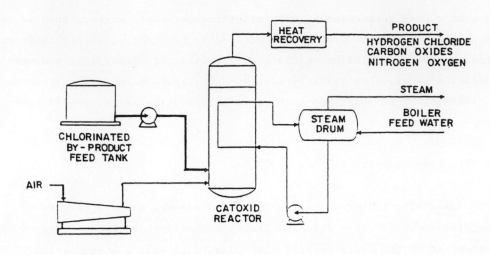

Figure 9.2.2.   Flow diagram of B. F. Goodrich CATOXID process. (Reference 7)

98 percent of the feed organic materials while operating at a combustion
temperature of approximately 820°F and a residence (catalyst contact) time of
150 milliseconds, as illustrated in Figure 9.2.3.   It was claimed that the
catalyst deactivation rate was low, as illustrated in Figure 9.2.4.
Unfortunately, the article failed to detail which catalyst materials were
involved in the test.

In 1981, Rockwell (Reference 11) announced that a two stage fluidized bed
process had achieved a destruction efficiency of greater than 99.9999 percent
for PCBs and PCB surrogates during a test burn of a sample of transformer
coolant (52 percent by weight PCB, 48 percent trichlorobenzene).   The
combustion temperatures during this test reportedly were at or below 1300°F.
The system combined a conventional fluidized bed reactor, which also contained
a sodium carbonate sorbent for HCl and a chromia-catalyzed secondary fluidized
bed reactor which oxidized any materials not combusted in the first unit.

Benson reported in 1979 (Reference 12) on the results achieved by the
largest scale catalytic oxidation process for chlorinated hydrocarbons, the
BF Goodrich CATOXID process.   At temperatures below 1000°F, a feedstock
containing several low molecular weight hydrocarbons was oxidized "essentially
to completion".   In this process, catalyst life was reported to be indefinite
and was not reported to have been significantly poisoned by the chlorinated
species.   Manning's own results, as shown in Figures 9.2.5 and 9.2.6,
indicated "effective oxidation for several chlorocarbons".   While the
poisoning of the catalyst clearly was shown in this study to be directly
proportional to chlorocarbon feed rate or concentration (see Figure 9.2.7),
Manning concluded that extended operation may be possible at only slightly
reduced rates if feed stoichiometry (specifically the H/Cl ratio in the mixed
feed) is controlled.   Manning further concluded that, while a catalyzed system
may not be best for a large-scale commercial waste disposal facility handling
diverse waste materials, it may be applicable to usage in the chemical process
industry, where much of the volume of HOC waste is presently being treated
onsite.   In this environment, where the waste streams are less variable in
composition and volume, the application of catalytic oxidation in fluidized
bed reactors offers more immediate potential.[7]   This observation is
substantiated by the pattern of usage of the technology to date.

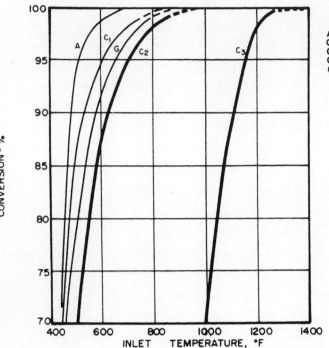

A - ACETONE  $CH_3COCH_3$  HHV  13,220  BTU/#
$C_1$ - CUMENE  $C_6H_5$ $CH(CH_3)_2$ LHV 17,712  BTU/#
G - GASOLINE  $C_8H_{18}$        LHV 19,070  BTU/#
$C_2$ CUMENE
$C_3$ CUMENE  THERMAL  REACTION

Figure 9.2.3.   Calculated conversion efficiencies for
several organic materials.   (Reference 10)

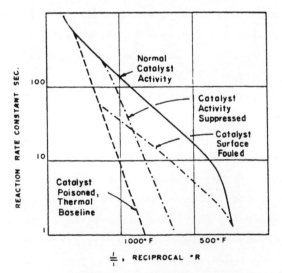

Figure 9.2.4.   Effect of fouling materials, poisons,
and suppressants on catalyst activity.
(Reference 10)

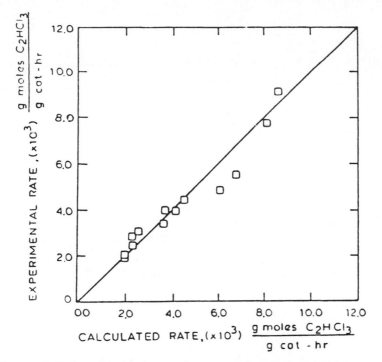

Figure 9.2.5.   Comparison of measured $C_2HCl_3$ oxidation rates and calculated rates from regression.   (Reference 7)

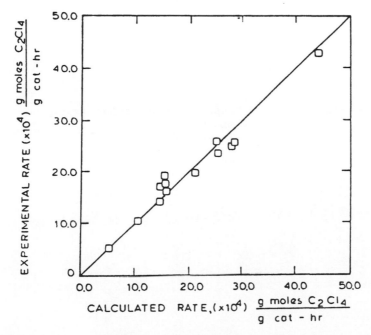

Figure 9.2.6.   Comparison of measured $C_2Cl_4$ oxidation rates and calculated rates from regression.   (Reference 7)

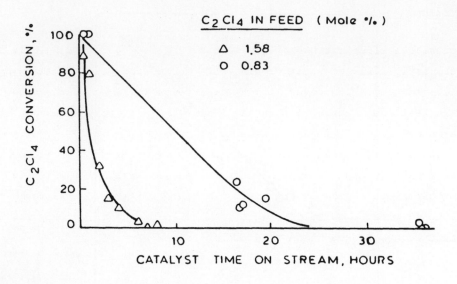

Figure 9.2.7. Catalyst deactivation during
oxidation of dry $C_2Cl_4$.
(Reference 7)

### 9.2.3  Costs

Because the fluidized bed catalytic incineration systems are still in a relatively early stage of their development, detailed cost information is not available.  In general, however, the capital costs would be expected to closely reflect the costs of conventional fluidized bed incinerators.  The catalyst materials used, typically made from exotic materials, would tend to add, perhaps significantly, to the overall capital cost.  The higher capital costs could be more than offset, however, by the significant savings in operating costs due to reduced fuel comsumption.  Such systems would be expected to involve somewhat higher maintenance costs, particularly for the destruction of HOC wastes which will poison the catalyst pellets.

### 9.2.4  Status of Development

The fluidized bed catalytic incineration system is primarily in the early stages of its development as a waste disposal technology.  Several units, however, appear to have been installed at industrial facilities.  Hardison and Dowd reported that two commercial units manufactured by Air Resources, Inc. were designed and placed into operation, including one destroying chlorinated organics at a chemical maunfacturing facility.  The BF Goodrich CATOXID process has also been implemented at several Goodrich plants, including one at the Calvert City VCM complex which handles all of the chlorinated by product streams at the plant.  Benson reported that several CATOXID units are in operation worldwide.

## 9.3   MOLTEN GLASS INCINERATION

Molten glass incinerators (MGI) are electric furnace reactors in which a pool of molten glass is used both as a means of destroying hazardous organic wastes and/or as a means for encapsulating the solid byproducts of hazardous waste treatment.  The system utilizes furnaces similar to those used extensively in the glass manufacturing industry.  Combustible hazardous wastes of virtually any physical form or chemical composition may be destroyed effectively in MGI systems.  The system is considered particularly attractive for the destruction of highly toxic organic wastes, wastes containing heavy metals, and contaminated soils.[1]  Solids introduced with the waste feed and many solid products of combustion become incorporated in a glass matrix, rendering them environmentally inert and land disposable.  Molten glass systems are being studied by two separate firms (Battelle Northwest and Penberthy Electromelt International) as hazardous waste treatment devices. The process is considered to have certain technological and economic advantages over other established incineration technologies.[1]

### 9.3.1   Process Description

The molten glass furnace is a tunnel-shaped reactor, lined with refractory brick, in which a pool of glass is maintained in a molten state by electric current passing through the glass between submerged electrodes.  Such furnaces are used extensively in the glass manufacturing industry.  The unit is designed to withstand temperatures as high as 1260°C (2300°F), and corrosion by acidic gases.  MGI systems, as designed, will be equipped with heat recovery and air pollution control systems, and can be combined with a preconditioning heater or primary incineration unit, as depicted in Figure 9.3.1.

In the absence of a primary incineration unit, wastes can be fed directly to the furnace chamber, above or into the pool of molten glass.  Solids, slurries, and highly viscous liquids are usually charged via a screw feeder. Liquids may also be sprayed into the chamber through nozzles located at the top of the unit.  Combustion air is fed to the system from two locations, one near the top (as shown in Figure 9.3.1), and the other nearer to the surface

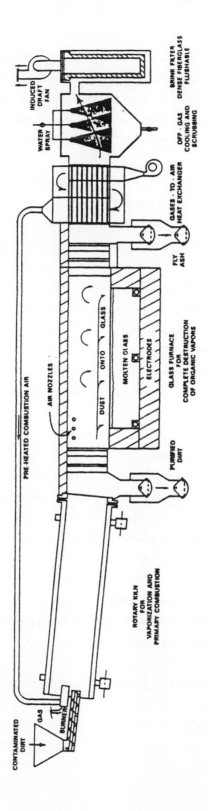

Figure 9.3.1.  Dirt purifier and hazardous waste incinerator.

Source:  Reference 13.

of the pool on the opposite side, in order to maximize the turbulence within
the reaction space.  The temperature within the chamber is maintained at
2300°F.  Residence time of gases within the chamber is about 2 seconds
although this can be increased if desired by reducing load.  Residence time of
solids within the glass will be appreciably longer, and is measured in terms
of hours.  Several furnace sizes accommodating various waste feed rates are
available.[13]

During operation, volatile waste materials mix with air, ignite, and
react in the space above, and at the surface of, the pool of molten glass.
The solid products of combustion, dirt, and other noncombustible materials
(e.g., heavy metal contaminants or the solid waste being treated) will be
incorporated into the glass bed.  Gaseous products flow out of the chamber
through a series of ceramic fiber filters which catch most of the particulate
matter.  The hot gases, consisting primarily of $CO_2$, water vapor, and HCl
(if chlorinated organics are incinerated), then pass through a heat exchanger
for heat recovery (heat is used to warm the combustion air, as shown in
Figure 9.3.1).  The exhaust gases flow next to a series of water spray-type
scrubbers.  The first spray chamber is designed to use a slightly alkaline
scrubbing liquor to capture acidic vapors.  Water is used in the other spray
chamber (or chambers) to remove remaining particulates and other scrubbable
vapors.  The gases are then reheated above the dewpoint, and passed through
charcoal and HEPA filters before being vented out the stack.  The entire
system is maintained under negative pressure by means of the exhaust
blower.[13]

After a period of usage, the molten glass bed, with the solid waste
materials incorporated, is tapped out of the chamber into metal canisters and,
after cooling, is sent to a disposal facility.  The ceramic filters, which
eventually become loaded with particulate matter, can be disposed of by
dissolving them in the molten glass bed.  The glass bed can also be used to
encapsulate the sludge from the spray chambers, the spent charcoal and HEPA
filters.

Major advantages of the molten glass incineration system are its
applicability to many forms of hazardous waste and the encapsulation of
residuals in a nonleachable glass matrix.  Performance testing, and data
generated from commercial usage of MGI units in the chemical processing

industry, while limited, have shown no significant difference in the effective operation of such systems for wastes of different physical forms and widely varying chemical composition.  However, preheating and chemical treatment of wastes are often used to aid combustion and reduce system maintenance and down time.  The waste related factors which may be of the greatest particular concern are moisture, metals and inorganics content.  The significance of these characteristics are discussed below.

A high concentration of water in waste will necessitate additional energy input to the system and may affect destruction efficiencies.  Penberthy has set a moisture content limit of 20 percent (by weight) for its systems.[13] Since many wastes contain water at levels higher than this, pretreatment of the waste will be needed.  Pretreatment systems which can be used include evaporation and sedimentation.  Dewatering options may be somewhat limited for certain wastes, due to characteristics such as volatility and miscibility with water.

Metals and minerals which are constituents of wastes pose a problem to the effective operation of molten glass incinerators.  Those materials which are denser than the molten glass will tend to accumulate near the bottom of the furnace.  (Battelle reports that its process, which involves intermixing of molten glass and waste, achieves 95 percent detention of nonvolatile heavy metals.)[1]  Eventually, due to their electrolytic properties, they may affect the operation of the metal electrodes.  Penberthy has recommended the usage of sumps to collect and localize settling metal particles.  Such systems have been found to be effective in reducing the effect of metals on furnace operation.[13]

## 9.3.2   Demonstrated Performance

No data, demonstrating DREs or quantifying exhaust gas emissions, are available for halogenated organic wastes or any other wastes.  These data are needed if this technology, which appears promising in concept, is to be utilized for hazardous waste treatment.

### 9.3.3  Cost of Treatment

Costs will depend to an appreciable extent upon the need for pretreatment and the demands placed on the system used to clean exhaust gases.

### 9.3.4  Status of Technology

Molten glass incinerators are available commercially from Penberthy Electromelt International Inc. for use as chemical processing units.  Battelle Northwest, another company involved in the development of MGI systems, has not yet produced equipment on a commercial scale.  The Penberthy system has not been sold or permitted specifically as a hazardous waste incinerator to date.  However, despite the lack of information concerning its application to hazardous wastes, the technology would appear to offer certain definite advantages.  Anticipated advantages are as follows:

- Able to achieve significant waste volume reduction

- Able to destroy almost all forms of hazardous waste, largely independent of physical state or chemical composition

- Heat recovery and air pollution control built into system

- Solid byproducts transformed via glass encapsulation to environmentally safe state.  The encapsulates are resistant if not inert to chemical reaction, leaching, and fracture.  They probably can be disposed of in landfills

- System is small in size, can be transportable

- Equipment used is relatively simple, representing basic technology that has been applied in heavy industry for 30 years

Limitations, also largely conjectural at this stage, include the following:

- Unproven technology.  There is no knowledge of long term operation and maintenance requirements, or how performance would be affected by long term usage with wastes

- Energy costs and capital costs are relatively high

- Control system as described may be inadequate for exhaust gases of the type anticipated from hazardous waste destruction

## 9.4    MOLTEN SALT DESTRUCTION

Molten salt incinerators involve the combustion of waste materials in a bed of molten salt.  Using the molten salt incineration process, "organic wastes may be burned while, at the same time scrubbing in situ any objectionable byproducts of that burning and thus preventing their emission in the effluent gas stream."[14]  Molten salt incinerators were developed by Rockwell International, specifically to burn hazardous organic wastes and, as designed, are applicable to both liquid and solid waste streams.  However, wastes with high ash content or a high percentage of water or noncombustible material are not good candidates for molten salt destruction.

### 9.4.1    Process Description

The molten salt destruction process has been under development by Rockwell International since 1969.[15]  The original intent was to use the process to gasify coal.  A variety of salts can be used, but the most recent studies have used sodium carbonate ($Na_2CO_3$) and potassium carbonate ($K_2CO_3$) in the 1450°F to 2200°F (790°C to 1200°C) temperature range.

As shown in a schematic of the Rockwell process (Figure 9.4.1), the waste is fed to the bottom of a vessel containing the liquid salt along with air or oxygen-enriched air.  The molten salt is maintained at temperatures of 800–1,000°C (1500 to 1850°F).[16]  The high rate of heat transfer to the waste causes rapid destruction.  Hydrocarbons are oxidized to carbon dioxide and water.  Constituents of the feed such as phosphorous, sulfur, arsenic, and the halogens react with the salt (i.e., sodium carbonate) to form inorganic salts, which are retained in the melt.[16]  The operating temperatures are low enough to prevent $NO_x$ emissions.[1,15]

Eventually, the build-up of inorganic salts must be removed from the molten bed to maintain its ability to absorb acidic gases.  Additionally, ash introduced by the waste must be removed to maintain the fluidity of the bed.  Ash concentrations in the melt must be below 20 percent to preserve fluidity.

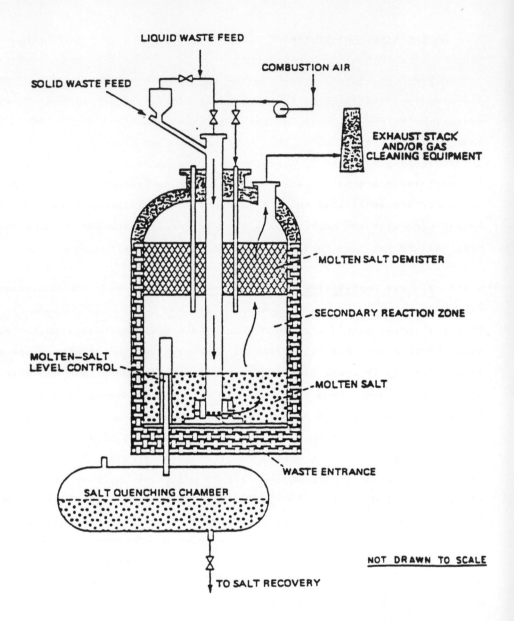

Figure 9.4.1.  Molten salt combustion system

Source:    Reference 14.

Melt removal can be performed continuously or in a batch mode.
Continuous removal is generally used if the ash feed rates are high.  The melt
can be quenched in water and the ash can be separated by filtration while the
salt remains in solution.  The salt can then be recovered and recycled.  Salt
losses, necessary recycle rates, and recycling process design are strongly
dependent on the waste feed characteristics.[1,15]

9.4.1.1  Waste Characteristics and Pretreatment Requirements--

Molten salt destruction (MSD) systems are limited in their applicability
to various hazardous wastes.  Although the system is capable of handling
hazardous wastes in both the liquid and solid state, MSD is in practice
limited to the incineration of hazardous organic wastes which have a
relatively low percentage of solids or inorganics.  Slurried wastes and most
"dry" solid wastes (e.g., contaminated soils) are not good candidates for
incineration by MSD.  When ash accumulates in the bed, it tends to form a
waste matrix which eventually affects bed fluidity, the overall transfer of
heat and capacity for waste byproduct neutralization within the molten mass.
Thus, 20 percent ash was determined to be the limit to which the system could
effectively operate.[16]

Wastes with high water content may limit the effectiveness of the molten
salt destruction process.  As moisture content increases, the waste will
require more fuel and combustion air, to the point where the reactor volume is
limited.  Thus, many wastes must be dewatered by pretreatment to ensure that
they are effectively destroyed in the MSD.

Discussion with Rockwell indicated that there is no established
pretreatment system designed as part of the MSD system.  However, separation
technology to remove solids and to dewater wastes prior to incineration in a
MSD unit must be considered.

9.4.1.2  Operating Parameters--

The operating parameters for a molten salt unit are:

- Temperature Range:          800-1000°C (1500-1850°F)

- Residence Time:
  Gas Phase                   Approximately 5 seconds
  Liquid or Solid Phase       Hours

- Energy Requirement:  Natural gas or oil to heat salt bed; Auxiliary fuel for noncombustible wastes; Power for exhaust

- Available Capacity:  Commercial units available at 2,000 lbs/hr; Pilot scale in use operating at 250 lbs/hr

- Operating Limitations:  Heat generation: MSD requires a cooling system to prevent operational failures

9.4.1.3 Post-Treatment Requirements--

Although post-treatment requirements have not yet been defined, it is likely that treatment will be required to remove products of combustion that are not scrubbed out of the exhaust gases by the molten salt. These products of combustion could include particulates, POHCs and PICs. Solid residues (i.e., used salt) must be reprocessed or disposed.

9.4.2 Demonstrated Performance

Rockwell International has built two bench scale combustors (0.5 to 2 lb/hr), a pilot plant (55 to 220 lb/hr), and a portable unit (500 lb/hr).[15] They have also built a 200 lb/hr coal gasifier based on the molten salt process. Destruction efficiency tests have been conducted at the bench and pilot scale levels. While no data were found to demonstrate the DRE of the molten salt destruction technology for the halogenated organics, data showing five nines to eleven nines DRE for certain organic compounds have been obtained.

Many wastes have been tested in the bench scale unit. Chemical warfare agents, including Mustard HD, a chlorinated organic, have been destroyed at efficiencies ranging from 99.999988 to 99.9999995 percent. Other chemicals that have been destroyed using the molten salt combustion process include: chlordane, malathion, Sevin, DDT, 2,4-D herbicide, tar, chloroform, perchloroethylene distillation bottoms, trichloroethane, tributyl phosphate, and PCBs.[16]

The PCB trial combustion data are presented in Table 9.4.1. The destruction efficiency at the lowest operating temperature, 700°C (1300°F), exceeded 99.99995 percent. The average residence time of the PCB in the

TABLE 9.4.1.   PCB COMBUSTION TESTS IN SODIUM-POTASSIUM-CHLORIDE-CARBONATE
               MELTS

| Temp (°C) | Stochiometric air (%) | Concentration of KCl, NaCl in melt (wt %) | Extent of PCB destruction[a] (%) | Concentration of PCB in off-gas[a] ($\mu g/m^3$) |
|---|---|---|---|---|
| 870 | 145 | 60 | 99.99995 | 52 |
| 830 | 115 | 74 | 99.99995 | 65 |
| 700 | 160 | 97 | 99.99995 | 51 |
| 895 | 180 | 100 | 99.99993 | 59 |
| 775 | 125 | 100 | 99.99996 | 44 |
| 775 | 90 | 100 | 99.99996 | 66 |

[a]PCBs were not detected in the off-gas, i.e., values shown are detection
limits.

Reference:   Reference 17.

melted salt was 0.25 to 0.50 seconds, based on gas velocities of 1 to 2 ft/sec through the 0.5 ft of melt.[15]

Hexachlorobenzene (HCB) and chlordane destruction were tested in the pilot plant facility.[17] Feed rates for HCB and chlordane were as high as 269 lb/hr and 72 lb/hr, respectively. Bed temperatures ranged from 1685° to 1805°F (920°C to 985°C) and residence times were in the 2 to 3 second range. HCB destruction efficiencies ranged from 99.9999999 to 99.999999999, and chlordane destruction efficiencies ranged from 99.99999 to 99.999999. The results of the pilot-scale tests are summarized in Table 9.4.2.

As shown in Table 9.4.2, very high DREs were noted for both compounds. HCl emissions were below 100 ppm, and no $Cl_2$ gas or phosgene gas was detected. Particulate emissions were measured, but were found to be quite low, and analysis showed that particulate matter was nonhazardous. The improved performance in the pilot scale reactor was attributed to greater residence times.

## 9.4.3   Costs of Treatment

Detailed estimates of costs for molten salt destruction have not been formulated. Based on the performance of the bench- and pilot-scale MSD units, it is speculated that general operating costs will be low, but that the initial capital costs will be high. Molten salt destruction operating costs should be lower than established technologies such as rotary kilns. Operating temperatures are low and the process does not require a complex air pollution control system and associated appurtenances, (although emission data are needed to verify this), or ash recovery and transport systems.

## 9.4.4   Status of Technology

Molten salt destruction systems are a proprietary design of the Rockwell International Corporation. Rockwell began development of the MSD system in 1969, obtaining several patents for the technology. By 1980, the system was made available for commercial-scale application, at a capacity of 2000 lbs/hr, for destruction of specific waste types. The company constructed, and currently maintains three different sized units, including a bench-scale (2 lbs/hr) unit, and a pilot-scale (200 lb/hr) unit, and a full scale

TABLE 9.4.2.   SUMMARY OF PILOT-SCALE TEST RESULTS

| | PCB | Chlordane |
|---|---|---|
| Combustor Feed Rate (lb/hr) | 20.9 - 122.0 | 12.1 - 32.7 |
| Combustor Off-gas | | |
| - mg/m$^3$ | $2.7 \times 10^{-4}$ - $7.1 \times 10^{-2}$ | $5.3 \times 10^{-3}$ - $6.8 \times 10^{-2}$ |
| - ppmv | $2.3 \times 10^{-5}$ - $6.1 \times 10^{-3}$ | $3.2 \times 10^{-4}$ - $4.1 \times 10^{-3}$ |
| Baghouse | | |
| - mg/m$^3$ | $6 \times 10^{-6}$ - $1.6 \times 10^{-4}$ | $3.6 \times 10^{-4}$ - $4.4 \times 10^{-3}$ |
| - ppmv | $5.2 \times 10^{-7}$ - $1.4 \times 10^{-5}$ | $2.1 \times 10^{-5}$ - $2.6 \times 10^{-4}$ |
| Spent Melt (ppmv) | 0.001 - 0.104 | 0.0044 - 1.2 |
| NO$_x$ (ppmv) | 70 - 125 | 0.5 - 630 |
| HC (ppmv) | 35 - 110 | 0.4 - 60 |
| Particulate (mg/m$^3$) | $6.2 \times 10^{-3}$ - 0.107 | $4.1 \times 10^{-3}$ - $1.75 \times 10^{-2}$ |
| DRE (%) | 11-9's - 9-9's | 8-9's - 7-9's |

Note:   The pH of the liquid in a small sampling scrubber in the off-gas line remained basic throughout the test indicating essentially no HCl emission.

Source:   Reference 1.

(2000 lbs/hr) unit, for demonstration of molten salt incineration capabilities. However, no commercial scale units have been sold by the company to date. Rockwell has indicated that development of this technology has been curtailed, due in part to the limited demand encountered. Rockwell will maintain their demonstration units and considers future development of MSD a possibility.[16]

As demonstrated in the molten salt destruction process performance tests, MSD systems have certain distinct advantages as an incineration technology alternative. The limitations of the system however, may prove to severely limit its further development.

Advantages--

- Achievement of high destruction efficiencies for many wastes, including highly toxic and highly halogenated wastes;

- Low $NO_x$ and heavy metal emissions

- Retention of halogens and metals in a manageable salt matrix;

- Compact size. The process has few moving parts and acts as its own, highly efficient scrubber for acid combustion gases;

- Especially well-suited to wastes whose combustion results in liberation of acids;

- Improved reliability due to simple design;

- Increased waste throughput possible

Limitations--

- Generally restricted to certain types of organic hazardous wastes;

- Sensitive to high ( 20 percent) ash content in wastes;

- Molten salt is corrosive to all but specific engineering alloys. Material and construction costs will therefore be high, and management of spent salt beds will be difficult;

- No commercial applications to date, thus, no existing record of long-term performance and operation and maintenance requirements

## 9.5   PYROLYSIS PROCESSES

Pyrolysis reactors are systems in which destruction of waste contaminants is accomplished by applying sufficient thermal input to bring about bond fracture and molecular decomposition.  No oxidation reactions are involved in these processes.  Pyrolysis reactors can achieve very high destruction efficiencies for wastes, including difficult to dispose wastes such as dioxins.  A variety of pyrolysis systems have been developed, including continuous and batch furnace pyrolyzers, the plasma arc reactor, and the high temperature fluid wall reactor.  These are described below.

### 9.5.1   Furnace Pyrolysis Systems

#### 9.5.1.1   Process Description--

The pyrolysis system shown in Figure 9.5.1 consists of three major components:  a continuous rotary furnace, a rich fume secondary combustion chamber, and a heat recovery unit.  The furnace is similar to furnaces employed to treat metals and other materials requiring controlled thermal treatment.  Waste is continuously fed to a rotating belt which passes through an indirect-fired, oxygen-free pyrolytic chamber.  The waste is heated to between 540°C (1000°F) and 870°C (1600°F).  Volatiles in the waste or resulting from pyrolysis are driven off leaving behind inert materials, metals, and other inorganics, which are continuously removed from the moving belt.  The volatile gases, containing organic compounds and products of pyrolysis such as $H_2$ and HCl, are combusted in the rich fume reactor to complete the destruction of any organic materials present.  These gases then flow through a waste heat boiler or a similar device used to recover energy.  Although some HCl formed by pyrolysis is removed through contact with alkaline components (either in the waste or added to the feed deliberately for that purpose), it is conceivable that some type of air pollution control device might be needed to control acid gas emissions.  Reportedly other pollutants such as particulates and nitrogen oxides will not be a problem because of the low turbulence level within the pyrolysis chamber and the reducing atmosphere of pyrolysis, respectively.

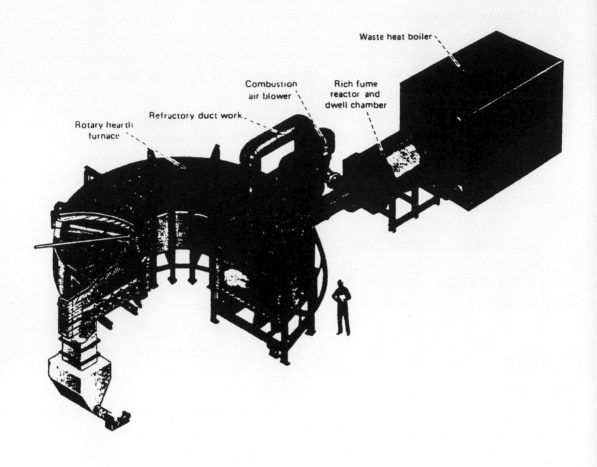

Figure 9.5.1.    Continuous pyrolyzer.

Source:    Reference 18.

Although wastes with a wide range of chemical characteristics may be treated in a pyrolytic incinerator, certain wastes are clearly better candidates than others.  As noted by Midland-Ross, the developer,[18] pyrolysis systems work best for wastes which fall into the following categories:

1.   Too viscous to atomize in liquid incinerators, yet too fluid for spreader-stoker incinerators.

2.   Low melting point materials that foul heat exchangers, spall refractories, and complicate residue discharge.

3.   High residue materials (ash), with easily entrained solids, that would generally require substantial stack gas cleanup.

4.   Material containing priority pollutants with excessive vapor pressure at incineration temperatures.

5.   Any material, drummed or loose bulk, where controlled thermal treatment is desired to make clean gases for heat recovery or for discharge to the atmosphere.

Operating conditions for the components of the two types of pyrolyzers (batch and continuous) produced by Midland-Ross are as follows:[1]

Pyrolyzer

- Temperature Range:  650°-870°C (1200°-1600°F)

- Residence Time Range:  15-30 minutes (continuous systems) 4-6 hours (batch systems)

- Auxiliary Fuel Requirements:  Natural gas, fuel oils, and/or electrically-fired

Rich Fume Incinerator (Reactor)

- Temperature Range:  980°-1200°C+ (1800°-2200°F+)

- Residence Time Range:  1.0-2.0 seconds

Commercial System

- PyroBatch  Systems:  1,000 lb/load to 30,000 lb/load

- PyroTherm  Systems:  500 lb/hr

9.5.1.2  Demonstrated Performance--

A Midland-Ross batch pyrolysis system, operated by the McDonnell Douglas Company in St. Charles, Missouri, was RCRA permitted in 1984 after a series of trial burns using wastes with five POHCs with 50-70 percent chlorinated hydrocarbons.  Average DREs for the five POHCs during the trial burns were 99.9999 percent.  Removal efficiency for HCl was 99.9 percent, and particulate emissions were 0.035 grains per dry standard cubic foot.[1]  No other data appear to be available.

9.5.1.3  Cost of Treatment--

As noted in Reference 1, the developer states the following with regard to cost.

> "Our cost estimates are proprietary information and are supplied only to customers with whom we have projects.  To date, most of our clients' wastes applications are different from one another, hence project capital costs are also different.  However, inherent benefits of pyrolytic incineration help our clients realize significant overall project cost reductions relative to direct incineration systems."

9.5.1.4  Status of Technology--

Both batch and continuous pyrolysis units are supplied commercially by the Midland-Ross Corporation.  The company also maintains a research facility and offers complete bench and pilot test facilities.

The pyrolysis systems are particularly suited for sludges and solid wastes because of the long residence times that can be employed to assist destruction.[19]  In addition to the potential to destroy all organics and to handle difficult waste types, pyrolysis systems, as noted by the developer, offer the following advantages:[1]

1.  Salts and metals (inert materials) with moderate melting points are not liquified because the pyrolyzer operates at a design temperature below the melting points of most salts and metals.

2.  Since the same salts and metals are normally not vaporized, refractory spalling, surface fouling, and formation of inert aerosol condensates are all greatly reduced.

3.  Particulate emissions with most types of pyrolyzers are greatly reduced because the waste is not agitated or contacted with turbulent gases during pyrolysis, so particulate cleanup devices in many cases are not needed to meet Federal standards.

4.  Waste-borne $NO_x$ is reduced in a pyrolysis atmosphere to $N_2$ and $H_2O$.  Hence, $NO_x$ emissions from the process are considerably lower.

5.  Chlorinated or halogenated materials (e.g., hydrochloric acid) typically liberated by thermal treatment of a waste can be adsorbed by caustics present in, or added to, the feed prior to pyrolysis. This often leads to a reduction in emissions of HCl and $SO_x$ from 50 to 90 percent.

6.  Leaching of metals and salts from the carbonaceous residue (char) is reduced because they are exposed to a reducing atmosphere throughout the process, and they tend to be physically or chemically tied up in the char.

7.  Overall, gas cleanup equipment is greatly reduced or not required to pyrolytically treat the same waste materials treated by direct incineration.

## 9.5.2   Plasma Arc Pyrolysis

9.5.2.1   Process Description--

In this process, under development by Pyrolysis Systems Inc. of Welland, Ontario, waste molecules are destroyed by the action of a thermal plasma field.  The field is generated by passing an electric charge through a low pressure air stream thereby ionizing the gas molecules and generating temperatures up to 10000°C.

A flow diagram of the plasma pyrolysis system is shown in Figure 9.5.2. The plasma device is horizontally mounted in a refractory-lined pyrolysis chamber with a length of approximately 2 meters and a diameter of 1 meter. The colinear electrodes of the plasma device act as a plug-flow atomization zone for the liquid waste feed, and the pyrolysis chamber serves as a mixing zone where the atoms recombine to form hydrogen, carbon monoxide, hydrogen chloride, and particulate carbon.  The approximate residence times in the atomization zone and the recombination zone are 500 microseconds and 1 second, respectively.  The temperature in the recombination zone is normally maintained at 900-1200°C (1650°F - 2190°F).[21]

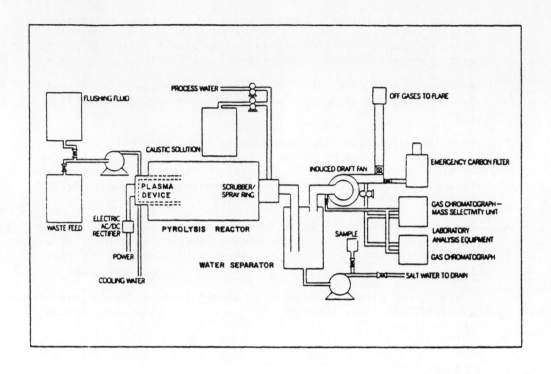

Figure 9.5.2.   Pyroplasma process flow diagram.

Source:   Reference 20.

After the pyrolysis chamber, the product gases are scrubbed with water and caustic soda to remove hydrochloric acid and particulate matter. The remaining gases, a high percentage of which are combustible, are drawn by an induction fan to the flare stack where they are electrically ignited. In the event of a power failure, the product gases are vectored through an activated carbon filter to remove any undestroyed toxic material.

The treatment system that is currently being used for testing purposes is rated at 4 kg/minute of waste feed or approximately 55 gal/hour. The product gas production rates are 5-6 $m^3$/minute prior to flaring. To facilitate testing, a flare containment chamber and 30 ft. stack have also been added to the system. The gas flow rate at the stack exit is approximately 36 $m^3$/minute.[21]

A major advantage of this system is that it can be moved from waste site to waste site as desired. The entire treatment system, including a laboratory, process control and monitoring equipment, and transformer and switching equipment, are contained on a 45 foot tractor-trailer bed.[21]

Two residual streams are generated by this process. These are the exhaust gases that are released up the stack as a flare, and the scrubber water stream. Since the product gas (after scrubbing) is mainly hydrogen, carbon monoxide, and nitrogen, it burns with a clean flame after being ignited. Analysis of the flare exhaust gases, presented in the following section, indicates virtually complete destruction of toxic constituents.

The scrubber water stream is composed mainly of salt water from neutralization of HCl and particulates, primarily carbon. Analyses of the scrubber water for the waste constituent of concern (e.g., carbon tetrachloride ($CCl_4$) and PCB in the feed material) have shown that the constituents were present at low ppb concentrations. The quality of scrubber water generated would depend on the water feed rate and corresponding product gas and scrubber waste flowrates. During a test in which 2.5 kg/min of waste containing 35 to 40 percent $CCl_4$ was fed to the reactor, a scrubber water effluent flowrate of 30 liters/minute was generated.[21]

The reactor as it is currently designed can only be used to treat liquid waste streams with viscosities up to that of 30 to 40 weight motor oils. Particulates are removed by a 200 mesh screen prior to being fed into the reactor. Contaminated soils and viscous sludges cannot be treated.

9.5.2.2   Demonstrated Performance--

The plasma arc system has been tested using several liquid feed materials, including carbon tetrachloride ($CCl_4$), polychlorinated biphenyls (PCBs), and methyl ethyl ketone (MEK).

Table 9.5.1 presents the results of three test burns conducted in Kingston, Ontario using carbon tetrachloride in the feed material. The carbon tetrachloride was fed to the reactor along with ethanol, methyl ethyl ketone, and water at a rate of 1 kg of $CCl_4$/minute. The duration of each of these tests was 60 minutes, and stack gas flowrates and temperatures averaged 32.5 dry standard cubic meter/minute (dscm/min) and 793°C (1460°F), respectively. As can be seen in the table, the destruction and removal efficiency (DRE) of $CCl_4$ in each of the tests was high, exceeding six nines. In addition, the concentration of HCl in exhaust gases was less than the upper limit of 1.8 kg/hr required by RCRA guidelines. As far as PCBs are concerned, the destruction and removal efficiency in each of the tests was greater than 6 nines, and in some cases reached 8 nines. Similar or better results can be anticipated for most halogenated organics.

9.5.2.3   Costs of Treatment--

The approximate capital cost of a unit similar to the one tested would be in the range of 1 to 1.5 million dollars.[21] More accurate figures will be available once a commercial unit has been built.

9.5.2.4   Status of Technology--

The construction and testing of the plasma arc system is being jointly sponsored by the New York State Department of Environmental Conservation (NYDEC) and the U.S. EPA Hazardous Waste Engineering Research Laboratory (HWERL). The project is comprised of four phases, which are:

Phase 1:   Design and construction of the mobile plasma arc system by the contractor, Pyrolysis Systems, Inc. (PSI).

Phase 2:   Performance testing of the plasma arc system at the Kingston, Ontario, Canada test site.

Phase 3:   Installation of the plasma arc system and additional performance testing at Love Canal, Niagra Falls, N.Y.

Phase 4:   Demonstration testing, as designated by NYDEC.

TABLE 9.5.1.   CARBON TETRACHLORIDE TEST RESULTS

| Parameter | Test 1 | Test 2 | Test 3 |
|---|---|---|---|
| Chlorine Mass Loading (%) | 35 | 40 | 35 |
| Scrubber Effluent | | | |
| $CCl_4$(ppb) | 1.27 | 5.47 | 3.26 |
| mg/hr | 2.29 | 9.85 | 5.87 |
| Flare Exhaust | | | |
| $CCl_4$(ppb) | 0.83 | 0.43 | 0.63 |
| mg/hr | 12.1 | 4.9 | 7.2 |
| $NO_x$ | | | |
| ppm(v/v) | 106 | 92 | 81 |
| lbs/hr | 1.02 | 0.69 | 0.02 |
| CO | | | |
| ppm(v/v) | 48 | 57 | 81 |
| lbs/hr | 0.28 | 0.26 | 0.37 |
| HCl | | | |
| mg/dscm | 1 | 137.7 | 247.7 |
| kg/hr | 1 | 0.25 | 0.44 |
| Destruction Removal Efficiency | 99.99998 | 99.99998 | 99.99998 |

Source:   Reference 20.

Phase 1 took place in 1982 and Phase 2, the results of which have been presented above, was completed in early 1986. Phase 3 will be initiated later in 1986.

The plasma technology is being jointly marketed by Westinghouse Electric Corporation Waste Technology Services Division and PSI. Once the system has been properly tested, they plan to lease these units to companies or organizations that require the system for waste clean up. The current system is only designed to handle liquid wastes. Future plans by PSI and Westinghouse include the design of units which could handle contaminated soil and other solid wastes.[22]

### 9.5.3 High Temperature Fluid Wall (HTFW) Destruction - Advanced Electric Reactor

#### 9.5.3.1 Process Description--

The HTFW reactor was originally developed by Thagard Research of Costa Mesa, California. However, the J.M. Huber Corp. of Borger, Texas has developed proprietary modifications to this original design. The reactor, called the Advanced Electric Reactor (AER), is shown in Figure 9.5.3. The reactor is a thermal destruction device which employs radiant energy provided by electrically heated carbon electrodes to heat a porous reactor core. The heated core then radiates heat to the waste materials. The reactor core is isolated from the waste by a blanket of gas formed by nitrogen flowing radially through the porous core walls.

The only feed streams to the reactor are the waste material and the inert nitrogen gas blanket. Therefore, the destruction is by pyrolysis rather than oxidation. Because of the low gas flow rate and the absence of oxygen, long gas phase residence times can be employed, and intensive downstream cleanup of off gases can be achieved economically.

Destruction via pyrolysis instead of oxidation significantly reduces the concentrations of typical incineration products such as carbon dioxide and oxides of nitrogen. The principal products formed during treatment of halogenated organics would be hydrogen, chlorine (if calcium oxide is added to the reactor, calcium chloride is formed instead), hydrochloric acid, elemental carbon, and free-flowing granular material.[23-25]

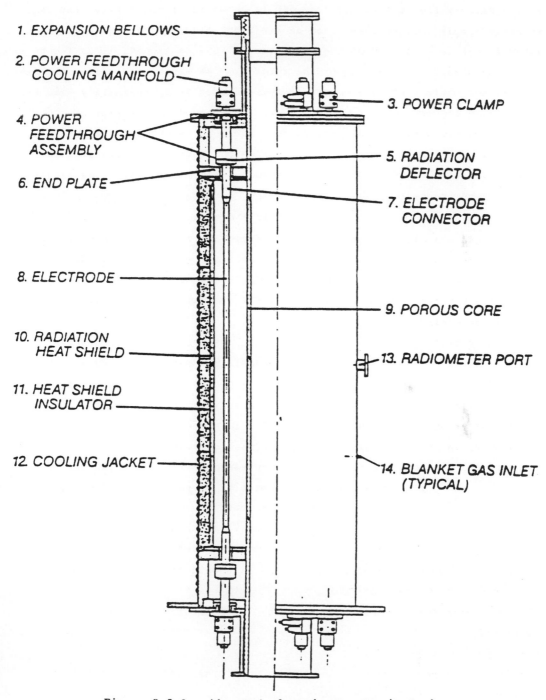

1. EXPANSION BELLOWS

2. POWER FEEDTHROUGH
   COOLING MANIFOLD

3. POWER CLAMP

4. POWER
   FEEDTHROUGH
   ASSEMBLY

5. RADIATION
   DEFLECTOR

6. END PLATE

7. ELECTRODE
   CONNECTOR

8. ELECTRODE

9. POROUS CORE

10. RADIATION
    HEAT SHIELD

13. RADIOMETER PORT

11. HEAT SHIELD
    INSULATOR

12. COOLING JACKET

14. BLANKET GAS INLET
    (TYPICAL)

Figure 9.5.3.   Advanced electric reactor [Huber].

A process flow diagram for the AER is shown in Figure 9.5.4.  The waste, if it is a solid, is released from an air tight feed bin through a metered screw feeder into the top of the reactor.  If it is a liquid, it is fed by an atomizing nozzle into the top of the reactor.  The waste then passes through the reactor where pyrolysis occurs at temperatures of approximately 4500°F (2480°C) in the presence of nitrogen gas.  Downstream of the reactor, the product gas and waste solids pass through two post-reactor treatment zones, the first of which is an insulated vessel which provides additional high temperature (2000°F or 1090°C) and residence time (5 seconds).  The second post-reactor treatment zone is water-cooled, and its primary purpose is to cool the gas prior to downstream particulate cleanup.

Off gas cleaning equipment includes a cyclone to collect particles which do not fall into the solids bin, a bag filter to remove fines, an aqueous caustic scrubber for acid gas and free chlorine removal, and two banks of five parallel activated carbon beds in series for removal of trace residual organics and chlorine.

The stationary pilot scale reactor which has been used for testing various wastes at Huber's Borger, Texas facility consists of a porous graphite tube, 1 foot in diameter and 12 feet high, enclosed in a hollow cylinder with a double wall cooling jacket.  This pilot unit is capable of processing 5,000 tons/yr of waste.  Huber also has a 3 inch diameter mobile unit which has been transported to hazardous waste sites for testing purposes.

The AER cannot currently handle two-phase materials (i.e., sludge); it can only burn single-phase materials consisting of solids, or liquids, or gases alone.[24,26]  Generally, a solid feed must be free flowing, nonagglomerating, and smaller than 100 mesh (less than 149 micrometers or 0.0059 inches).[24]  However, depending on the required destruction, solids smaller than 10 mesh may be suitable.  Soils should be dried and sized before being fed into the reactor.

The operating parameters as described by References 25 and 27 are as follows:

- Residence Time                         0.1 seconds
  (100 mesh solids)

- Gas Flow Rate                          500 scfm for 150 ton/day

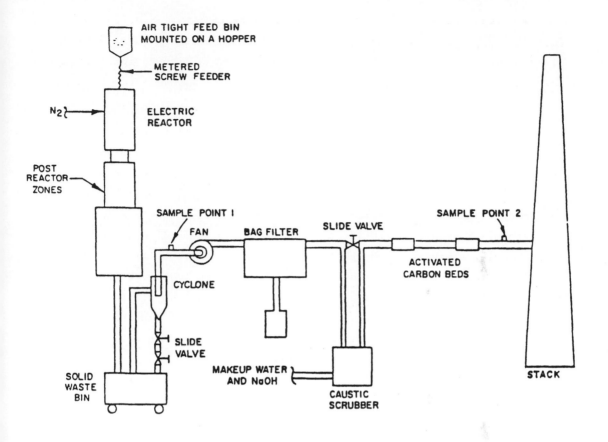

Figure 9.5.4.   High temperature fluid wall process configuration for
the destruction of carbon tetrachloride [Huber].

- Gas Phase                                 5 seconds
  Residence Time
  (at 2500°F or 1370°C)

9.5.3.2  Demonstrated Performance--

In 1983, Thagard conducted a series of tests on PCB-contaminated soils using a 3-inch diameter research reactor.[27] The results of these tests showed an average DRE of 99.9997 percent. The destruction efficiency was found to be independent of the feed rate in the 50 to 100 g/m range at 2343°C. Pyrolysis products other than carbon and hydrogen chloride were not detected using a GC with electron capture detection. It was concluded that the method for dispersing the feed into the reactor needed improvement. Problems with slagging in the reactor occurred that were believed to be related to the small diameter of the reactor and also to the design of the fluid wall flow. After modifications, additional tests on a 6-inch prototype reactor were conducted by Thagard using hexachlorobenzene dispersed on carbon particles; 99.99991 percent destruction efficiency was achieved.[27]

J. M. Huber Corporation purchased the patent rights and made further improvements to the process.[26] The J.M. Huber Corporation then began tests in its stationary reactor system which has a diameter of 12 inches. Included in this system are: an insulated post-reactor vessel, a water-jacketed cooling vessel, a cyclone, a baghouse, a wet scrubber, and an activated carbon bed. Several research burns have been conducted with this system. Results and operating parameters for pertinent burns are summarized in Table 9.5.2.

A series of four trial PCB-burns were conducted during September 1983 using a synthesized mixture of Aroclor 1260 and locally available sand to obtain a total concentration of 3000 ppm PCBs.[1,24] After treatment, the sand had a PCB content ranging from 0.0001 to 0.0005 ppm (0.1 to 0.5 ppb). The destruction and removal efficiency was measured to be 99.99960 to 99.99995 percent. Additional studies were conducted with the 12 inch diameter reactor using soils contaminated with octachlorodibenzo-p-dioxin (OCDD) and carbon tetrachloride. Seven nines DRE (99.99999 percent) were reportedly achieved at feed rates up to 2500 lbs/hr.

TABLE 9.5.2.   SUMMARY OF OPERATING PARAMETERS AND RESULTS
FOR HUBER AER RESEARCH/TRIAL BURNS

| Condition | PCBs (Sept. 1983) | $CCl_4$ (May 1984) | Dioxins (Oct/Nov 1984) |
|---|---|---|---|
| Reactor Core Temperature (°F) | 4100 | 3746-4418 | 3500-4000 |
| Waste Feed Rate (lb/min) | 15.5-15.8 | 1.1-40.8 | 0.4-0.6 |
| Nitrogen Feed Rate (scfm) | 147.2 | 104.3-190.0 | 6-10 |
| %-DRE | 99.99999 | 99.9999 | 99.999 |

Source:   References 24 and 28.

9.5.3.3  Cost of Treatment--

Operating costs will vary depending on the quantity of material to be processed and the characteristics of the waste feed.  Pretreatment may be necessary for bulky wastes having a high moisture content.  Typical energy requirements for contaminated soils range from 800 to 1000 kwh/ton.

The Huber process is not cost competitive with standard thermal destruction techniques (such as the rotary kiln) for materials with a high Btu content.[23,26]  It is cost-effective for wastes with a low Btu content (e.g., highly chlorinated compounds) because unlike standard thermal destruction techniques, the Huber process does not require supplementary fuels to obtain the necessary Btu content for incineration.

Cost estimates for processing contaminated soil at a site containing more than 100,000 tons of waste material were approximately $365 to $565/ton in 1985.  The cost breakdown for this estimate was 12 percent for maintenance, 7 percent labor, 29 percent energy, 18 percent depreciation and 34 percent for other costs (permitting, setup, post-treatment, etc.).[29,30]  These costs have recently been updated.  The new costs are expected to be released in 1986.[26]

9.5.3.4  Status of Technology--

Huber maintains two fully equipped reactors at their pilot facility in Borger, Texas.[24]  The smaller reactor, which is equipped for mobile operation, has a 3-inch core diameter and a capacity of 0.5 lb/min. The larger reactor is commercial scale with a 12-inch core diameter and a capacity of 50 lb/min.  Both of these reactors are used primarily for research purposes. In May 1984, the Huber reactor was certified by the EPA under TSCA to burn PCBs wastes.  Recently, the U.S. EPA and the Texas Water Commission jointly issued J.M. Huber Corporation a RCRA permit which authorizes the incineration of any non-nuclear RCRA hazardous waste (including dioxin-containing wastes) in the Huber Advanced Electric Reactor.[31]  This was the first commercial permit issued under RCRA for treating dioxin-containing wastes.  The J.M. Huber Corporation intends to use the permit for research and development of a full-scale transportable AER.  Huber does not intend to operate a hazardous waste disposal operation, but rather to construct and market stationary and/or mobile units for use by companies or organizations involved in hazardous waste destruction.[26]

9.6    IN SITU VITRIFICATION

In situ vitrification (ISV) was originally developed by Battelle Pacific
Northwest Laboratories as a means of stabilizing in-place high level nuclear
waste.  More recently, however, ISV has been studied as a means of destroying
soils contaminated with chlorinated organic wastes, including PCBs and dioxin
wastes, and heavy metals.  The system was patented in 1983.

In situ vitrification converts contaminated soils, or sludges, into a
solid glassy matrix through melting by electrical heating.  As depicted in
Figure 9.6.1, the process begins when graphite electrodes are placed into the
ground in a square array.  A conductive path is established by placing
graphite over the soil between the electrodes.  Electrical current is passed
between the electrodes, creating high temperatures (1700°C or 3100°F) which
melt the soil, and pyrolyze the organic waste constituents.  Gaseous effluents
which are produced are collected by a hood over the area and are exhausted to
off-gas treatment systems.  When pyrolysis is complete, current is shut off
and the mass cools to form a glass-like material.  A picture of the system is
presented in Figure 9.6.2, showing the enclosed hood.

Battelle engineers have developed 30, 500, and 3,750 kw size units.  The
small unit produces up to a ton of vitrified mass per setting, the 500 kw unit
produces approximately 10 tons per setting, and the large unit produces 400 to
800 tons per setting.

The cost estimates reported by PNL, and discussed below for transuranic
(TRU) wastes treated by the ISV process, account for charges associated with
site preparation, consumable supplies such as electrical power, and
operational costs such as labor and annual equipment charges.[33]
Specifically, for variations in manpower levels, power source costs, and
degree of heat loss, it was determined that the costs for TRU waste
vitrification ranges from 160 to 360 $/m$^3$ to vitrify to a depth of
5 meters.  These costs are a function of many variables, but are most
sensitive to variations in the amount of moisture in the soil and the cost of
electrical power in the vicinity of the process.  Figure 9.6.3, developed by
PNL, illustrates the variation in total costs as a function of both electrical
power costs and the moisture content of the TRU soil which was experimentally
treated.  The vertical line represents the value beyond which it is more cost
effective to lease a portable generator.[34]

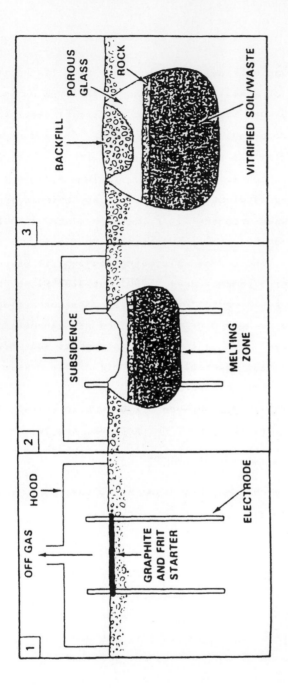

Figure 9.6.1.   Operating sequence of in situ vitrification.

Source:   Reference 32.

Figure 9.6.2.  Off-gas containment and electrode support hood.

Source:   Reference 32.

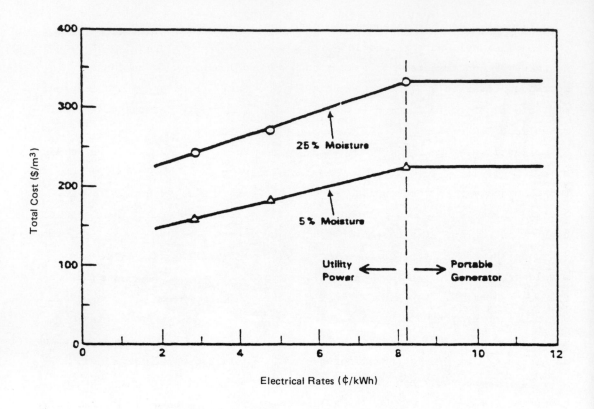

Figure 9.6.3.   Cost of in situ vitrification for TRU wastes as functions
of electrical rates and soil moisture [Fitzpatrick, 1984].

Recently, PNL has assessed the cost implications for ISV treatment of three additional waste categories; i.e., industrial sludges and hazardous waste (PCB) contaminated soils at both high and low moisture contents.[35] Representatives at PNL indicated that for industrial sludges with moisture contents of 55 to 75 percent (classified as a slurry), the total costs would range from $70 to $130/m$^3$. Additionally, treatment of high (greater than 25 percent) moisture content hazardous waste-PCB contaminated soil would cost approximately $150 to$ 250/m$^3$ versus costs of $128 to $230/m$^3$ for low (approximately 5 percent) moisture content PCB contaminated soil.

As these recent data and past TRU waste cost data suggest, the moisture content of the contaminated material treated is particularly important in influencing treatment costs; high moisture content increases both the energy and length of time required to treat the contaminated material. Furthermore, PNL representatives suggest that treatment costs are also influenced by the degree of off-gas treatment required for a given contaminated material, i.e., ISV application to hazardous chemical wastes will likely not require as sophisticated an off-gas treatment system as would TRU waste treatment.

PNL has recently assessed the treatment of and costs associated with hazardous waste contaminated soils. Specifically, during the summer of 1985, tests were conducted for the Electric Power Research Institute (EPRI) on PCB contaminated soil. While the draft report on these tests has been completed, it has not been published and/or made available to date. However, an EPRI project summary publication, dated March 1986, entitled "Proceedings: 1985 EPRI PCB Seminar" (EPRI CS/EA/EL 4480), has recently been made available to EPRI members. Preliminary results suggest that a destruction/removal efficiency (DRE) of six to nine nines was achieved from the off-gas treatment system and that a vitrification depth of 2 feet was achieved. Additional information will soon be available to the public. PNL expects to continue with research in the area of hazardous waste soils.

## REFERENCES

1. Freeman, H. M. Innovative Hazardous Waste Treatment. EPA-600/2-85-049. U.S. Environmental Protection Agency, Hazardous Waste Engineering Research Laboratory. Cincinnati, OH. April 1985.

2. Breton, M. A., et al. Technical Resource Document: Treatment Technologies for Solvent-Containing Wastes. Prepared for U.S. EPA, HWERL, Cincinnati, OH under Contract No. 68-03-3243, Work Assignment No. 2. August 1986.

3. Arienti, M., et al. Technical Resource Document: Treatment Technologies for Dioxin-Containing Wastes. Prepared for U.S. EPA, HWERL, Cincinnati, OH under Contract No. 68-03-3243, Work Assignment No. 2. August 1986.

4. GA Technologies, Inc. San Diego, CA. Material received in correspondence with M. Kravett, GCA Technology Division, Inc. April 1986.

5. Rickman, W. Telephone conversation with M. Kravett, GCA Technology Division, Inc., GA Technologies, Inc., San Diego, CA. April 1986.

6. Chang, D. and N. Sorbo. Evaluation of a Pilot Scale Circulating Bed Combustor with a Surrogate Hazardous Waste Mixture. In: Incineration and Treatment of Hazardous Waste: Proceedings of the 11th Annual Research Symposium. EPA-600/9-85-028. U.S. EPA, HWERL, Cincinnati, OH. September 1985.

7. Manning, M. P. Fluid Bed Catalytic Oxidation: An Underdeveloped Hazardous Waste Disposal Technology. In: Hazardous Waste. Volume 1, Number 1, 1984. pp. 41-65.

8. Conklin, J. H., D. M. Sowards, J. H. Kroehling. Exhaust Control, Industrial. In: Kirkothmer Encyclopedia of Chemical Technology. Volume 9. John Wiley & Sons, New York, NY. 1980. pp. 512-518.

9. Radanof, R. VOC Incineration and Heat Recovery - Systems and Economics. In: Third Conference on Advanced Pollution Control for the Metal Finishing Industry. EPA-600/2-81-028. U.S. Environmental Protection Agency, Office of Research and Development, Industrial Environmental Research Laboratory, Cincinnati, OH. February 1981. pp. 84-91.

10. Hardison, L. C. and E. J. Dowd. Air Pollution Control: Emission Control Via Fluidized Bed Oxidation. In: Chemical Engineering Progress. Volume 73, Number 7. August 1977. pp. 31-35.

11.   Chemical Engineering News.  News Bulletin.  Chemical Engineering News.
      Volume 34.  July 20, 1981.

12.   Benson, J. S.  Catoxid for Chlorinated Byproducts.  B. F. Goodrich
      Chemical Division, Cleveland, OH.  In:  Hydrocarbon Processing.  October
      1979.  pp. 107-109.

13.   Penberthy Electromelt Corporation.  Seattle, WA.  Material received in
      correspondence with M. Kravett, GCA Technology Division, Inc.  February
      1986.

14.   U.S. EPA.  Assessment of Incineration as a Treatment Method for Liquid
      Organic Hazardous Wastes (Background Report Series).  U.S. Environmental
      Protection Agency, Office of Policy, Planning, and Evaluation,
      Washington, D.C.  March 1985.

15.   Edwards, B. H., J. N. Paullin, K. C. Jordan.  Emerging Technologies for
      the Control of Hazardous Wastes.  Noyes Data Corporation, Park Ridge, New
      Jersey.  1983.

16.   Kohl, A.  Rockwell International Corporation, Canoga Park, CA.  Telephone
      conversation with M. Kravett, GCA Technology Division, Inc. February 1986.

17.   Johanson, J. G., S. J. Yosim, L. G. Kellog, and S. Sudar.  Elimination of
      Hazardous Waste by the Molten Salt Destruction Process.  In:
      Incineration and Treatment of Hazardous Waste, Proceedings of the Eighth
      Annual Research Symposium.  EPA-600/9-83-003.  April 1983.

18.   Midland-Ross Corporation.  Surface Combustion Division, Idaho, OH.
      Material received in correspondence with M. Kravett, GCA Technology
      Division, Inc.  March 1986.

19.   Daiga, V.  Telephone conversation with M. Kravett, GCA Technology
      Division, Inc., Midland-Ross Corporation, Surface Combustion Division,
      Toledo, OH.  March 1986.

20.   Kolak, Nichlos P., Thomas G. Barton, C.C. Lee and Edward F. Peduto.
      Trial Burns - Plasma Arc Technology.  EPA Twelfth Annual Research
      Symposium on Land Disposal, Remedial Action.  Incinerator and Treatment
      of Hazardous Waste.  Cincinnati, Ohio.  April 21-23, 1986

21.   Barton, Thomas G. Mobile Plasma Pyrolysis.  Hazardous Waste.  1 (2)
      pp. 237-247.  1984.

22.   Haztech News.  Plasma Arc Technology Used to Atomize Liquid Organics.  1
      (5)  pp. 33-34.  1986.

23.   Boyd, J., H.D. Williams, and T.L. Stoddard.  Destruction of Dioxin
      Contamination By Advanced Electric Reactor.  Preprinted Extended Abstract
      of Paper Presented Before the Division of Environmental Chemistry,
      American Chemical Society, 191st National Meeting,  New York, New York:
      Vol 26, No 1.  April 13-18, 1986.

24.  Schofield, William R., Oscar T. Scott, and John P. DeKany.  Advanced Waste Treatment Options: The Huber Advanced Electric Reactor and The Rotary Kiln Incinerator.  Presented HAZMAT Europa 1985 and HAZMAT Philadelphia 1985.

25.  GCA Technology Division, Inc.  Draft Report:  Identification of Remedial Technologies.  Prepared for U.S. EPA, Office of Waste Programs Enforcement, under EPA Contract No. 68-01-6769, Work Assignment No. 84-120.  GCA-TR-84-109-G(0).  March 1985.

26.  Boyd, James. J.M. Huber Corporation.  Telephone Conversations with Lisa Farrell, GCA Technology Division, Inc.  January 28, 1986; April 3, 1986; May 1, 1986.

27.  Horning, A.W., and H. Masters.  Rockwell International, Newbury Park, California.  Destruction of PCB-Contaminated Soils With a High-Temperature Fluid-Wall (HTFW) Reactor.  Prepared for U.S. EPA, Office of Research and Development, Municipal Environmental Research Laboratory, Cincinnati, Ohio.  EPA-600/D-84-072.  1984.

28.  Roy F. Weston, Inc. and York Research Consultants.  Times Beach, Missouri:  Field Demonstration of the Destruction of Dioxin in Contaminated Soil Using the J.M. Huber Corporation Advanced Electric Reactor.  February 11,1985.

29.  Lee, Kenneth W., William R. Schofield, and D. Scott Lewis.  Mobile Reactor Destroys Toxic Wastes in "Space".  Chemical Engineering.  April 2, 1984.

30.  Freeman, Harry M.  Hazardous Waste Destruction Processes.  Environmental Progress.  Volume 2, Number 4.  November 1983.

31.  Hazardous Materials Intelligence Report (HMRI).  First Commercial Dioxin Incineration Permit Granted to J.M. Huber.  January 24, 1986.

32.  Buelt, J. L. and S. T. Freim.  Demonstration of In-Situ Vitrification for Volume Reduction of Zirconia/Lime Sludges.  Battelle Northwest Laboratories.  April 1986.

33.  Oma, K. H. et al.  1983.  In-Situ Vitrification of Transuranic Wastes:  Systems Evaluation and Applications Assessment.  PNL-4800, Pacific Northwest Laboratory, Richland, Washington.

34.  Fitzpatrick, V. F., et al.  1984.  In Situ Vitrification - A Potential Remedial Action Technique for Hazardous Wastes.  Presented at the 5th National Conference on Management of Uncontrolled Hazardous Waste Sites, Washington, DC

35.  Buelt, J. L.  Battelle Memorial Institute, Pacific Northwest Laboratories.  Telephone conversation with Michael Jasinski.  GCA Technology Division, Inc.  1986.

# 10. Land Disposal of Residuals

All treatment processes leave residuals that, at a minimum, must be characterized to ensure that land disposal is achievable without risk to human health and the environment.

Even processes such as incineration that are carried out to achieve essentially total destruction of organic constituents can produce residuals such as ash and scrubber waters. These residuals can be hazardous; e.g., as a result of concentration of EP toxicity metals in the ash or POHCs and PICs in scrubber waters. In such cases, measures such as fixation and encapsulation may be required for the ash to eliminate the characteristics of EP toxicity and further treatment (e.g., carbon adsorption) may be needed to provide an aqueous stream suitable for discharge. The discussion here will concentrate on examining the status of solidification and fixation processes and other techniques designed to isolate and contain wastes within a stable, non-leachable medium.

As noted in Reference 1, chemical fixation involves the chemical interaction of the hazardous waste constituents with the fixation medium. Solidification is a process in which the waste is physically entrapped within a solid, essentially continuous, nonporous matrix. A third process, microencapsulation, relies on containment of the waste within a coating or outer enclosure; e.g. a sealed glass vessel.

Despite the interest shown in immobilization techniques and some generalizations made concerning their applicability to organic containing wastes, there are little, if any, data provided in the literature. Most techniques described must be considered physical processes and few can be considered to represent chemical fixation. Even if fixation could be demonstrated, little is known concerning the long term stability of the matrix

367

and the possible breakdown products over time.  The processes described below,
will require further study to demonstrate their effectiveness for halogenated
organic wastes.  The extent of their future use will depend upon the
regulatory criteria now being established by EPA for fixation/encapsulation
processes.

## 10.1   SOLIDIFICATION/CHEMICAL FIXATION

Solidification can be used to chemically fix or structurally isolate
halogenated organic wastes to a solid, crystalline, or polymeric matrix.  The
resultant monolithic solid mass can then be safely handled, transported, and
disposed of using established methods of landfilling or burial.
Solidification technologies are usually categorized on the basis of the
principal binding media, and include such additives as: cement-based
compounds, lime-based pozzolanic materials, thermoplasts, and organic polymers
(thermosets).  The resulting stable matrix produces a material that contains
the waste in a nonleachable form, is nondegradable, and does not render the
land it is disposed in unusable for other purposes.  A brief summary of the
compatibility and cost data for selected waste solidification/
stabilization systems is presented in Tables 10.1 and 10.2.

### Cement Based Systems

These systems utilize type I Portland cement, water, proprietary
additives, possibly fly ash, and waste sludges to form a monolithic, rock-like
mass.  In an EPA publication,[3] several vendors of cement based systems
reported problems with organic wastes containing oils, solvents, and greases
not miscible with an aqueous phase.  Although the unreactive organic wastes
become encased in the solids matrix, their presence can retard setting, cause
swelling, and reduce final strength.[4]  These systems are most commonly used
to treat inorganic wastes such as incinerator generated wastes and heavy metal
sludges.

TABLE 10.1.  COMPATIBILITY OF SELECTED WASTE CATEGORIES WITH DIFFERENT WASTE SOLIDIFICATION/STABILIZATION TECHNIQUES

| Waste component | Treatment Type | | | | | | |
|---|---|---|---|---|---|---|---|
| | Cement based | Lime based | Thermoplastic solidification | Organic polymer (UF)* | Surface encapsulation | Self-cementing techniques | Classification and synthetic mineral formation |
| **Organics:** | | | | | | | |
| 1. Organic solvents and oils | May impede setting, may escape as vapor | Many impede setting, may escape as vapor | Organics may vaporize on heating | May retard set of polymers | Must first be absorbed on solid matrix | Fire danger on heating | Wastes decompose at high temperatures |
| 2. Solid organics (e.g., plastics, resins, tars) | Good--often increases durability | Good--often increases durability | Possible use as binding agent | May retard set of polymers | Compatible--many encapsulation materials are plastic | Fire danger on heating | Wastes decompose at high temperatures |
| **Inorganics:** | | | | | | | |
| 1. Acid wastes | Cement will neutralize acids | Compatible | Can be neutralized before incorporation | Compatible | Can be neutralized before incorporation | May be neutralized to form sulfate salts | Can be neutralized and incorporated |
| 2. Oxidizers | Compatible | Compatible | May cause matrix break down, fire | May cause matrix break down | May cause deterioration of encapsulating materials | Compatible if sulfates are present | High temperatures may cause undesirable reactions |
| 3. Sulfates | May retard setting and cause spalling unless special cement is used | Compatible | May dehydrate and rehydrate causing splitting | Compatible | Compatible | Compatible | Compatible in many cases |
| 4. Halides | Easily leached from cement, may retard setting | May retard set, most are easily leached | May dehydrate | Compatible | Compatible | Compatible if sulfates are also present | Compatible in many cases |
| 5. Heavy metals | Compatible | Compatible | Compatible | Acid pH solubilizes metal hydroxides | Compatible | Compatible if sulfates are present | Compatible in many cases |
| 6. Radioactive materials | Compatible | Compatible | Compatible | Compatible | Compatible | Compatible if sulfates are present | Compatible |

*Urea-Formaldehyde resin.

Source:  Reference 2.

TABLE 10.2.    PRESENT AND PROJECTED ECONOMIC CONSIDERATIONS FOR WASTE SOLIDIFICATION/STABILIZATION SYSTEMS

| Type of treatment system | Major materials required | Unit cost of material | Amount of material required to treat 100 lbs of raw waste | Cost of material required to treat 100 lbs of raw waste | Trends in price | Equipment costs | Energy use |
|---|---|---|---|---|---|---|---|
| Cement-based | Portland Cement | $0.03/lb | 100 lb | $ 3.00 | Stable | Low | Low |
| Pozzolanic | Lime Flyash | $0.03/lb | 100 lb | $ 3.00 | Stable | Low | Low |
| Thermoplastic (bitumen-based) | Bitumen Drums | $0.05/lb $27/drum | 100 lb 0.8 drum | $18.60 | Keyed to Oil Prices | Very high | High |
| Organic polymer (polyester system) | Polyester Catalyst Drums | $0.45/lb $1.11/lb $17/drum | 43 lb of polyester-catalyst mix | $27.70 | Keyed to Oil Prices | Very high | High |
| Surface encapsulation (polyethylene) | Polyethylene | Varies | Varies | $ 4.50* | Keyed to Oil Prices | Very high | High |
| Self-cementing | Gypsum (from waste) | ** | 10 lb | ** | Stable | Moderate | Moderate |
| Glassification/mineral synthesis | Feldspar | $0.03/lb | Varies | -- | Stable | High | Very high |

*Based on the full cost of $91/ton.

**Negligible but energy cost for calcining are appreciable.

Source:  Reference 2.

## Lime Based (Pozzolanic) Techniques

Pozzolanic concrete is the reaction product of fine-grained aluminous siliceous (pozzolanic) material, calcium (lime), and water.  The pozzolanic materials are wastes themselves and typically consist of fly ash, ground blast furnace slag, and cement kiln dust.  The cementicious product is a bulky and. heavy solid waste used primarily in inorganic waste treatment such as the solidification of flue gas desulfurization sludge.  However, biological and paint sludges have been treated, although high concentrations (greater than 20 percent) of organics tend to prevent the formation of a high strength product.[5]

## Thermoplastic Material

In a thermoplastic stabilization process, the waste is dried, heated (260-450°F), and dispersed through a heated plastic matrix.  Principal binding media include asphalt, bitumen, polypropylene, polyethylene, or sulfur.  The resultant matrix is resistant to leaching and biodegradation, and the rates of loss to aqueous contacting fluids are significantly lower than those of cement or lime based systems.  However this process is not suited to wastes that act as solvents for the thermoplastic material.  Also there is a risk of fire or secondary air pollution with wastes that thermally decompose at high temperature.[2]

## Organic Polymers (Thermosets)

Thermosets are polymeric materials that crosslink to form an insoluble mass as a result of chemical reaction between reagents, with catalysts sometimes used to initiate reaction.  Waste constituents could conceivably enter into the reaction, but most likely will be merely physically entrapped within the crosslinked matrix.  The crosslinked polymer or thermoset will not soften when heated after undergoing the initial set.  Principal binding agents or reactants for stabilization include ureas, phenolics, epoxides, and polyesters.  Although the thermosetting polymer process has been used most frequently in the radioactive waste management industry, there are formulations that may be applicable to certain organic contaminants.  It is important to note that the concept of thermoset stabilization, like thermoplastic stabilization, does not require that chemical reaction take place during the solidification process.  The waste materials are physically trapped in an organic resin matrix that, like thermoplastics, may biodegrade and release much of the waste as a leachate.[6]  It is also an organic material that will thermally decompose if exposed to a fire.

## New Technology

An EPA sponsored study recently indicated that most solidification processes in current use (silicates, lime, and cement), including those described above, stabilize contaminants through microencapsulation rather than chemical fixation.[7]  Microencapsulation is a process that entraps micro and macroscopic particles individually as the fixative solidifies.  An inorganic polymer that is a candidate for true chemical fixation is the HWT product series marketed by International Waste Technologies.  The HWT series is a set of inorganic, irreversible colloidal polymers which improve on a successful Japanese approach which has been used in Japan for over 10 years.

In the HWT fixation process, there is a two-step reaction in which the toxic elements and compounds are complexed first in a rapid reaction and then permanently complexed in the building of macromolecules which continue to grow over a long period of time.  Step one of the detoxification reaction is the blending of contaminants and HWT chemicals to achieve a homogeneous state so that all the toxic compounds are exposed.  This blending generates irreversible colloidal structures and ion exchanges with toxic metals and organics.  Step two is the generation of an irreversible, three-dimensional, macromolecule which provides the crosslinking framework.  The vendor claims that both inorganic and organic wastes are treatable in either concentrated or dilute form, although pretreatment may be necessary.  Table 10.3 shows the effect of the inorganic  polymer on samples of PCB and PCP.  The levels of toxic compounds before and after treatment were determined by EPA approved laboratory testing.  A company spokesman indicated that data on the effectiveness of HWT on concentrated trichloroethylene still bottoms will be available in the near future.[8]

TABLE 10.3.   SUMMARY OF TEST RESULTS ON TOXIC ORGANICS

| Toxic organic | HWT - 20 weight percent | Concentration (µg/L) | |
|---|---|---|---|
| | | Untreated | Treated |
| PCB | 15 | 1,140 | 0.006 |
| | 15 | 1,800 | 0.069 |
| | 15 | 9,200 | 0.337 |
| PCP | 15 | 11,000 | 450 |

Source:   International Waste Technology.


International Waste Technology has estimated average treatment levels by
HWT compounds run between 8-15 percent by weight of waste with HWT compounds
costing between 12-25¢/lb.   The company estimates that heavy metal electric
arc furnace dust could be treated for $19/ton while chemical still bottoms
(halogenated hydrocarbons, benzene compounds, phenols in pure state) would
cost $90-100/ton in materials costs for low volumes of waste.   The bases for
these cost estimates are not entirely clear.   As a fixant for low molecular
weight organics, it would appear that HWT amounts far greater than 8 to
15 percent by weight of waste would be required.   At an assumed level of 50:50
HWT/waste, costs would range from $120-250/ton for HWT material with
additional costs required for transportation, processing, and disposal.

## 10.2   MACROENCAPSULATION

Encapsulation is often used to describe any stabilization process in
which the waste particles are enclosed in a coating or jacket of inert
material.   A number of systems are currently available utilizing
polybutadiene, inorganic polymers (potassium silicates), portland concrete,
polyethylene, and other resins as macroencapsulation agents for wastes that
have or have not been subjected to prior stabilization processes.   Several
different encapsulation schemes have been described in Reference 7.   The
resulting products are generally strong encapsulated solids, quite resistant

to chemical and mechanical stress, and to reaction with water.  Wastes (nonsolvent) successfully treated by these methods and their costs are summarized in Tables 10.4.

TABLE 10.4.   ESTIMATED COSTS OF ENCAPSULATION

| Process Option | Estimated Cost |
|---|---|
| Resin Fusion: | |
| Unconfined waste | $110/dry ton |
| 55-gallon drums | $0.45/gal |
| Resin spray-on | Not determined |
| Plastic Welding | $253/ton = $63.40/drum (80,000 55-gal drums/year) |

Source:   Reference 7.

The technologies could be considered for stabilizing organic wastes but are dependent on the compatibility or the organic waste and the encapsulating material.  Additional research is needed concerning the interaction of organic wastes and stabilization materials and the durability of the matrix, if the safe disposal of wastes and treatment residuals is to be realized through these processes.  EPA is now in the process of developing criteria which stabilized/solidified wastes must meet in order to make them acceptable for land disposal.[9]

REFERENCES

1.    Breton, M., et al.  Technical Resource Document:  Treatment Technologies
      for Solvent-Containing Wastes.  Prepared for U.S. EPA, HWERL, Cincinnati,
      OH under Contract No. 68-03-3243, Work Assignment No. 2.   August 1986.

2.    Guide to the Disposal of Chemically Stabilized and Solidified Waste,
      EPA SW-872, Sept. 1980.

3.    Environmental Laboratory U.S. Army Engineer Waterways Experiment Station,
      Survey of Solidification/Stabilization Technology for Hazardous
      Industrial Wastes, EPA-600/2-79-056.

4.    McNeese, J.A., Dawson, G.W., and Christensen, D.C., Laboratory studies of
      fixation of Kepone contaminated sediments, in "Toxic and Hazardous Waste
      Disposal", Vol. 2 Pojasek, R.B. Ed., Ann Arbor Science, Ann Arbor,
      Michigan.  1979.

5.    Stabilizing Organic Wastes:  How Predictable are the Results?  Hazardous
      Waste Consultant.  May 1985 pg. 18.

6.    Thompson, D.W. and Malone P.G., Jones, L.W., Survey of available
      stabilization technology in Toxic and Hazardous Waste Disposal, Vol. 1,
      Pojasek, R.B., Ed.  Ann Arbor Science, Ann Arbor Michigan.  1979.

7.    Lubowitz, H.R.  Management of Hazardous Waste by Unique Encapsulation
      Processes.  Proceedings of the Seventh Annual Research Symposium.
      EPA-600/9-81-002b.

8.    Newton, Jeff, International Waste Technology, Personal communication with
      Steve Palmer, GCA/Technology Division.  1986.

9.    C. Wiles, Hazardous Waste Engineering Research Laboratory, U.S. EPA,
      Private Communication; and Critical Characteristics and Properties of
      Hazardous Waste Solidification/Stabilization, HWERL, U.S. EPA, Contract
      No. 68-03-3186 (in publication).

# 11. Considerations for System Selection

Waste management options consist of three basic alternatives: source reduction, recycling/reuse, and use of a treatment/disposal processing system or some combination of these waste handling practices (see Figure 11.1). Recovery, treatment, and disposal may be performed onsite in new or existing processes or through contract with a licensed offsite firm which is responsible for the final disposition of the waste. Selection of the optimal waste management alternative will ultimately be a function of regulatory compliance and economics, with additional consideration given to factors such as safety, public and employee acceptance, liability, and uncertainties in meeting cost and treatment objectives.

Many of the technologies discussed in previous sections can be utilized to achieve high levels of halogenated organic removal or destruction; however, practicality will limit application to waste streams possessing specific characteristics. Since many processes yield large economies of scale, waste volume will be a primary determinant in system selection. The physical and chemical nature of the waste stream and pertinent properties of its constituents, including many of those properties identified in Appendix A, will also determine the applicability of waste treatment processes. Treatment will often involve the use of more than one technology in a system designed to progressively recover or destroy hazardous constituents in the most economical manner. Incremental costs of hazardous waste constituent removal will increase rapidly as low concentrations are attained.[1]

## 11.1 GENERAL APPROACH

All generators of hazardous wastes will be required to undertake certain steps to characterize regulated waste streams and to identify potential

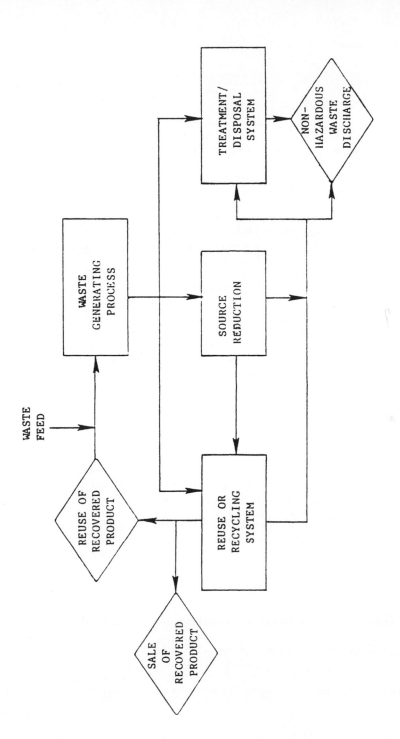

Figure 11.1.   Halogenated organic waste management options.

treatment options.  Treatment process selection should involve the following fundamental steps:

1.  Characterize the source, flow, and physical/chemical properties of the waste.

2.  Evaluate the potential for source reduction.

3.  Evaluate the potential for reuse or sale of recycled halogenated organics and other valuable waste stream constituents.

4.  Identify potential treatment and disposal options based on technical feasibility of meeting the required extent of waste constituent removal or destruction.  Give consideration to waste stream residuals and fugitive emissions to air.

5.  Determine the availability of potential options.  This includes the use of offsite services, access to markets for recovered products, and availability of commercial equipment and existing onsite systems.

6.  Estimate total system cost for various options, including costs of residual treatment and/or disposal and value of recovered organic product.  Cost will be a function of items 1 through 5.

7.  Screen candidate management options based on preliminary cost estimates.

8.  Use mathematical process modeling techniques and lab/pilot scale testing as needed to generate detailed treatment system design characteristics and processing capabilities.  The latter will define product and residual properties and identify need for additional treatment.

9.  Perform process trials of recovered product in its anticipated end use applications or determine marketability based on projected stream characteristics.

10.  Perform a detailed cost analysis based on modeling and performance results.

11.  Final system selection based on relative cost and other considerations; e.g., safety, acceptance, liability, and risks associated with data uncertainties.

Key system selection steps are discussed in more detail below.

## 11.2   ASSESSMENT OF ALTERNATIVES

### Waste Characterization

The first step in identifying appropriate waste management alternatives
to land disposal involves characterizing the origin, flow, and quality of
generated wastes.  An understanding of the processing or operational practices
which result in generation of the waste forms the basis for evaluating waste
minimization options.  Waste flow characteristics include quantity and rate.
Waste quantity has a direct impact on unit treatment costs due to economies of
scale in both treatment costs and marketability of recovered products.  Flow
rate can be continuous, periodic, or incidental (e.g., spills) and can be
relatively constant or variable.  This will have a direct impact on storage
requirements and treatment process design; e.g., continuous or batch flow.

Waste physical and chemical characteristics are generally the primary
determinant of waste management process selection for significant volume
wastes.  Of particular concern is whether the waste is pumpable, inorganic or
organic, and whether it contains recoverable materials, interfering compounds
or constituents which may foul heat or mass transfer surfaces.  Waste
properties such as corrosivity, reactivity, ignitability, heating value,
viscosity, concentrations of specific chemical constituents, biological and
chemical oxygen demand, and solids, oil, grease, metals and ash content need
to be determined to evaluate applicability of certain treatment processes.
Individual constituent properties such as solubility, vapor pressure, partition
coefficients, thermal stability, reactivity with various biological and
chemical (e.g., oxidants and reductants) reagents, and adsorption coefficients
are similarly required to assess treatability.  Finally, variability in waste
stream characteristics will necessitate overly conservative treatment process
design and additional process controls.  Variability will adversely affect
processing economics and marketability of recovered products.

### Source Reduction Potential

Source reduction potential is highly site specific, reflecting the
diversity of industrial waste generating processes and product requirements.
Source reduction alternatives which should be investigated include raw

material substitution, product reformulation, process redesign and waste segregation. The latter may result in additional handling and storage requirements, while differential processing cost and impact on product quality may be more important considerations for the other alternatives. Source reduction should be considered a highly desireable waste management alternative. In the wake of increasing waste disposal and liability costs, it has repeatedly proven to be cost effective while at the same time providing for minimal adverse health and environmental impact.

## Recycling Potential

As part of the waste characterization step, the presence of potentially valuable waste constituents should be determined. Economic benefits from recovery and isolation of these materials may result if they can be reused in onsite applications or marketed as saleable products. In the former case, economic benefits result from decreased consumption of virgin raw materials. This must be balanced against possible adverse effects on process equipment or product quality resulting from buildup or presence of undesirable contaminants. Market potential is limited by the lower value of available quantity or demand. Market potential will be enhanced with improved product purity, availability, quantity, and consistency.

## Identifying Potential Treatment and Disposal Options

Following an assessment of the potential for source reduction and recycling, the generator should evaluate treatment systems which are technically capable of meeting the necessary degree of halogenated organic removal or destruction. Guideline considerations for the investigation of treatment technologies are summarized in Table 11.1. The treatment objectives for a waste stream at a given stage of treatment will define the universe of candidate technologies. Possible restrictive waste characteristics (e.g., concentration range, flow, interfering compounds) may further reduce the number of candidate technologies. Consideration must be given to pretreatment options, for eliminating restrictive waste characteristics, to

TABLE 11.1.  GUIDELINE CONSIDERATIONS FOR THE INVESTIGATION OF
WASTE TREATMENT TECHNOLOGIES

A.  Objectives of Treatment:

- Primary function (pretreatment, treatment, residuals treatment)
- Primary mechanisms (destruction, removal, conversion, separation)
- Recover waste for reuse (fuel, process feed)
- Recovery of specific chemicals, group of chemicals
- Polishing for effluent discharge
- Immobilization or encapsulation to reduce migration
- Overall volume reduction of waste
- Selective concentration of hazardous constituents
- Detoxification of hazardous constituents

B.  Waste Applicability and Restrictive Waste Characteristics:

- Acceptable concentration range of primary & restrictive waste constituents
- Acceptable range in flow parameters
- Chemical and physical interferences

C.  Process Operation and Design:

- Batch versus continuous process design
- Fixed versus mobile process design
- Equipment design and process control complexity
- Variability in system designs and applicability
- Spatial requirements or restrictions
- Estimated operation time (equipment down-time)
- Feed mechanisms (wastes and reagents; solids, liquids, sludges, slurries)
- Specific operating temperature and pressure
- Sensitivity to fluctuations in feed characteristics
- Residuals removal mechanisms
- Reagent requirements
- Ancillary equipment requirements (tanks, pumps, piping, heat transfer equipment)
- Utility requirements (electricity, fuel and cooling, process and make-up water)

D.  Reactions and Theoretical Considerations:

- Waste/reagent reaction (destruction, conversion, oxidation, reduction)
- Competition or suppressive reactions
- Enhancing conditions (specify chemicals)
- Fluid mechanics limitations (mass, heat transfer)
- Reaction kinetics (temperature and pressure effects)
- Reactions thermodynamics (endothermic/exothermic/catalytic)

(continued)

TABLE 11.1 (continued)

---

E.   Process Efficiency:

- Anticipated overall process efficiency
- Sensitivity of process efficiency to:

   • feed concentration fluctuations
   • reagent concentration fluctuations
   • process temperature fluctuations
   • process pressure fluctuations
   • toxic constituents (biosystems)
   • physical form of the waste
   • other waste characteristics

   Acceptable range of fluctuations

F.   Emissions and Residuals Management:

- Extent of fugitive and process emissions and potential sources (processing equipment, storage, handling)
- Ability (and frequency) of equipment to be "enclosed"
- Availability of emissions data/risk calculations
- Products of incomplete reaction
- Relationship of process efficiency to emission data
- Air pollution control device requirements
- Process residuals (cooling and scrubber water, bottom ash, fly ash, fugitive/residual reagents, recovered products, filter cakes, sludges)
- Residual constituent concentrations and leachability
- Delisting potential

G.   Safety Considerations:

- Safety of storing and handling wastes, reagents, products and residuals
- Special materials of construction for storage and process equipment
- Frequency and need for use of personnel protection equipment
- Requirements for extensive operator training
- Hazardous emissions of wastes or reagents
- Minimization of operator contact with wastes or reagents
- Frequency of maintenance of equipment containing hazardous materials
- High operating temperatures or pressures
- Difficult to control temperatures or pressures
- Resistance to flows or residuals buildup
- Dangerously reactive wastes/reagents
- Dangerously volatile wastes/reagents

---

required treatment of process emissions and residuals, and to opportunities for by-product recovery.  System design will be based on the most difficult compound to remove or destroy.

A number of approaches to selecting potential treatment technologies for halogenated solvent and halogenated organic waste streams have been proposed[1-11].  Many of these references also provide cost information to assist the user in making a final determination of the cost effectiveness of a process.  The distinction between halogenated solvents and other halogenated organics as related to the applicability of recovery/treatment processes is obscure in many cases.  Physical and chemical properties can exhibit a high degree of similarity and both solvent and nonsolvent compounds coexist as significant constituents of many specific waste streams, including many of the K type wastes included in the halogenated organic category.  One scheme that specifically addresses the management of solvent bearing wastes is also directly applicable to nonsolvent halogens.[3]  In the Reference 3 scheme, management alternatives, including recycle/reuse, destructive treatments such as those resulting from thermal oxidations, and treatments for the removal of solvent constituents prior to land disposal, are reviewed.  The reference discusses the applicability of these waste management alternatives to waste streams having various physical characteristics.  Several waste treatment techniques are described including incineration, agitated thin film evaporation, fractional distillation, steam stripping, wet oxidation, carbon adsorption, and activated sludge biological treatment.

For the purposes of discussing treatment approaches, wastes can be divided into  three broad categories:  1) aqueous and mixed aqueous/organic liquids, 2) organic liquids, and 3) sludges.[3]  As defined, aqueous streams have water contents of 95 percent of higher, while organic streams are described as containing 50 percent or more organic liquids.  Mixed aqueous/organic streams fall in between.  Sludges are streams with solids content greater than 2 percent.  Decision charts for aqueous and mixed aqueous/organic liquids and for organic liquid waste stream treatment are provided in Figures 11.2 and 11.3.  Discussion of these charts in Reference 3 identifies some possible treatment options and stresses the importance of the possible need for treatment of residuals.

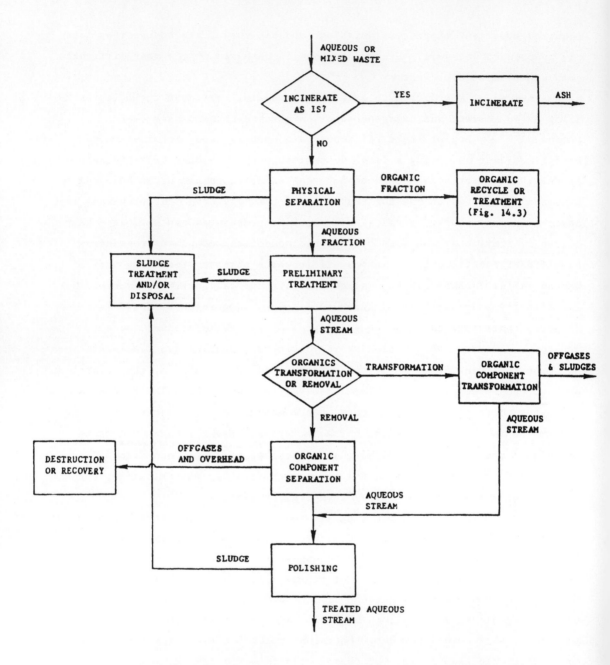

Figure 11.2.   Simplified decision chart for aqueous and mixed aqueous/organic
               waste stream treatment.

Source:   Reference 3.

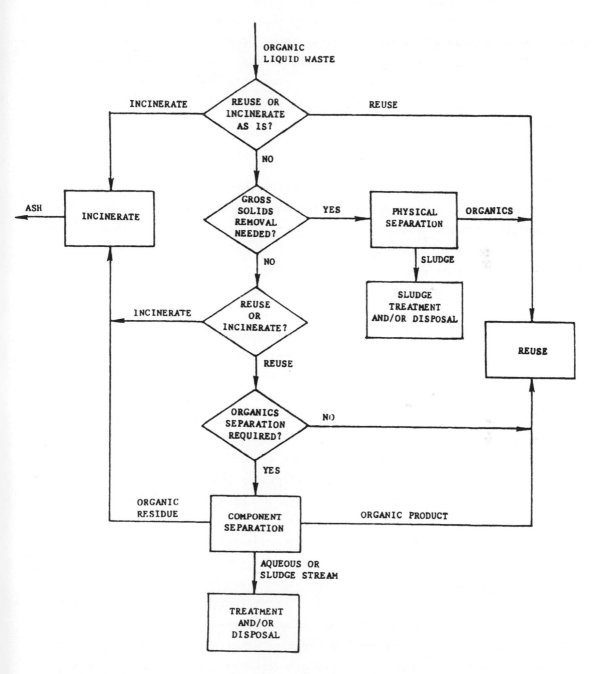

Figure 11.3.  Simplified decision chart for organic liquid
              waste stream treatment.

Source:  Reference 3.

The treatment processes potentially applicable to the three broad categories of waste are shown in Table 11.2. The identification of potentially applicable treatment processes should be considered as tentative since the treatments used will depend upon specific waste stream characteristics (not fully defined by the three general waste categories) and the purpose of the treatment. In addition, other innovative and emerging technologies described in previous sections of this document (or in the solvent and dioxin TRDs) could also be considered as applicable processes for some of these waste categories.

In addition to physical form, concentration of organic constituents within the waste is a principal determinant in assessing the applicability of a treatment process. Concentration ranges for which treatment processes are generally applicable are shown in Figure 11.4. Generally, techniques used for wastes with organic concentrations over 10 percent are applicable to lower concentrations as well, but other processes are generally more economical. Other waste characteristics which affect process selection are waste viscosity, solids content, volatility, solubility and contaminant type. Viscosity is important in that it indicates whether the waste stream is sufficiently fluid to undergo treatment. If not, high temperature to improve flow properties or treatment such as incineration in a kiln may be required. The presence of excess solids can cause plugging of certain equipment (e.g., packed towers) and necessitate solids removal prior to treatment. Dissolved solids may also require removal if they precipitate or otherwise interfere with process performance. Solubility and volatility are indicators of the ease of removal of a volatile compound by processes such as distillation or stripping. Finally, the type of contaminant will play a role in process selection. Certain types of compounds may be susceptible to reaction and degradation, and may, as in the case of many halogens, produce corrosive byproducts and be inherently low in Btu value.

As discussed previously, halogen content is a major factor in determining the extent to which an organic compound can be reacted or destroyed. Halogen contents of nonsolvent halogenated organics are listed in Table 2.2. The extent of halogenation will affect many of the key physical and chemical properties of the halogen compounds (see Appendix A), and thereby, determine the relative ease and practicality of thermal destruction, biodegradability, dehalogenation, and other approaches to chemical detoxification and destruction.

TABLE 11.2.   TREATMENT PROCESSES POTENTIALLY APPLICABLE TO HALOGENATED WASTES

| Process | Aqueous and mixed aqueous/ organic wastes | Organic wastes | Sludges |
|---|---|---|---|
| **Preliminary Treatment** | | | |
| pH adjustment | Y | NA | NA |
| Dissolved solid precipitation | Y | NA | NA |
| **Phase Separation** | | | |
| Solids removal | Y | Y | NA |
| Drying | NA | Y | Y |
| Organic fraction | Y | Y | Y |
| **Organic Component Separation** | | | |
| Stream stripping | Y | Y | Y |
| Carbon adsorption | Y | NA | NA |
| Fractional distillation | Y | Y | Y |
| Resin adsorption | Y | Y | NA |
| Solvent extraction | Y | Y | Y |
| **Organic Compound Destruction** | | | |
| Incineration | Y | Y | Y |
| Biological degradation | Y | NA | NA |
| Chemical oxidation | Y | NA | NA |
| Wet air oxidation | Y | NA | Y |
| Supercritical water | Y | NA | NA |
| Supercritical water oxidation | Y | NA | NA |
| Stabilization/Solidification | NA | NA | Y |

Y = Yes

NA = Generally not applicable.

Source:   Adapted from Reference 3.

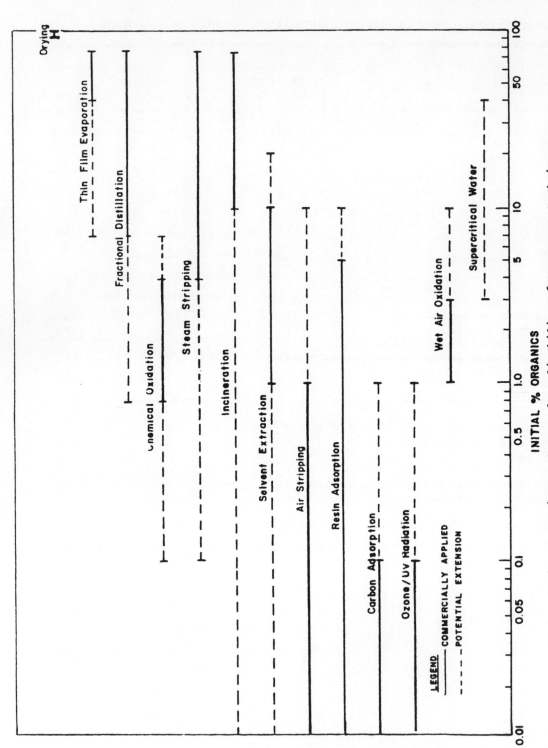

Figure 11.4. Approximate ranges of applicability of treatment techniques as a function of organic concentration in liquid waste streams.

Source: Reference 12.

The general susceptibility of halogenated solvents to biological, chemical, and thermal treatment has been summarized in Reference 12.  As noted therein, other researchers have provided similar qualitative assessments of the applicability of treatment processes for specific compounds. Reference 11, for example, provides a numerical rating assessing the applicability of many of the waste treatment processes considered here to various W-E-T model streams and their constituents.  Although this rating system was developed for assessing the treatment of volatile components within the waste stream, it contains information concerning the treatability of many of the nonsolvent halogenated organics addressed in this TRD.

The volativity of solvent and nonsolvent halogenated organics is often a key distinction between these two categories of halogenated compounds. Although volatilities (and other properties) are similar for many halogens, the nonsolvent category contains many high molecular weight compounds, (e.g., most of the pesticides) which exist as solids at 25°C.  Many of these will not be amenable to recovery by distillation and similar processes or will appear as constituents of the bottoms product resulting from such processing operations.  In many cases, further recovery may not be possible because of volativity or thermal stability considerations and ultimate disposal by incineration may be required.  Solidification/encapsulation may be another disposal option for such residuals.

The advantages and limitations of the treatment processes discussed in this document are summarized in Table 11.3.  Incineration and other thermal destruction processes are discussed first in the table because of their general applicability to the treatment of halogenated organic wastes.  As noted by Blaney and others, incineration may well prove to be the ultimate disposal method, at least for sludges for which halogenated organic recovery is impractical.  Incineration will also be the major method used to dispose of still bottoms following recovery operations.  However, the extent to which incineration will be used for these difficult to treat wastes will depend to some extent on the technical and regulatory requirements that will be imposed on performance of solidification/stabilization technologies.

Some of the technologies discussed in Table 11.3 are not generally intended to be used as final treatment processes.  Agitated thin film evaporation and distillation, for example, are concerned primarily with

TABLE 11.3.  SUMMARY OF HALOGENATED ORGANIC TREATMENT PROCESSES

| Process | Applicable waste streams | Stage of development | Performance | Residuals generated |
|---|---|---|---|---|
| **Incineration** | | | | |
| Liquid injection incineration | All pumpable liquids provided wastes can be blended to Btu level of 8500 Btu/lb. Some solids removal may be necessary to avoid plugging nozzles. | Estimated that over 219 units are in use. Most widely used incineration technology. | Excellent destruction efficiency (>99.99%). Blending can avoid problems associated with residuals, e.g., HCl. | TSP, possibly some PICs, and HCl. Little ash if solids removed in pre-treatment processes. |
| Rotary kiln incineration | All wastes provided Btu level is maintained. | Over 40 units in service; most versatile for waste destruction. | Excellent destruction efficiency (>99.99%). | Requires APCDs. Process residuals should be acceptable if charged properly and treated for acid gas removal. |
| Fluidized bed incineration | Liquids or nonbulky solids. | Nine units reportedly in operation-circulating bed units under development. | Excellent destruction efficiency (>99.99%). | As above. |
| Fixed/multiple hearths | Can handle a wide variety of wastes. | Approximately 70 units in use. Old technology for municipal waste combustion. | Performance may be marginal for halogenated wastes. | As above. |
| Industrial kilns | Generally all wastes, but Btu level, chlorine content, and other impurity content may require blending to control charge characteristics and product quality. | Only a few units now burning hazardous waste. | Usually excellent destruction efficiency (>99.99%) because of long residence times and high temperatures. | As above. |
| **Other Thermal Technologies** | | | | |
| Circulating bed combustor | Liquids or nonbulky solids. | Only one U.S. manufacturer. No units treating hazardous waste. | Manufacturer reports high efficiencies (>99.99%). | Bed material additives can reduce HCl emissions. Residuals should be acceptable. |
| Molten glass incineration | Almost all wastes, provided moisture and metal impurity levels are within limitations. | Technology developed for glass manufacturing. Not available yet as a hazardous waste unit. | No performance data available, but DREs should be high (>99.99%). | Will need APC device for HCl and possibly PICs; solids retained (encapsulated) in molten glass. |
| Molten salt destruction | Not suitable for high (>20%) ash content wastes. | Technology under development since 1969, but further development on hold. | Very high destruction efficiencies for organics (six nines for PCBs). | Needs some APC devices to collect material not retained in salt. |

(continued)

TABLE 11.3 (continued)

| Process | Applicable waste streams | Stage of development | Performance | Residuals generated |
|---|---|---|---|---|
| Furnace pyrolysis units | Most designs suitable for all wastes. | One pyrolysis unit RCRA permitted. Certain designs available commercially. | Very high destruction efficiencies possible (>99.99%). Possibility of PIC formation. | TSP emissions lower than those from conventional combustion; will need APC devices for HCl. Certain wastes may produce an unacceptable tarry residual. |
| Plasma arc pyrolysis | Present design suitable only for liquids. | Commercial design appears imminent, with future modifications planned for treatment of sludges and solids. | Efficiencies exceeded six nines in tests with sol···· | Requires APC devices for HCP and TSP, needs flare for $H_2$ and CO destruction. |
| Fluid wall advanced electric reactor | Suitable for all wastes if solids pretreated to ensure free flow. | Ready for commercial development. Test unit permitted under RCRA. | Efficiencies have exceeded six nines. | Requires APC devices for TSP and HCl. |
| In situ vitrification | Technique for treating contaminated soils, could possibly be extended to slurries. Also use as solidification process. | Not commercial, further work planned. | No data available, but DREs of over six nines reported. | Off gas system needed to control emissions to air. Ash contained in vitrified soil. |
| **Physical Treatment Methods** | | | | |
| Distillation | This is a process used to recover and separate volatile organics. Fractional distillation will require solids removal to avoid plugging columns. | Technology well developed and equipment available from many suppliers; widely practiced technology. | Separation depends upon reflux (99+ percent achievable). This is a recovery process. | Bottoms will usually contain levels of volatiles in excess of 1,000 ppm; condensate may require further treatment. |
| Evaporation | Agitated thin film units can tolerate higher levels of solids and higher viscosities than other types of stills. | Technology is well developed and equipment is available from several suppliers; widely practices technology. | This is a volatile organic recovery process. Typical recovery of 60 to 70 percent. | Bottoms will contain volatiles. Generally suitable for incineration. |
| Steam Stripping | A simple distillation process to remove volatile organics from aqueous solutions. Preferred for low concentrations and organics with low solubilities. | Technology well developed and available. | Not generally considered a final treatment, but can achieve low residual organic levels. | Aqueous treated stream will probably require polishing. Further concentration of overhead steam generally required. |
| Liquid-Liquid Extraction | Generally suitable only for liquids of low solid content. | Technology well developed for industrial processing. | Can achieve high efficiency separation for certain organic/waste combinations. | Organic compound solubility in aqueous phase should be monitored. |

(continued)

TABLE 11.3 (continued)

| Process | Applicable waste streams | Stage of development | Performance | Residuals generated |
|---|---|---|---|---|
| Carbon Adsorption | Suitable for low solid, low concentration aqueous waste streams. | Technology well developed; used as polishing treatment. | Can achieve low levels of organics in effluent. | Adsorbate must be processed during regeneration. Spent carbon and wastewater may also need treatment. |
| Resin Adsorption | Suitable for low solid waste streams. Consider for recovery of valuable compounds. | Technology well developed in industry for special resin/organic compound combinations. Applicability to waste streams not demonstrated. | Can achieve low levels of organics in effluent. | Adsorbate must be processed during regeneration. |
| **Chemical Treatment Processes** | | | | |
| Wet air oxidation | Suitable for aqueous liquids, also possible for slurries. Organic concentrations up to 15%. | High temperature/ pressure technology, widely used as pretreatment for municipal sludges, only one manufacturer. | Pretreatment for biological treatment. Some compounds resist oxidation. | Some residues likely which need further treatment. |
| Supercritical water oxidation | For liquids and slurries containing optimal concentrations of about 10% organics. | Supercritical conditions may impose demands on system reliability. Commercially available in 1986. | Supercritical conditions achieve high destruction efficiencies (>99.99%) for all constituents. | Residuals not likely to be a problem. Halogens can be neutralized in process. |
| UV/Ozonation | Oxidation with ozone (assisted by UV) suitable for low solid, dilute aqueous solutions. | Now used as a polishing step for wastewaters. | Not likely to achieve residual levels in the low ppm range for most wastes. | Residual contamination likely; will require additional processing of off gases. |
| Dechlorination | Dry soils and solids. | Not fully developed. | Destruction efficiency of over 99% reported for dioxin. | Residual contamination seems likely. |
| **Biological Treatment Methods** | | | | |
| | Aerobic technology suitable for dilute wastes although some constituents will be resistant. | Conventional treatments have been used for years. | May be used as final treatment for specific wastes, may be pretreatment for resistant species. | Residual contamination likely; will usually require additional processing. |

recovery/reuse.   Others like wet air oxidation and liquid-liquid extraction
are pretreatment processes than can be used to make a waste amenable to a
finishing step such as biological treatment.

Ultimately, the selection of a specific treatment system from the list of
potentially applicable processes will depend on cost, availability, and site
specific factors.   These considerations are discussed below.

## Management System Cost Estimation

The relative economic viability of candidate waste management systems
will be the primary determinant of ultimate system selection.   This must be
evaluated on the basis of total system costs which includes the availability
of onsite equipment, labor and utilities, net value of recovered products and
treatment/disposal processing costs.   Costs for a given management system will
also be highly dependent on waste physical, chemical, and flow
characteristics.   Thus, real costs are very site specific and limit the
usefulness of generalizations.   The reader is referred to the sections on
specific technologies for data on costs and their variability with respect to
flow and waste characteristics.   Major cost centers which should be considered
are summarized in Table 11.4.

## Modeling System Performance and Pilot Scale Testing

Following a preliminary cost evaluation which will enable the generator
to narrow his choice of waste management options, steps must be taken to
further finalize the selection process.   These could involve the use of
mathematical models to predict design and operating requirements.   However,
models often sacrifice accuracy for convenience and are not always adequate
for complex waste streams.   Laboratory data, or pilot plant and full-scale
data, may ultimately be needed to confirm predicted performance.   In fact,
some data may be needed as model inputs for predicting system behavior.

Processes which rely on Henry's Law constant are a good example of the
need for experimentally documented data.   Removal efficiency approximations
using Henry's Law constant based on a ratio of pure compound vapor pressure to

TABLE 11.4.   MAJOR COST CENTERS FOR WASTE MANAGEMENT ALTERNATIVES

A.   Credits

  - Material/energy recovery resulting in decreased consumption of
    purchased raw materials
  - Sales of waste products

B.   Capital Costs*

  - Processing equipment
  - Ancillary equipment (storage tanks, pumps, piping)
  - Pollution control equipment
  - Vehicles
  - Buildings, land
  - Site preparation, installation, start-up

C.   Operating and Maintenance Costs

  - Overhead, operating, and maintenance labor
  - Maintenance materials
  - Utilities (electricity, fuel, water)
  - Reagent materials
  - Disposal, offsite recovery and waste brokering fees
  - Transportation
  - Taxes, insurance, regulatory compliance, and administration

D.   Indirect Costs and Benefits

  - Impacts on other facility operations; e.g., changes in product
    quality as a result of source reduction or use of recycled materials
  - Use of processing equipment for mangement of other wastes

*Annual costs derived by using a capital factor:

$$CRF = \frac{i(1+i)^n}{(1+i)^n - 1}$$

Where:   i = interest rate and n = life of the investment.  A CRF of 0.177 was
used to prepare cost estimates in this document.  This corresponds to
an annual interest rate of 12 percent and an equipment life of
10 years.

its solubility often overestimate stripping by as much as two orders of magnitude.[2]  However, if Henry's Law constant is obtained experimentally using headspace analysis and batch stripping methods, it can be effectively used to estimate equilibrium partitioning behavior.

Many models are useful for predicting constituent behavior in separation processes.  These models are based on thermodynamic equilibrium partitioning and may also include kinetic factors to establish separation performances. Perry's Chemical Engineers' Handbook and other Chemical Engineering textbooks are sources of information about such models.[14-18]  Standard analytical packages are also available to predict the fate of waste stream contaminants as they are exposed to unit operations such as stripping and distillation.

The need for experimental data will depend upon the complexity of waste stream/process interactions.  Equipment manufacturers are often able to provide experimental equipment and models to establish process parameters and cost, including the costs required for disposal of residuals.

REFERENCES

1.   Allen, C. C., and B. L. Blaney. Techniques for Treating Hazardous Waste
     to Remove Volatile Organic Constituents. Research Triangle Institute for
     EPA HWERL.  EPA-600/2-85-127 PB85-218782/REB.  March 1985.

2.   Allen, C. C., and B. L. Blaney.  Techniques for Treating Hazardous Waste
     to Remove Volatile Organic Constituents.  JAPCA, Vol. 35, No. 8.
     August 1985.

3.   Blaney, B. L.  Alternative Techniques for Managing Solvent Wastes.
     Journal of the Air Pollution Control Association, 36(3):  275-285.
     March 1986.

4.   Ehrenfeld, J., and J. Bass, Arthur D. Little, Inc.  Evaluation of
     Remedial Action Unit Operations at Hazardous Waste Disposal Sites.
     Cambridge, MA, Noyes Publication.

5.   Bee, R.W., et al.  The Aerospace Corporation.  Evaluation of Disposal
     Concepts for Used Solvents at DOD Bases.  Report No. TDR-0083(3786)-01.
     February 1983.

6.   U. S. EPA Technologies and Management Strategies for Hazardous Waste
     Control.  U. S. EPA Office of Technology Assessment.  1983.

7.   U. S. EPA  Superfund Strategy. OTA-ITE-252, U. S. EPA Office of
     Technology Assessment.  April 1985.

8.   White, R. E., Busman, T., and J. J. Cudahy, et al.,IT Enviroscience, Inc.
     New Jersey Industrial Waste Study (Waste Projection and Treatment).
     Knoxville, TN.  EPA/600/6-85/003.  May 1985.

9.   Michigan Department of Commerce.  Hazardous Waste Management in the Great
     Lakes:  Opportunities for Economic Development and Resource Recovery.
     September 1982.

10.  Spivey, J. J. et al., Research Triangle Institute.  Preliminary
     Assessment of Hazardous Waste Pretreatment as an Air Pollution Control
     Technique.  U. S. EPA/IERL.  15 March 1984.

11.  Engineering-Science.  Supplemental Report on the Technical Assessment of
     Treatment Alternatives for Waste Solvents.  Washington, D. C.:  U. S.
     Environmental Protection Agency.  1985.

12.  Breton, M., et al.  Technical Resource Document:  Treatment Technologies for Solvent-Bearing Wastes.  Reports prepared for U.S. EPA, HWERL, Cincinnati, OH under Contract No. 68-03-3243, Work Assignment No. 2. August 1986.

13.  Arienti, M., et al.  Technical Resource Document:  Treatment Technologies for Dioxin-Containing Wastes.  Report prepared for U.S. EPA, HWERL, Cincinnati, OH under Contract No. 68-03-3243, Work Assignment No. 2. August 1986

14.  Perry, J. H. et.al.  Chemical Engineers' Handbook.  Sixth Edition, McGraw Hill.  1984.

15.  Henley, E., and J. D. Seader.  Equilibrium-Stage Separation Operations in Chemical Engineering, John Wiley and Sons, Inc., New York.  1981.

16.  McCabe, W. L., and J. C. Smith.  Unit Operations of Chemical Engineering (Third Edition), McGraw-Hill Book Company, New York, 1979.

17.  Treybal, R. E.  Liquid Extraction, Second Edition, McGraw-Hill Book Company, New York, 1963, pp. 359, 376.

18.  Holland, C. D.  Fundamentals and Modeling of Separation Processes, Prentice Hall, New York.  1975.

# Appendix A—Chemical and Physical Properties of Halogenated Organic Compounds

TABLE A-1.   CHEMICAL AND PHYSICAL PROPERTIES OF HALOGENATED ORGANIC COMPOUNDS [a]

| Compound | CAS No. | Molecular formula | Molecular weight | Melting point (°C) | Boiling point @ 760 torr (°C) | Vapor pressure @ 20-25°C [b] (torr) | Vapor density (air$^{-1}$) | Liquid density (g/ml) | Solubility in water (mg/l) | Log octanol water partition coefficient | Henry's Law constant (atm-m$^3$/mole) |
|---|---|---|---|---|---|---|---|---|---|---|---|
| **A. HALOGENATED ALIPHATICS** | | | | | | | | | | | |
| **Alkanes** | | | | | | | | | | | |
| Methylene chloride | 75-09-2 | $CH_2Cl_2$ | 84.9 | -95 | 39.8 | 362.4 | 2.9 | $1.33^{20}$ | $20,000^{25}$ | 1.25 | $3.19 \times 10^{-3}$ |
| 1,1,1-trichloroethane | 71-55-6 | $C_2H_3Cl_3$ | 133.4 | -30.4 | 74.1 | 96.0 | 4.6 | $1.34^{20}$ | $950^{25}$ | 2.17 | $4.92 \times 10^{-3}$ |
| Carbon tetrachloride | 56-23-5 | $CCl_4$ | 153.8 | -22.9 | 76.8 | 89.55 | 5.3 | $1.59^{20}$ | $785^{20}$ | 2.64 | $3.02 \times 10^{-2}$ |
| 1,1,2-trichloro-1,2,2-trifluoroethane | 76-13-1 | $C_2Cl_3F_3$ | 187.4 | -36.4 | 48 | 270 | 6.5 | $1.56^{25}$ | $10^{25}$ | 2.00 | $4.3 \times 10^{-2}$ |
| Trichlorofluoromethane | 75-69-4 | $CCl_3F$ | 137.4 | -111 | 23.8 | 667.4 | 4.7 | $1.49^{17}$ | $1,100^{25}$ | 2.53 | $5.83 \times 10^{-2}$ |
| Chloroform | 67-66-3 | $CHCl_3$ | 119.4 | -64 | 62 | 160 | 4.12 | 1.49 | 8,000 | 1.97 | $3.39 \times 10^{-3}$ |
| Methylene bromide | 74-95-3 | $CH_2Br_2$ | 173.9 | -52.7 | 97 | 40 | 6.05 | 2.5 | $11.7^{15}$ | | |
| Methyl chloride | 74-87-3 | $CH_3Cl$ | 50.5 | -97.7 | -24 | 3,800 | 1.8 | $0.991^{-25}$ | 4,000 cc/l (vapor) | 0.91 | $8.14 \times 10^{-3}$ |
| Hexachloroethane | 67-72-1 | $C_2Cl_6$ | 237 | 187 (subl.) | 187 (subl.) | 0.4 | 8.2 | $2.09^{20/4}$ | $50^{22}$ | 3.34 | $4.85 \times 10^{-3}$ |
| 1,2,3-trichloropropane | 96-18-4 | $C_3H_5Cl_3$ | 147.4 | -14 | 156 | 2 | 5.0 | $1.42^{15/4}$ | 213 | | $2.8 \times 10^{-2}$ |
| 1,2-dichloroethane | 107-06-2 | $C_2H_4Cl_2$ | 99 | -35 | 84 | 62 | 3.35 | $1.25^{20/4}$ | 8,000 | 1.5 | $1.35 \times 10^{-3}$ |
| 1,1,2-trichloroethane | 79-00-5 | $C_2H_3Cl_3$ | 133.4 | -35 | 113.7 | 19 | 4.6 | $1.44^{20/4}$ | 4,500 | 2.17 | $7.42 \times 10^{-4}$ |
| 1,1,2,2-tetrachloroethane | 79-34-5 | $C_2H_2Cl_4$ | 167.8 | -43 | 146 | 5 | 5.8 | $1.60^{20/4}$ | 2,900 | 2.56 | $3.8 \times 10^{-4}$ |
| 1,1,1,2-tetrachloroethane | 630-18-4 | $C_2H_2Cl_4$ | 167.8 | -70 | 138 | 13.9 | | 1.60 | 236 | 3.0 | $1.3 \times 10^{-2}$ |
| Trichloromethanethiol | | $CHCl_3S$ | 151.5 | | | | | | | | |

(continued)

## TABLE A-1 (continued)

| Compound | CAS No. | Molecular formula | Molecular weight | Melting point (°C) | Boiling point @ 760 torr (°C) | Vapor pressure @ 20-25°C[b] (torr) | Vapor density (air=1) | Liquid density (g/ml) | Solubility in water (mg/l) | Log octanol water partition coefficient | Henry's law constant (atm-m³/mole) |
|---|---|---|---|---|---|---|---|---|---|---|---|
| Methyl bromide | 74-83-9 | $CH_3Br$ | 95 | -93.6 | 3.6 | 1,420 | 3.27 | 1.73 | $900^{20}$ | 1.1 | $2.2 \times 10^{-1}$ |
| 1,2-dibromo-3-chloropropane[c] | 96-12-8 | $C_3H_5Br_2Cl$ | 236.4 | 10 | 196 | 0.8 | | $2.08^{20/20}$ | $1,000^{25}$ | 3.0 | $2.45 \times 10^{-4}$ |
| Ethylene dibromide[c] | 106-93-4 | $C_2H_4Br_2$ | 187.9 | 10 | 131 | 11 | 6.5 | $2.725/4$ | $4,300^{30}$ | | $8.82 \times 10^{-4}$ |
| Dichlorodifluoromethane | 75-71-8 | $CCl_2F_2$ | 120.9 | -158 | -29.8 | 4,250 | 4.18 | 1.33 | $280^{25}$ | 2.16 | 3.88 |
| Ethylidene dichloride | 75-34-3 | $C_2H_4Cl_2$ | 99 | -97 | 57 | 182 | 3.4 | $1.17^{20/4}$ | $5,500^{20}$ | 1.79 | $2.9 \times 10^{-2}$ |
| 1,2-dichloropropane | 78-87-5 | $C_3H_6Cl_2$ | 113 | -100 | 96.8 | 42 | 3.9 | $1.16^{20/20}$ | $2,700^{20}$ | 2.28 | $2.8 \times 10^{-3}$ |
| Methyl iodide | 74-88-4 | $CH_3I$ | 142 | -66.5 | 42.5 | 400 | 4.9 | $2.28^{20/4}$ | $1420$ | | $2.1 \times 10^{-2}$ |
| Pentachloroethane | 76-01-7 | $C_2HCl_5$ | 202.3 | -29 | 162 | 3.4 | 7.2 | $1.67^{25/4}$ | 55 | 3.7 | $2.1 \times 10^{-2}$ |
| Bromoform | 75-25-2 | $CHBr_3$ | 252.8 | 6 | 149 | 5.6 | 8.7 | $2.9^{20/4}$ | $3,190^{30}$ | 2.30 | $5.3 \times 10^{-4}$ |
| **Alkenes** | | | | | | | | | | | |
| Allyl chloride | 107-05-1 | $C_3H_5Cl$ | 76.5 | -136 | 45 | 340 | 2.64 | 0.94 | 100 | | |
| Chloroprene | 126-99-8 | $C_4H_5Cl$ | 88.5 | -130 | 59.4 | 215 | 3.06 | 0.96 | slightly | | |
| 1,3-dichloropropene | 542-75-6 | $C_3H_4Cl_2$ | 111 | | 108 | 25 | 3.8 | 1.22 | 2,700 | 1.98 | $3.55 \times 10^{-3}$ |
| Tetrachloroethylene | 127-18-4 | $C_2Cl_4$ | 165.9 | -22.7 | 121 | 14 | 5.8 | 1.63 | 150 | 2.88 | $2.82 \times 10^{-2}$ |
| Trichloroethylene | 79-01-6 | $C_2HCl_3$ | 131.4 | -73 | 87.2 | 57.9 | 4.53 | 1.46 | 1,000 | 2.29 | $1.17 \times 10^{-2}$ |
| Hexachlorobutadiene | 87-68-3 | $C_4Cl_6$ | 260.8 | -19 to -22 | 210 to 220 | 0.15 | | $1.68^{15/15}$ | $2^{20}$ | 3.74 | $1.03 \times 10^{-2}$ |
| Vinyl chloride | 75-01-4 | $C_2H_3Cl$ | 62.5 | -153 | -13.9 | 2,600 | 2.15 | $0.91^{15/4}$ | $1,100^{25}$ | 0.6 | $1.6 \times 10^{-2}$ |

(continued)

## TABLE A-1 (continued)

| Compound | CAS No. | Molecular formula | Molecular weight | Melting point (°C) | Boiling point @ 760 torr (°C) | Vapor pressure @ 20-25°C[b] (torr) | Vapor density (air=1) | Liquid density (g/ml) | Solubility in water (mg/l) | Log octanol water partition coefficient | Henry's Law constant (atm-m³/mole) |
|---|---|---|---|---|---|---|---|---|---|---|---|
| Vinylidene chloride | 75-35-4 | $C_2H_2Cl_2$ | 97 | -122.5 | 31.9 | 500 | 3.25 | $1.22^{20/4}$ | $5,000^{20}$ | 1.48 | $1.5 \times 10^{-2}$ |
| 1,4-dichloro-2-butene | 764-47-0 | $C_4H_6Cl_2$ | 125 | 3.5 | 158 | | | $1.19^{25/4}$ | Insoluble | | |
| 1,2-trans-dichloroethylene | 540-59-0 | $C_2H_2Cl_2$ | 97 | -50 | 48 | 315 | 3.34 | 1.26 | $600^{20}$ | 1.48 | $5.32 \times 10^{-3}$ |
| Hexachloropropene | 1888-71-7 | $C_3Cl_6$ | 248.8 | | | | | | | | |
| **Cyclic Compounds** | | | | | | | | | | | |
| Lindane | 58-89-9 | $C_6H_6Cl_6$ | 290.9 | 113 | 323 | 0.03 | 10.0 | $1.87^{20/4}$ | $17^{24}$ | 3.72 | $7.8 \times 10^{-6}$ |
| Hexachlorocyclopentadiene | 77-47-4 | $C_5Cl_6$ | 272.8 | -9 | 234 | .08 | 9.4 | | $0.805^{25}$ | 3.99 | $1.6 \times 10^{-2}$ |
| Uracil mustard[c] | 66-75-1 | $C_8H_{11}Cl_2N_3O_2$ | 252.1 | 206 (dec.)[?] | | | | | Sparingly | | |
| **Aldehydes** | | | | | | | | | | | |
| Chloroacetaldehyde | 107-20-0 | $C_2H_3ClO$ | 78.5 | -16 | 85 | 60 | 2.7 | $1.19^{25/25}$ | miscible | -0.4 | $2.6 \times 10^{-5}$ |
| Trichloroacetaldehyde | 75-87-6 | $C_2HCl_3O$ | 147.4 | -57.5 | 97.8 | 35 | 5.1 | $1.5^{20/4}$ | Freely Soluble | | |
| **Epoxides** | | | | | | | | | | | |
| Epichlorohydrin | 106-89-8 | $C_3H_5ClO$ | 92.5 | -26 | 116 | 13 | $3.19^{116}$ | $1.18^{20/4}$ | 64,000 | -0.4 | $3.23 \times 10^{-5}$ |
| **Acids** | | | | | | | | | | | |
| Fluoroacetic acid, sodium salt | 144-49-0 | $C_2H_2FNaO_2$ | 100 | 200 (dec.) | | | | | Very Soluble | | |
| **Ketones** | | | | | | | | | | | |
| Bromoacetone | 598-31-2 | $C_3H_5BrO$ | 137 | -36.5 | 137 | | | $1.63^{23}$ | Sparingly | | |

(continued)

## TABLE A-1 (continued)

| Compound | CAS No. | Molecular formula | Molecular weight | Melting point (°C) | Boiling point @ 760 torr (°C) | Vapor pressure @ 20-25°C [b] (torr) | Vapor density (air=1) | Liquid density (g/ml) | Solubility in water (mg/l) | Log octanol water partition coefficient | Henry's law constant (atm-m$^3$/mole) |
|---|---|---|---|---|---|---|---|---|---|---|---|
| **Ethers** | | | | | | | | | | | |
| Bis(chloromethyl)ether | 542-88-1 | $C_2H_4Cl_2O$ | 115.0 | -41.5 | 104 | 30.0 | 3.97 | $1.32^{20/4}$ | 22,000; dec. | 0.38 | $2.1 \times 10^{-4}$ |
| Bis(2-chloroethyl)ether | 111-44-4 | $C_4H_8Cl_2O$ | 143 | -25 | 178 | 0.71 | 4.93 | $1.22^{20/4}$ | 10,200 | 1.58 | $2.16 \times 10^{-5}$ |
| Bis(2-chloroisopropyl)ether | 108-60-1 | $C_6H_{12}Cl_2O$ | 171.1 | -97 | 189 | 0.85 | 6.0 | 1.11 | 1,700 | 2.58 | $1.53 \times 10^{-4}$ |
| 2-chloroethyl vinyl ether | 110-75-8 | $C_4H_7ClO$ | 106.6 | -70.3 | 109 | 26.8 | | $1.06^{10/4}$ | 15,000; dec. | 1.28 | $2.16 \times 10^{-3}$ |
| Chloromethoxymethane | 107-30-2 | $C_2H_5ClO$ | 80.5 | -103.5 | 59.5 | | | 1.06 | dec. | | |
| Bis(2-chloroethoxy)methane | 111-91-1 | $C_5H_{10}Cl_2O_2$ | 173.1 | -32.8 | 218 | <0.1 | | | 81,000 | 1.26 | $3.78 \times 10^{-7}$ |
| **Nitrogen Compounds** | | | | | | | | | | | |
| 3-chloropropionitrile | 542-76-7 | $C_3H_4ClN$ | 89.5 | -51 | 176 | $6^{50}$ | 3.1 | $1.14^{25}$ | 45,000 | | |
| Cyanogen chloride | 506-77-4 | CClN | 61.5 | -6.5 | 13.1 | 1,000 | 1.98 | $1.24^{4/4}$ | $30,000^{25}$ | 0.64 | |
| Fluoroacetamide [c] | 640-19-7 | $C_2H_4FNO$ | 77 | 108 | | | | | Freely | -1.05 | |
| Cyanogen bromide | 506-68-3 | CBrN | 105.9 | 52 | 61 | 100 | | $2.02^{20/4}$ | Slowly dec. in CW | | |
| Uracil mustard [c] | 66-75-1 | $C_8H_{11}Cl_2N_3O_2$ | 252.1 | 206 (dec.) | | | | | Sparingly | | |
| **B. HALOGENATED AROMATICS** | | | | | | | | | | | |
| **Benzenes** | | | | | | | | | | | |
| Chlorobenzene | 108-90-7 | $C_6H_5Cl$ | 112.6 | -45.6 | 132 | 8.8 | 3.88 | 1.11 | 488 | 2.84 | $3.93 \times 10^{-3}$ |
| o-dichlorobenzene | 95-50-1 | $C_6H_4Cl_2$ | 147 | -17.6 | 180.5 | 1 | 5.05 | 1.30 | 145 | 3.38 | $1.94 \times 10^{-3}$ |

(continued)

## TABLE A-1 (continued)

| Compound | CAS No. | Molecular formula | Molecular weight | Melting point (°C) | Boiling point @ 760 torr (°C) | Vapor pressure @ 20-25°C[b] (torr) | Vapor density (air=1) | Liquid density (g/ml) | Solubility in water (mg/l) | Log octanol water partition coefficient | Henry's law constant (atm-m³/mole) |
|---|---|---|---|---|---|---|---|---|---|---|---|
| Hexachlorobenzene | 118-74-1 | $C_6Cl_6$ | 284.8 | 228 | 325 | $1.089 \times 10^{-5}$ | 9.86 | $2.04^{23}$ | $0.11^{24}$ | 6.18 | $1.70 \times 10^{-2}$ |
| Pentachlorobenzene | 608-93-5 | $C_6HCl_5$ | 250.3 | 85 | 275 | <1 | 8.6 | 1.61 | $0.24^{22}$ | 5.3 | |
| Dichlorophenylarsine | 696-28-6 | $C_6H_5AsCl_2$ | 228.9 | | | | | | | | |
| Benzene sulfonyl chloride | 98-09-9 | $C_6H_5ClO_2S$ | 176.6 | 14.5 | 246 (dec.) | | | $1.38^{23}$ | Insoluble | | |
| 1-bromo-4-phenoxy benzene | 101-55-3 | $C_{12}H_9BrO$ | 249 | 18.7 | 310 | 0.015 | | | | 5.15 | |
| m-dichlorobenzene | 541-73-1 | $C_6H_4Cl_2$ | 147 | -24.8 | 172 | 2.28 | 5.08 | $1.29^{20/4}$ | $123^{25}$ | 3.38 | $2.63 \times 10^{-3}$ |
| p-dichlorobenzene | 106-46-7 | $C_6H_4Cl_2$ | 147 | 53 | 173.4 | 0.6 | 5.07 | $1.46^{20/4}$ | $80^{25}$ | 3.39 | $2.72 \times 10^{-1}$ |
| 3,3'-dichlorobenzidine[c] | 91-94-1 | $C_{12}H_{10}Cl_2N_2$ | 253.1 | 132 | | | | | $4.0^{22}$ (pH 6.9) | 3.02 | |
| Pentachloronitrobenzene | 82-68-8 | $C_6Cl_5NO_2$ | 295.4 | 144 | 328 | 0.013 | 10.2 | 1.72 | Pract. Insol. | | |
| 1,2,4,5-tetrachlorobenzene | 95-94-3 | $C_6H_2Cl_4$ | 215.9 | 138 | 245 | 0.1 | 7.4 | $1.86^{21/4}$ | $0.32^{22}$ | 4.72 | |
| 1,2,4-Trichlorobenzene | 120-82-1 | $C_6H_3Cl_3$ | 181.5 | 17 | 213 | 0.29 | 6.25 | 1.57 | 19 | 4.28 | $2.3 \times 10^{-3}$ |
| **Arenes** | | | | | | | | | | | |
| Benzyl chloride | 100-44-7 | $C_7H_7Cl$ | 126.6 | -39 | 179 | 1 | 4.3 | $1.10^{20/20}$ | Insol. CW; Dec. HW | 2.30 | $5.22 \times 10^{-2}$ |
| Benzotrichloride | 98-07-7 | $C_7H_5Cl_3$ | 195.5 | -5 | 213 | 145.8 | 6.77 | $1.38^{20/4}$ | Reacts with $H_2O$ | 4.26 | $3.1 \times 10^{-4}$ |
| Benzal chloride | 98-87-3 | $C7H6Cl2$ | 161 | -16 | 205 | 135.4 | | $1.26^{14/14}$ | Insoluble | | |
| Chlorambucil[c] | 305-03-3 | $C_{14}H_{19}Cl_2NO_2$ | 304.2 | 65 | | | | | Insoluble | | |
| 4-chloro-2-methyl benzenamine[c] | 95-69-2 | $C_7H_8ClN$ | 141.6 | 27 | 241 | | | | Sparingly | | |

(continued)

## TABLE A-1 (continued)

| Compound | CAS No. | Molecular formula | Molecular weight | Melting point (°C) | Boiling point @ 760 torr (°C) | Vapor pressure @ 20-25°C[b] (torr) | Vapor density (air=1) | Liquid density (g/ml) | Solubility in water (mg/l) | Log octanol water partition coefficient | Henry's Law constant (atm-m³/mole) |
|---|---|---|---|---|---|---|---|---|---|---|---|
| Melphalan[c] | 148-82-3 | $C_{13}H_{18}Cl_2N_2O$ | 305 | 182 (dec.) | | | | | Pract. Insol. | | |
| 4,4'-methylene bis(2-chloroaniline)[c] | 101-14-4 | $C_{13}H_{12}Cl_2N_2$ | 267.2 | 99-107 | | | | 1.44 | | | |
| Pronamide[c] | 23950-58-5 | $C_{12}H_{11}Cl_2NO$ | 256.1 | 155 | | $8.5 \times 10^{-5}$ | | | $15^{25}$ | | |
| o-toluidine hydrochloride[c] | 636-21-5 | $C_7H_{10}ClN$ | 143.6 | 215 | 242 | | | | Very soluble | | |
| **Phenols** | | | | | | | | | | | |
| Pentachlorophenol | 87-86-5 | $C_6HCl_5O$ | 266.4 | 190 | 310 (dec.) | $1.1 \times 10^{-4}$ | 9.2 | 1.98 | $14^{20}$ | 5.01 | $8.82 \times 10^{-6}$ |
| 2-chlorophenol | 95-57-8 | $C_6H_5ClO$ | 128.6 | 9.3 | 175 | 2.2 | | 1.26 | 28,500 | 2.15/2.19 | $8.28 \times 10^{-6}$ |
| p-chloro-m-cresol | 59-50-7 | $C_7H_7ClO$ | 142.6 | 66 | 235 | | | | 3,850 | 2.95/3.10 | |
| 2,4-dichlorophenol | 120-83-2 | $C_6H_4Cl_2O$ | 163 | 45 | 210 | 0.12 | 5.62 | $1.38^{60/25}$ | $4,600^{20}$ | 2.75 | $6.66 \times 10^{-6}$ |
| 2,6-dichlorophenol | 87-65-0 | $C_6H_4Cl_2O$ | 163 | 65 | 219 | 1.0 | 5.6 | | 340 | 2.9 | $4.0 \times 10^{-5}$ |
| 2,4,6-trichlorophenol | 88-06-2 | $C_6H_3Cl_3O$ | 197.5 | 68 | 246 | $1^{76.5}$ | | $1.5^{75/4}$ | $800^{25}$ | 3.38 | $7.2 \times 10^{-6}$ |
| Hexachlorophene[c] | 70-30-4 | $C_{13}H_6Cl_6O_2$ | 406.9 | 164 | | | | | Pract. Insol. | | |
| 2,3,4,6-tetrachlorophenol | 58-90-2 | $C_6H_2Cl_4O$ | 231.9 | 69 | 288 | $1^{100}$ | 8.0 | $1.6^{60/4}$ | Insoluble | 4.3 | |
| 2,4,5-trichlorophenol | 95-95-4 | $C_6H_3Cl_3O$ | 197.5 | 62 | 253 | $1^{72}$ | 6.8 | $1.5^{75}$ | 1,190 | 3.72 | |
| **Bi-Phenyls** | | | | | | | | | | | |
| Methoxychlor[c] | 72-43-5 | $C_{16}H_{15}Cl_3O_2$ | 345.7 | 98 | | | | | $1.41^{25}$ | $0.04^{24}$ | |
| Ethyl 4,4'-dichlorobenzilate (Chlorobenzilate) | 510-15-6 | $C_{16}H_{14}Cl_2O_3$ | 325.2 | 36 | 147 | $2.2 \times 10^{-6}$ | | 1.57 | Slightly | | |

(continued)

## TABLE A-1 (continued)

| Compound | CAS No. | Molecular formula | Molecular weight | Melting point (°C) | Boiling point @ 760 torr (°C) | Vapor pressure @ 20-25°C[b] (torr) | Vapor density (air=1) | Liquid density (g/ml) | Solubility in water (mg/l) | Log octanol water partition coefficient | Henry's Law constant (atm-m³/mole) |
|---|---|---|---|---|---|---|---|---|---|---|---|
| DDD[c] | 72-54-8 | $C_{14}H_{10}Cl_4$ | 320.1 | 112 | | $1.02 \times 10^{-6}$ | 11.0 | | $0.160^{24}$ | 5.98 | $2.16 \times 10^{-5}$ |
| DDT[c] | 50-29-3 | $C_{14}H_9Cl_5$ | 354.5 | 108 | 185 | $1.9 \times 10^{-7}$ | | | $0.0034^{25}$ | 4.98 | |
| Hexachlorophene[c] | 70-30-4 | $C_{13}H_6Cl_6O_2$ | 406.9 | 164 | | | | | Pract. insol. | | |
| **Polynuclear Compounds** | | | | | | | | | | | |
| Chlornaphazine | 494-03-1 | $C_{14}H_{15}Cl_2N$ | 268.2 | 55 | 210 | | | | Sparingly | | |
| 2-chloronaphthalene | 91-58-7 | $C_{10}H_7Cl$ | 162.6 | 59.5 | 256 | 0.017 | | 1.61 | $6.7^{25}$ | 4.12 | $6.12 \times 10^{-4}$ |
| **Amines/Amides** | | | | | | | | | | | |
| p-chloroaniline | 106-47-8 | $C_6H_6ClN$ | 127.6 | 72.5 | 232 | 0.015 | 4.41 | $1.43^{19/4}$ | Solub. in HW | 1.83 | |
| 1-(o-chlorophenyl)thiourea | | $C_7H_7ClN_2S$ | 187 | 146 | | | | | | | |
| Chlorambucil[c] | 305-03-3 | $C_{14}H_{19}Cl_2NO_2$ | 304.2 | 65 | | | | | Insoluble | | |
| 4-chloro-2-methyl benzenamine[c] | 95-69-2 | $C_7H_8ClN$ | 141.6 | 27 | 241 | | | | Sparingly | | |
| 3,3'-dichlorobenzidine[c] | 91-94-1 | $C_{12}H_{10}Cl_2N_2$ | 253.1 | 132 | | | | | $4.0^{22}$ (pH 6.9) | 3.02 | |
| Melphalan[c] | 148-82-3 | $C_{13}H_{18}Cl_2N_2O_2$ | 305 | 182 (dec.) | | | | | Pract. insol. | | |
| 4,4'-methylene bis(2-chloroaniline)[c] | 101-14-4 | $C_{13}H_{12}Cl_2N_2$ | 267.2 | 99-107 | | | | 1.44 | | | |
| Pronamide[c] | 23950-58-5 | $C_{12}H_{11}Cl_2NO$ | 256.1 | 155 | | $8.5 \times 10^{-5}$ | | | $15^{25}$ | | |
| o-toluidine hydrochloride[c] | 636-21-5 | $C_7H_{10}ClN$ | 143.6 | 215 | 242 | | | | Very soluble | | |

(continued)

## TABLE A-1 (continued)

| Compound | CAS No. | Molecular formula | Molecular weight | Melting point (°C) | Boiling point @ 760 torr (°C) | Vapor pressure @ 20-25°C [b] (torr) | Vapor density (air=1) | Liquid density (g/ml) | Solubility in water (mg/l) | Log octanol water partition coefficient | Henry's Law constant (atm-m³/mole) |
|---|---|---|---|---|---|---|---|---|---|---|---|
| **C. HALOGENATED PESTICIDES** | | | | | | | | | | | |
| **Hydrocarbon Insecticides** | | | | | | | | | | | |
| Lindane[c] | 58-89-9 | $C_6H_6Cl_6$ | 290.9 | 113 | 323 | 0.03 | 10.0 | $1.87^{20/4}$ | $17^{24}$ | 3.72 | $7.8 \times 10^{-6}$ |
| Methoxychlor[c] | 72-43-5 | $C_{16}H_{15}Cl_3O_2$ | 345.7 | 98 | | | | $1.41^{25}$ | $0.04^{24}$ | | |
| DDD[c] | 72-54-8 | $C_{14}H_{10}Cl_4$ | 320.1 | 112 | | $1.02 \times 10^{-6}$ | 11.0 | | $0.16^{24}$ | 5.98 | $2.16 \times 10^{-5}$ |
| DDT[c] | 50-29-3 | $C_{14}H_9Cl_5$ | 354.5 | 108 | 185 | $1.9 \times 10^{-7}$ | | | $0.0034^{25}$ | 4.98 | |
| 1,2-dibromo-3-chloropropane[c] | 96-12-8 | $C_3H_5Br_2Cl$ | 236.4 | | 196 | 0.8 | | $2.08^{20/20}$ | $1,000^{25}$ | 3.0 | $2.45 \times 10^{-4}$ |
| Ethylene dibromide[c] | 106-93-4 | $C_2H_4Br_2$ | 187.9 | 10 | 131 | 11 | 6.5 | $2.7^{35/4}$ | $4,300^{30}$ | | $8.82 \times 10^{-4}$ |
| **Cyclodiene Insecticides** | | | | | | | | | | | |
| Endrin | 72-20-8 | $C_{12}H_8Cl_6O$ | 380.9 | 226-230 | | $2 \times 10^{-7}$ | | | 0.26 | 5.6 | $5 \times 10^{-7}$ |
| Chlordane | 57-74-9 | $C_{10}H_6Cl_8$ | 409.8 | 105 | 175 @ 2 mm Hg | $1 \times 10^{-5}$ | 14.2 | $1.6^{15/15}$ | .009 | 2.78 | $4.8 \times 10^{-5}$ |
| Heptachlor | 76-44-8 | $C_{10}H_5Cl_7$ | 373.4 | 95 | 140 @ 2 mm Hg | $3 \times 10^{-4}$ | 12.9 | 1.57 | $.056^{25-29}$ | 5.05 | $1.48 \times 10^{-3}$ |
| Aldrin | 309-00-2 | $C_{12}H_8Cl_6$ | 364.9 | 104 | | $2.3 \times 10^{-5}$ | | | $0.27^{27}$ | 5.00 | $4.96 \times 10^{-4}$ |
| Dieldrin | 60-57-1 | $C_{12}H_8Cl_6O$ | 380.9 | 176 | | $3.1 \times 10^{-6}$ | 13.2 | 1.75 | $0.186^{25-29}$ | 4.56 | $5.8 \times 10^{-5}$ |
| Isodrin | 465-73-6 | $C_{12}H_8Cl_6$ | 365 | 241 | | | | | | | |
| Endosulfan | 115-29-7 | $C_9H_6Cl_6O_3S$ | 406.9 | 70-100 | | $1.0 \times 10^{-5}$ | | | Pract. Insol. | | |
| **Carbamate Insecticides** | | | | | | | | | | | |
| S(2,3-dichloroallyl)diisopropyl thiocarbamate | 2303-16-4 | $C_{10}H_{17}ClNOS$ | 270.2 | 25-30 | 150 @ 9 mm Hg | $1.5 \times 10^{-4}$ | | 1.18 | 14 | | |

(continued)

## TABLE A-1 (continued)

| Compound | CAS No. | Molecular formula | Molecular weight | Melting point (°C) | Boiling point @ 760 torr (°C) | Vapor pressure @ 20-25°C[b] (torr) | Vapor density (air=1) | Liquid density (g/ml) | Solubility in water (mg/l) | Log octanol water partition coefficient | Henry's Law constant (atm-m³/mole) |
|---|---|---|---|---|---|---|---|---|---|---|---|
| Dimethyl carbamoyl chloride | 79-44-7 | $C_3H_6ClNO$ | 107.6 | -33 | 166 | | 3.73 | 1.68 | | | |
| Phenoxy Acids and Salts | | | | | | | | | | | |
| 2,4-D | 94-75-7 | $C_8H_6Cl_2O_3$ | 221 | 140.5 | 160 | 0.4[160] | 7.63 | 1.42[25] | 890[25] | 2.81 | |
| 2,4,5-TP (Silvex) | 93-72-1 | $C_9H_7Cl_3O_3$ | 269.5 | 180 | 293 | $0.1 \times 10^{-3}$ | 9.3 | | 140[25] | 2.4 | |
| 2,4,5-T | 93-76-5 | $C_8H_6Cl_3O_3$ | 255.5 | 158 | | | | | 278[25] | | |
| Camphenes | | | | | | | | | | | |
| Toxaphene | 8001-35-2 | $C_{10}H_{10}Cl_8$ | 413.8 | 65-90 | 120 (dec.) | 0.2-0.4 | 14.3 | 1.63 | 3 | 3.3 | $4.89 \times 10^{-3}$ |
| Organophosphorus Compounds | | | | | | | | | | | |
| Diisopropyl fluorophosphate | | $C_6H_{14}FO_3P$ | 184 | -82 | 46 @ 5 mm | 0.579 | 5.24 | 1.07 | 15,400 | 1.38 | |
| Miscellaneous Pesticides | | | | | | | | | | | |
| Fluoroacetamide[c] | 640-19-7 | $C_2H_4FNO$ | 77 | 108 | | | | | Freely | -1.05 | |
| Kepone | 143-50-0 | $C_{10}Cl_{10}O$ | 490.7 | 350 (sublimes) | 350 (sublimes) | $3 \times 10^{-7}$ | | | 7.6 | | |
| Pronamide[c] | 23950-58-5 | $C_{12}H_{11}Cl_2NO$ | 256.1 | 155 | | $8.5 \times 10^{-5}$ | | | 15[25] | | |

(continued)

## TABLE A-1 (continued)

| Compound | CAS No. | Molecular formula | Molecular weight | Melting point (°C) | Boiling point @ 760 torr (°C) | Vapor pressure @ 20-25°C[b] (torr) | Vapor density (air=1) | Liquid density (g/ml) | Solubility in water (mg/l) | Log octanol water partition coefficient | Henry's Law constant (atm-m³/mole) |
|---|---|---|---|---|---|---|---|---|---|---|---|
| **D. MISCELLANEOUS COMPOUNDS** | | | | | | | | | | | |
| Phosgene (carbonyl chloride) | 75-44-5 | $CCl_2O$ | 98.9 | -118 | 8.1 | 1,216 | 3.42 | $1.39^{19/4}$ | Very slightly sol.; dec. | | |
| Acetyl chloride | 75-36-5 | $C_2H_3ClO$ | 78.5 | -112 | 51 | 299 | 2.70 | $1.11^{20/4}$ | dec. | -0.17 | |
| Carbonyl fluoride | 353-50-4 | $CF_2O$ | 66 | -114 | -83.1 | | | $1.14^{-114}$ | | | |
| Methyl chlorocarbonate | 79-22-1 | $C_2H_3ClO_2$ | 94.5 | | 71 | | 3.26 | $1.22^{20/4}$ | slightly sol., dec. | | |
| Tris(2,3-dibromopropyl)phosphate | 126-72-7 | $C_9H_{15}BR_6PO_4$ | 697.7 | | | | | 2.24 | $8.0^{24}$ | | |

[a]Source: Arienti, M., et al (see Reference 2-Section 2).

[b]Note: 1 torr = 1 mm Hg.

[c]Indicates compounds which are listed in more than one functional group category.

*Other Noyes Publications*

# TREATMENT TECHNOLOGIES FOR SOLVENT CONTAINING WASTES

by

## M. Breton, P. Frillici, S. Palmer
## C. Spears, M. Arienti, M. Kravett
## A. Shayer, N. Surprenant

Alliance Technologies Corporation

*Pollution Technology Review No. 149*

This book provides technical information describing management options for solvent containing wastes. These options include treatment and disposal of waste streams as well as waste minimization procedures such as source reduction, reuse, and recycling.

Emphasis is placed on proven technologies such as incineration, use as fuel, distillation, steam stripping, biological treatment, and activated carbon adsorption; however, a full range of waste minimization processes and treatment recovery technologies which can be used to manage solvent wastes is covered in the book.

Potentially viable technologies are described in terms of performance in removal of regulated constituents, associated process residuals and emissions, and restrictive waste characteristics which affect the ability of a given technique to effectively treat the wastes under consideration.

Approaches to the selection of treatment/recovery options are reviewed, and pertinent properties of organic solvents which impact treatment technology/waste interactions are provided.

## CONTENTS

**ISBN 0-8155-1158-2 (1988)**

753 pages

# STORAGE AND TREATMENT OF HAZARDOUS WASTES IN TANK SYSTEMS

## U.S. Environmental Protection Agency
## Office of Solid Waste

*Pollution Technology Review No. 146*

This book provides guidance to owners and operators of hazardous waste storage tanks regarding storage and treatment methods for compliance with U.S. Environmental Protection Agency standards and regulations. Types of wastes covered in the book include liquid hazardous wastes; non-liquid wastes such as solid hazardous wastes, residues, and dried sludges; and gaseous hazardous wastes.

The overall goal of the EPA regulations is the protection of human health and the environment from the risks posed by hazardous waste storage and treatment tank system facilities. This is to be accomplished by the prevention of migration of hazardous waste constituents to ground and surface waters. A prime element in prevention is the rapid detection of waste release so that appropriate response action can be taken. The expense of cleaning up soil contaminated by a leaking tank can be enormous.

Six key features of EPA's regulatory approach are covered: (1) maintaining the integrity of the primary containment system, (2) proper installation of tank systems, (3) secondary containment with monitoring to detect leaks in the primary system, (4) response methods for hazardous waste releases, (5) proper operaton and inspection, and (6) adequate closure and post-closure care.

The condensed contents below includes **chapter titles and selected subtitles.**

1. **INTRODUCTION**

2. **BACKGROUND**

3. **THE PERMITTING PROCESS**

4. **WRITTEN ASSESSMENT OF TANK SYSTEMS**
   Tank System Design and Testing
   Internal Inspections
   Ancillary Equipment Assessment

5. **DESIGN OF NEW TANK SYSTEMS OR COMPONENTS**
   Description of Feed Systems, Safety Cutoff, Bypass Systems, and Pressure Controls
   External Corrosion Protection

Protection from Vehicular Traffic

6. **INSTALLATION OF NEW TANK SYSTEMS**
   Proper Handling Procedures
   Backfilling
   Ancillary Equipment Installation

7. **SECONDARY CONTAINMENT SYSTEMS AND RELEASE DETECTION**
   Design Parameters
   Types of Secondary Containment
   Liner Requirements
   Vault Requirements
   Double-Walled Tank Requirements

8. **VARIANCES FROM SECONDARY CONTAINMENT**
   Technology-Based Variance
   Risk-Based Variance

9. **CONTROLS AND PRACTICES TO PREVENT SPILLS AND OVERFILLS**
   Underground Tanks
   Aboveground/Inground/Onground Tanks
   Uncovered Tanks—Freeboard

10. **INSPECTIONS**
    Fiberglass-Reinforced Plastic (FRP) Tanks
    Concrete Tanks
    Inspection Tools and Electromechanical Equipment
    Reporting Requirements

11. **RESPONSE TO LEAKS OR SPILLS AND DISPOSITION OF LEAKING OR UNFIT-FOR-USE TANK SYSTEMS**

12. **CLOSURE AND POST-CLOSURE REQUIREMENTS**
    Decontamination/Removal Procedures for Closure
    Closure Plan and Closure Activities
    Closure of Tank System
    Closure/Post-Closure Cost Estimates

13. **PROCEDURES FOR TANK SYSTEMS THAT STORE OR TREAT IGNITABLE, REACTIVE, OR INCOMPATIBLE WASTES**
    Ignitable or Reactive Wastes, General Precautions
    Incompatible Wastes

**APPENDICES**

ISBN 0-8155-1138-8 (1987)

428 pages

# MINIMIZING EMPLOYEE EXPOSURE TO TOXIC CHEMICAL RELEASES

by

## Ralph W. Plummer, Terrence J. Stobbe, James E. Mogensen, Luanne K. Jeram
West Virginia University

*Pollution Technology Review No. 145*

This book describes procedures for minimizing employee exposure to toxic chemical releases and suggested personal protective equipment (PPE) to be used in the event of such chemical release. How individuals, employees, supervisors, or companies perceive the risks of chemical exposure (risk meaning both probability of exposure and effect of exposure) determines to a great extent what precautions are taken to avoid risk.

The objective of this study was to obtain information from chemical manufacturers and facilities which use chemicals in other processes, to determine what types of procedures are currently in practice, and to develop a set of recommended procedures and suggested PPE which can be used to minimize the possibility and the risk of toxic chemical releases. In Part I, the authors develop an approach which divides the project into three phases: kinds of procedures currently being used; the types of toxic chemical release accidents and injuries that occur; and, finally, integration of this information into a set of recommended procedures which should decrease the likelihood of a toxic chemical release and, if one does occur, will minimize the exposure and its severity to employees. Part II covers the use of personal protective equipment. It addresses the questions: what personal protective equipment ensembles *are used* in industry in situations where the release of a toxic or dangerous chemical may occur or has occurred; and what personal protective equipment ensembles *should be used* in these situations.

A condensed table of contents listing **part titles, chapter titles and selected subtitles** is given below.

ISBN 0-8155-1131-0 (1987)

257 pages

# UNDERGROUND TANK LEAK DETECTION METHODS

by

### Shahzad Niaki and John A. Broscious
IT Corporation

*Pollution Technology Review No. 139*

This state-of-the-art review details available and developing methods for detecting leaks in underground storage tanks. Thirty-six volumetric, non-volumetric, inventory monitoring, and leak effects monitoring detection methods are described as used for underground tanks used primarily for liquid hydrocarbons and products. The book provides general engineering comments on the methods and discusses variables which may affect their accuracy.

In recent years, the increase in leaks from underground fuel storage tanks has had a significant adverse impact on the United States. Current estimates are that there are from 1.5 to 3.5 million of these storage tanks in the U.S. As many as 100,000 may have some leakage already, and it is anticipated that up to 350,000 may develop leaks within the next five years.

Corrosion is one of the primary causes of leakage. Product loss from leaking tanks may adversely affect the environment, endanger lives, reduce income, and require expensive cleanup procedures. To prevent, reduce, or avoid such problems, accurate methods must be used to determine whether or not an underground tank is leaking. This book identifies existing and developing techniques for leak detection. In so doing, each method, its capabilities, its claimed precision and accuracy, are reviewed.

The condensed table of contents given below lists **chapter titles and selected subtitles.**

ISBN 0-8155-1117-5 (1987)

Printed and bound by CPI Group (UK) Ltd, Croydon, CR0 4YY

19/10/2024

01776427-0001